T0141695

Advances in Intelligent Systems and Computing

Volume 884

Series editor

Janusz Kacprzyk, Polish Academy of Sciences, Warsaw, Poland
e-mail: kacprzyk@ibspan.waw.pl

The series "Advances in Intelligent Systems and Computing" contains publications on theory, applications, and design methods of Intelligent Systems and Intelligent Computing. Virtually all disciplines such as engineering, natural sciences, computer and information science, ICT, economics, business, e-commerce, environment, healthcare, life science are covered. The list of topics spans all the areas of modern intelligent systems and computing such as: computational intelligence, soft computing including neural networks, fuzzy systems, evolutionary computing and the fusion of these paradigms, social intelligence, ambient intelligence, computational neuroscience, artificial life, virtual worlds and society, cognitive science and systems, Perception and Vision, DNA and immune based systems, self-organizing and adaptive systems, e-Learning and teaching, human-centered and human-centric computing, recommender systems, intelligent control, robotics and mechatronics including human-machine teaming, knowledge-based paradigms, learning paradigms, machine ethics, intelligent data analysis, knowledge management, intelligent agents, intelligent decision making and support, intelligent network security, trust management, interactive entertainment, Web intelligence and multimedia.

The publications within "Advances in Intelligent Systems and Computing" are primarily proceedings of important conferences, symposia and congresses. They cover significant recent developments in the field, both of a foundational and applicable character. An important characteristic feature of the series is the short publication time and world-wide distribution. This permits a rapid and broad dissemination of research results.

More information about this series at http://www.springer.com/series/11156

Miguel Botto-Tobar · Lida Barba-Maggi
Javier González-Huerta · Patricio Villacrés-Cevallos
Omar S. Gómez · María I. Uvidia-Fassler
Editors

Information and Communication Technologies of Ecuador (TIC.EC)

Springer

Editors
Miguel Botto-Tobar
Department of Mathematics and Computer Science
Eindhoven University of Technology
Eindhoven, Noord-Brabant, The Netherlands

Lida Barba-Maggi
Facultad de Ingeniería
Universidad Nacional de Chimborazo
Riobamba, Ecuador

Javier González-Huerta
Department of Software Engineering
Blekinge Tekniska Högskola
Karlskrona, Blekinge Län, Sweden

Patricio Villacrés-Cevallos
Facultad de Ingeniería
Universidad Nacional de Chimborazo
Riobamba, Ecuador

Omar S. Gómez
Escuela Superior Politécnica de Chimborazo
Riobamba, Ecuador

María I. Uvidia-Fassler
Facultad de Ingeniería
Universidad Nacional de Chimborazo
Riobamba, Ecuador

ISSN 2194-5357 ISSN 2194-5365 (electronic)
Advances in Intelligent Systems and Computing
ISBN 978-3-030-02827-5 ISBN 978-3-030-02828-2 (eBook)
https://doi.org/10.1007/978-3-030-02828-2

Library of Congress Control Number: 2018958314

This Springer imprint is published by the registered company Springer Nature Switzerland AG
The registered company address is: Gewerbestrasse 11, 6330 Cham, Switzerland

Preface

The sixth conference on Information and Communication Technologies "TIC-EC" was held in Riobamba—Ecuador from November 21 until 23, 2018. This academic event is considered as one of the most important conferences about ICT in Ecuador, as it brings scholars and practitioners from the country and abroad to discuss the development, issues, and projections of the use of information and communication technologies in multiples fields of application. In 2018, the "TIC-EC" conference was organized by Universidad Nacional del Chimborazo (Unach) and its Engineering School, and the Ecuadorian Corporation for the Development of Research and Academia (CEDIA). The content of this volume is related to the following subjects:

- Communication Networks
- Software Engineering
- Computer Sciences
- Architecture
- Intelligent Territory Management
- IT Management
- Web Technologies
- Engineering, Industry, and Construction with ICT Support
- Entrepreneurship and Innovation at the Academy: a business perspective

In its 2018 edition, the TIC-EC conference received 87 submissions in English from 234 authors coming from nine different countries. All these papers were peer-reviewed by the TIC-EC 2018 Program Committee consisting of 50 high-quality researchers coming from 12 different countries. To assure a high-quality and thoughtful review process, we assigned each paper at least three reviewers. Based on the results of the peer reviews, 27 full papers were accepted, resulting in a 31% acceptance rate, which was within our goal of less than 40%.

We would like to express our sincere gratitude to the invited speakers for their inspirational talks, to the authors for submitting their work to this conference, and the reviewers for sharing their experience during the selection process.

November 2018

Miguel Botto-Tobar
Lida Barba-Maggi
Javier González-Huerta
Patricio Villacrés-Cevallos
Omar S. Gómez
María I. Uvidia-Fassler

Organization

Honorary Committee

Nicolay Samaniego Erazo

Presidente de CEDIA/Rector Unach

Juan Pablo Carvallo Vega

Director Ejecutivo CEDIA

Patricio Villacrés-Cevallos

Decano Facultad de Ingeniería, Unach

Organizing Committee

Lida Barba-Maggi, Unach
Ciro Radicelli García, Unach
María Isabel Uvidia, Unach
Gabriela Jimena Dumancela Nina, Unach
Galia Rivas Toral, CEDIA
Andrea Daniela Morales Rodríguez, CEDIA
Ximena Lazo Álvarez, CEDIA

Program Committee

Miguel Botto-Tobar	Eindhoven University of Technology, The Netherlands
Angela Díaz Cadena	Universitat de Valencia, Spain
Andrés Robles	Edinburgh Napier University, UK
Andrés José Cueva Costales	Yachay EP, Ecuador
Yan Pacheco	Universidad de las Américas, Ecuador
Fadloun Samiha	University of Montpellier, France
Guillermo Pizarro	Universidad Politécnica Salesiana, Ecuador
Orlando Erazo	Universidad Técnica Estatal de Quevedo, Ecuador
María L. Montoya Freire	Aalto University, Finland
Erick Cuenca	University of Montpellier, France
David Rivera Espín	Interamerican Center of Tax Administrations, Panamá
Yuliana Jimenez	Universidad Técnica Particular de Loja, Ecuador
Juan Fernando Balarezo Serrano	Radical Alternativas de Avanzada, Ecuador
Luis Felipe Urquiza Aguiar	Escuela Politécnica Nacional, Ecuador
Gustavo Andrade-Miranda	Universidad de Guayaquil, Ecuador
Wayner Xavier Bustamante Granda	Universidad Internacional del Ecuador, Ecuador
Janneth Chicaíza	Universidad Técnica Particular de Loja, Ecuador
Diego Vallejo-Huanga	Universidad Politécnica Salesiana, Ecuador
Danilo Jaramillo Hurtado	Universidad Técnica Particular de Loja, Ecuador
Pablo Palacios Játiva	Universidad de las Américas, Ecuador
Marlon Navia Mendoza	ESPAM-MFL, Ecuador
Pablo Saa	Universidad Tecnológica Equinoccial, Ecuador
Jeffery Alex Naranjo Cedeño	Universidad Politécnica Estatal del Carchi, Ecuador
Jefferson Ribadeneira Ramírez	Escuela Superior Politécnica de Chimborazo (ESPOCH), Ecuador
Julio Proaño	Universidad Politécnica Salesiana, Ecuador
Maikel Leyva Vázquez	Universidad de Guayaquil, Ecuador
Alex Cazañas	Universidad de Coimbra, Portugal
Jaime Jarrín	AndeanTrade, Ecuador
Washington Velásquez	Universidad Politécnica de Madrid, Spain
Marco Fabricio Falconi Noriega	Corporación Nacional de Telecomunicaciones, Ecuador
José Luis Carrera Villacrés	University of Bern, Switzerland
Germania Rodríguez Morales	Universidad Técnica Particular de Loja, Ecuador
Patricia Ludeña González	Universidad Técnica Particular de Loja, Ecuador
Marcia M. Bayas Sampedro	Universidad Estatal Península de Santa Elena, Ecuador

Sponsoring Institutions

FACULTAD DE
INGENIERÍA

Universidad Nacional del Chimborazo
http://www.unach.edu.ec/

CEDIA
https://www.cedia.edu.ec/es/

Contents

Communication Networks

Millimeter-Wave Channel Estimation Using Coalitional Game

Pablo Palacios[1(✉)], José Julio Freire[2(✉)], and Milton Román-Cañizáres[2(✉)]

[1] Departamento de Ingeniería Eléctrica, Universidad de Chile, Santiago, Chile
pablo.palacios@ug.uchile.cl

[2] Departamento de Redes y Telecomunicaciones, Universidad De Las Américas, Quito, Ecuador
{jose.freire,milton.roman}@udla.edu.ec

Abstract. In millimeter-wave (mm-wave) massive MIMO systems, the channel estimation (CE) is a crucial component to set the mm-wave links. Unfortunately, acquiring channel knowledge is a source of training overhead. In this paper, we propose a CE method leveraging measurements at sub 6-Ghz frequencies in order to reduce the training overhead. This solution extracts spatial information from a sub 6-Ghz channel using a virtual channel transformation, such as the searching space is reduced to the information provided by the low frequency channel. In a second stage, a multicell system and its interference between cells is analyzed, proposing a coalitional game to deal with the intercell interference. In the single cell case, we analyze the proposed method in different SNR scenarios, the computational complexity and over user equipment (UE) mobility environment. Finally, we analyze how the coalitional game improves the throughput and its performance over UE in mobility cases.

Keywords: mm-wave · High-speed · Coalitional game
Channel estimation

1 Introduction

Large antenna arrays (i.e. Massive MIMO) at both sides eNodeB (eNB) and UE is a promising technology to achieve high-throughput services [1]. Large antenna arrays at the same time deal with the high path-loss in millimeter frequencies. By the other hand, channel states information (CSI) in terms of channel matrix or beam alignment are needed at the eNB to point the beams in the UE direction. Both strategies are usually acquired by a training sequence [2]. The sequence is used to measure every beamformer and combiner to estimate the pair of beams that are closer to the desired angles, but this exhaustive-search method need a large number of measurements to estimate the best beam-pair. Additionally, this fact could leads to lower channel rate. In vehicular or train scenarios where due to the UE speed, the channel coherence time becomes shorter and the training period could occupies all the coherence time, leaving no time for data transmission [3].

M. Botto-Tobar et al. (Eds.): TICEC 2018, AISC 884, pp. 3–17, 2019.
https://doi.org/10.1007/978-3-030-02828-2_1

The document is structured as follows: Sect. 2 describes the current work on millimeter waves, radio cognitive and work proposal. Then, Sect. 3 introduces the model of the system used for the evaluation of the proposed method. After that, Sect. 4 reports the channel estimation method for the proposed work. Section 5 includes the multicell analysis in which the proposed coalition game model is described with its respective algorithm. Finally, in Sect. 6 the numerical results obtained in the simulation are shown, to finalize with the conclusions that are described in Sect. 7

2 Related Work

In the literature we can find different mechanism to estimate the Angles of Arrival (AoA) and Angle of Departure (AoD) or the complete CSI with lower training, for example in [4] a beam alignment method is carried out taken advantage of UE location. Beamforming focused on wireless backhaul in small cell networks and wind effect on beam misalignment is studied at [5], in the other hand a typical mm-wave channel estimation (CE) process is carried out based on compressive sensing (CS) framework [6–8] that is a useful technique to decrease the training overhead although could not be appropriate for environments where frequent updates are required to estimate the channel, e.g. railway and vehicular scenarios.

Therefore some analysis in models with high mobility have been done in order to provide a better understanding about mm-wave propagation for vehicular and train environments, in [9] an analysis focused on urban areas was deployed, motivated by measurements and ray tracing, researchers concluded that interference from a NLOS parallel street is negligible, in [10] was found an optimal beam-width different than zero which maximizes the coherence time, the paper also includes the beam misalignment due to motion in the receiver.

In [11] a location-aided mm-wave channel estimation method was proposed exploiting the eNB and vehicle position to infer the LOS path's AoA and AoD. Unfortunately, the adaptive approach is likely to fail if the LOS path is blocked. Keeping in mind that beam training due to channel estimation process is a source of overhead, another point of view to resolve the problem is by using spatial information from different frequencies, specifically measurements from sub-6 Ghz frequencies which are being broadly used currently, although the number of common paths decreases with larger frequency separation, there is still a strong spatial information congruency among sub-6 Ghz and mm-wave frequencies as was proved in [12], where the researchers provided a mm-wave channel estimation method using two transform process to relate the spatial correlation matrix from sub-6 Ghz to mm-wave frequencies. While [13] took advantage of out-of-band information for beam-selection in a OFDM system, leveraging data from all active subcarriers to decide the best beam-pair.

In this paper, we propose a channel estimation method using out-of-band measurements. It is based on training and we assume the microwave channel have been already estimated. This assumption is taken based on several studies about channel estimation for static and high speed environments have been done for single carrier and multi-carrier systems in micro-wave frequencies, such that this is a practical assumption, this solution applies virtual channel decomposition [14] to the lower frequency channel in order to extract the dominant paths, and reminding the fact that due to higher path loss and shorter wavelength, high frequency systems are expected to use a larger antenna array than sub-6 Ghz [15], this implies a narrower beam at mm-wave system.

Notation: $\mathcal{A}$ is a set, $|\mathcal{A}|$ is the cardinality of set $\mathcal{A}$. lower-case a is a scalar, $\mathbf{a}$ is a vector, $\mathbf{A}$ is a matrix. $\mathbf{A}^T, \mathbf{A}^H$ denote the transpose and Hermitian of matrix $\mathbf{A}$.

3 System Model

We consider a mm-wave MIMO downlink system with uniform linear arrays (ULAs) conformed by N_t transmitter antennas in the UE and N_r receiver antennas in the eNB, as show in Fig. 1. We consider that both the transmitter and the receiver have only one RF chain, hence, only analog beamforming/combining can be applied.

We use $\mathbf{f}$, and $\mathbf{q}$ to denote the beamformer and combiner vector, respectively:

$$\mathbf{f} = \frac{1}{\sqrt{N_t}}[1, ..., \ e^{j(N_r-1)\frac{2\pi}{\lambda}d\cos\phi}]^T, \tag{1}$$

where $\phi \in [-\pi/2, \pi/2]$, is a quantized angle of departure, besides $\mathbf{f}$ has constant modulus entries and only phase can varying, in similar fashion the combiner:

$$\mathbf{q} = \frac{1}{\sqrt{N_r}}[1, ..., \ e^{j(N_r-1)\frac{2\pi}{\lambda}d\cos\theta}]^T, \tag{2}$$

where $\theta \in [-\pi/2, \pi/2]$, is a quantized angle of arrival, the AoAs and AoDs can be taken following regular or non regular sampling strategies, the detail about how we choose this angles is discussed later in Sect. 3.2. Then considering a narrowband channel model $\mathbf{H} \in \mathbb{C}^{N_r \times N_T}$, the received signal in the eNB can be modeled as:

$$y = \sqrt{\rho}\mathbf{q}^H\mathbf{H}\mathbf{f}x + \mathbf{q}^H\mathbf{v}, \tag{3}$$

where $\sqrt{\rho}$ is the average transmit power in the training phase, x is the training symbol, and $\mathbf{v}$ is the vector of i.i.d. $\sim \mathcal{CN}(0, \sigma_0^2\mathbf{I})$ noise.

3.1 Millimeter-Wave Channel Model

We adopt a geometric channel model with L scatterers, where each scatterer contributes to one propagation path. Accordingly, the channel matrix $\mathbf{H}$, can be expressed as:

$$\mathbf{H} = \sqrt{N_tN_r}\sum_{l=1}^{L}\alpha_l\mathbf{a}_r(\theta_l)\mathbf{a}_t^H(\phi_l), \tag{4}$$

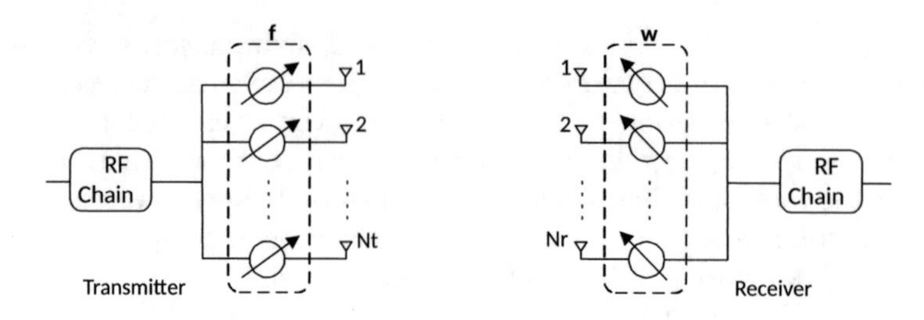

Fig. 1. Transmitter and receiver architecture.

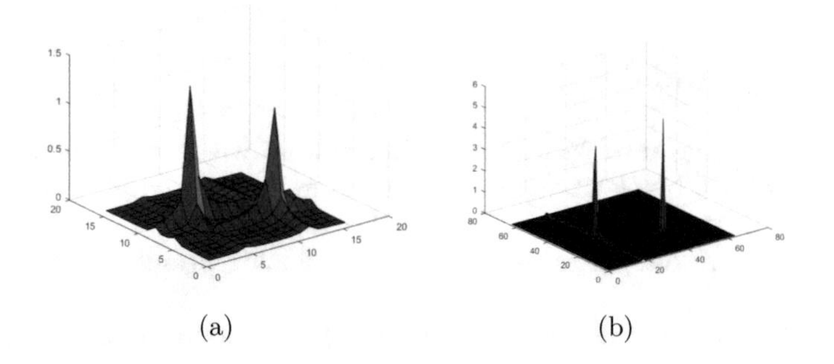

(a) (b)

Fig. 2. Illustration of virtual channel matrix for: (a) Sub-6 Ghz with Nr = Nt = 16, (b) Mmwave with Nr = Nt = 64.

where L is the number of paths, α_l represents the complex path gain of the l-th propagation path, $\theta_l \in [-\pi/2, \pi/2]$ and $\phi_l \in [-\pi/2, \pi/2]$ denote the AoA and AoD of the L-th path at transmitter and receiver, respectively. The vectors $\mathbf{a}_t(\cdot)$ and $\mathbf{a}_r(\cdot)$ denote the array response vectors for transmitting and receiving antenna arrays. The array response vector of $\mathbf{a}_r(\theta_l)$ is given by:

$$\mathbf{a}_r[\theta_l] = \frac{1}{\sqrt{N_r}}[1,\ e^{j\frac{2\pi}{\lambda}d\cos\theta_l},\ ...,\ e^{j(N_r-1)\frac{2\pi}{\lambda}d\cos\theta_l}]^T, \tag{5}$$

where λ is the transmission wave length and d is the antenna spacing. Furthermore the array response vector in (5) has a unit norm and the factor $\sqrt{N_t N_r}$ in (4) reflects this normalization. The array response vector $\mathbf{a}_t(\phi_l)$, can be written in a similar fashion.

3.2 Beam-Codebook Design

Different codebook models have been designed according to the channel property for instance Grassmannian codebooks, or to satisfy an hybrid architecture for example Multi-Resolution hierarchical codebook In this work due to full-analogous architecture we denote the codebooks $\mathcal{F}$ and $\mathcal{Q}$ at the transmitter and receiver respectively as:

$$\mathcal{F} = \{\mathbf{f}_1, \mathbf{f}_2, ..., \mathbf{f}_{N_t}\} \tag{6}$$

$$\mathcal{Q} = \{\mathbf{q}_1, \mathbf{q}_2, ..., \mathbf{q}_{N_r}\} \tag{7}$$

where every beamforming vector has the same form of the array response vector, such that:

$$\mathbf{f}_m = \mathbf{a}_t(\tilde{\phi}_m),\ m \in \{1, ..., N_t\} \tag{8}$$

$$\mathbf{q}_n = \mathbf{a}_r(\tilde{\theta}_n),\ n \in \{1, ..., N_r\} \tag{9}$$

The angles are chosen as in [14] such that every beam has same magnitude but different width, that is to say narrower at broadside direction and broader at endfire direction, if we set the inter-antenna spacing as $\lambda/2$, the positive and negative angles of departure are given by:

$$\tilde{\phi}_m^{(+)} = \arcsin(\frac{2m}{N_t}),\ m \in \{1, ..., \frac{N_t}{2}\} \tag{10}$$

$$\tilde{\phi}_m^{(-)} = \arcsin(\frac{2m}{N_t}),\ m \in \{-1, ..., \frac{-N_t}{2}\} \tag{11}$$

The AoAs design follow the same rule than AoDs.

4 Channel Estimation Method

4.1 Extracting Spatial Information

Here thanks to a matrix transformation we propose an easy methodology to obtain the spatial information from sub-6 Ghz channel. Considering a geometric channel model $\mathbf{H}_{6G}$ already estimated, whose structure doesn't provide clear information about AoAs and AoDs in the different paths, therefore a channel representation that provides a simpler geometric interpretation of the scattering environment is needed.

A virtual channel representation (VCR) of $\mathbf{H}_{6G}$ will provide spatial information uniformly spaced over the virtual angles, which are determined by the spatial resolution of the array. Thus, VCR characterizes the MIMO channel via beamforming in the direction of fixed virtual transmit and receive angles [14], that is:

$$\mathbf{H}_{6G} = \mathbf{U}_r \hat{\mathbf{H}}_{6G} \mathbf{U}_t^H \tag{12}$$

where $\mathbf{U}_r = [\mathbf{u}_r(\bar{\theta}_{-k}), ..., \mathbf{u}_r(\bar{\theta}_k)]$, $-N_r/2 \geq k \leq N_r/2$, is a matrix $N_r \times N_r$ which carries the receiver response vector in the virtual directions that satisfy $\bar{\theta}_k = \arcsin(\frac{2k}{N_r})$, likewise $\mathbf{U}_t = [\mathbf{u}_t(\bar{\phi}_{-i}), ..., \mathbf{u}_t(\bar{\phi}_i)]$, $-N_t/2 \geq i \leq N_t/2$ is a matrix $N_t \times N_t$ that carries the transmitter response vector in the virtual directions that satisfy $\bar{\phi}_t = \arcsin(\frac{2i}{N_t})$, consequently $\mathbf{U}_t$ and $\mathbf{U}_r$ are unitary discrete Fourier transform (DFT) matrices reflecting the fixed virtual receive and transmit angles, and $\hat{\mathbf{H}}_{6G} \in \mathbb{C}^{N_r \times N_t}$ is the virtual channel matrix and its entries reveals the desired channel parameters, virtual AoAs, AoDs and path gain [14]. In addition, typical sub-6 Ghz MIMO systems carry with lower number of antennas than mm-wave, this fact will lead to broader virtual angles in the low frequency channel compared with the high frequency channel, as shown in Fig. 2 several mm-wave virtual AoAs and AoDs overlap to those at virtual sub-6 Ghz, reducing the searching space to those overlapped mm-wave angles, then the candidate beam list provided by the spatial information from the sub-6 Ghz channel, is stored in the set $\mathcal{S}$. At this point the task is select the pair of virtual angles in the mm-wave virtual representation that provide higher gain in order in the receiver in order to estimate the mm-wave channel.

Considering the set $\mathcal{S}$ is carrying the bunch of possibles arrival and departure virtual angles $\bar{\theta}_p$ and $\bar{\phi}_r$ respectively, where $\bar{\theta}_p, p = 1, 2, ..., P; P < N_r$ and $\bar{\phi}_r, r = 1, 2, ..., N_t; R < N_t$ decreasing the searching space and training overhead to $P \times R$. Then using the set of arrival virtual angles to construct the combiner at the receiver i.e. $\mathbf{Q} = [\mathbf{q}_1(\bar{\theta}_1), \mathbf{q}_2(\bar{\theta}_2), ..., \mathbf{q}_P(\bar{\theta}_P)]$. The departure virtual angles are feedback to the transmitter in order to build its beamformer $\mathbf{F}$, where $\log_2 P$ bits are needed, that could be sent using the sub-6 Ghz channel. The new received signal in mm-wave systems is

$$\mathbf{Y} = \sqrt{\rho}\mathbf{Q}_{\mathcal{S}}^H \mathbf{H}\mathbf{F}_{\mathcal{S}} + \mathbf{V}, \tag{13}$$

After a vectorization step:

$$\mathbf{y} = \sqrt{\rho}(\mathbf{Q}_{\mathcal{S}}^T \otimes \mathbf{F}_{\mathcal{S}}^H)\mathbf{h} + \mathbf{v} \tag{14}$$

where $\mathbf{y} = vec(\mathbf{Y})$, $\mathbf{h} = vec(\mathbf{H})$, and $\mathbf{v} = vec(\mathbf{V})$. The largest absolute value entry in (15) determines the best beam-pair, that is to say $i_{opt} = \arg\max |[\mathbf{y}]_i|$, will match the best pair of AoA and AoD.

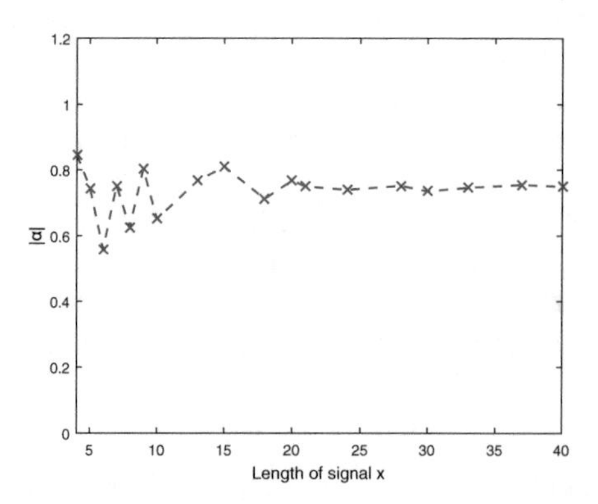

Fig. 3. Complex path gain estimation for a typical value.

4.2 Path Gain Estimation

At this point the AoAs and AoDs for different paths have been estimated by the $\mathbf{q}_{\hat{i}}$ and $\mathbf{f}_{\hat{i}}$ vectors, therefore the final step is to estimate the nonzero values entries of the channel, to simplify the analysis, we assume a single path channel, then re-writing Eq. (3) with a vector training symbol such that $\mathbf{x} \in \mathbb{C}^{1 \times s}$, the received signal is:

$$\mathbf{y} = \sqrt{\rho}\mathbf{q}_{\hat{i}}^H \mathbf{H}\mathbf{f}_{\hat{i}}\mathbf{x} + \mathbf{v}, \tag{15}$$

We apply the linear LS estimator to calculate the complex path value associated to the given AoA and AoD, given by

$$z = (\mathbf{x}\mathbf{x}^H)^{-1}\mathbf{x}\mathbf{y}^H \tag{16}$$

where $z = \sqrt{\rho}\mathbf{q}_{\hat{i}}^H\mathbf{H}\mathbf{f}_{\hat{i}}$, that carries the path gain estimation. About the vector training symbol length the Fig. 3 shows the variability of the path gain estimated α_l according to the signal length, we can observe there is a convergence with a length further than 20.

Additionally we focus on analyze the rate between the eNB and an UE in downlink transmission. Then the rate R is affected by changing the channel coherence time due to MS velocity, as

$$R_{Single} = \frac{T_o - T_\tau}{T_o}\log_2\left(1 + \frac{\rho\mathbf{q}_{\hat{i}}^H\mathbf{H}\mathbf{f}_{\hat{i}}\mathbf{f}_{\hat{i}}^H\mathbf{H}^H\mathbf{q}_{\hat{i}}}{\sigma_o^2}\right) \tag{17}$$

where the pre log-factor takes account the training overhead necessary to estimates the channel, here $T_o = \frac{\lambda B_o}{2v_o}$, λ, B_o, v_o, are mm-wave carrier wavelength, bandwidth and mobile station speed respectively, additionally T_τ is the number of $P \times R$ training blocks needed to estimate the channel.

Another metric choosed to evaluate the performance of the proposed method is the computational complexity compared with Fast Channel Estimation (FCE) method described in [16]. The complexity cost is given by $O(N_rN_tL + N_rL)$ counting the beam and gain selection, by the other hand the FCE computational complexity is $O(N_rT + LN_tT + LT)$.

5 Multicell Analysis

From now, considering an Orthogonal Frequency Division Multiple Access (OFDMA) multicell system as is shown in Fig. 4. We assume in every cell there is a small cell microwave station and millimeter-wave small station located in the same position. Also, we assume a microwave channel estimation have been already done in every small cell microwave station, such that a CE training-based method is applied in every millimeter-wave small station. We will analyze the performance of this mm-wave CE proposed method. Therefore, we focus on this frequency band. Additionally we assume no interference among the subchannels.

Let's consider the Fig. 4 where the UE1 is located in the eNB2 coverage area border, such that this user must deal with handoff management and interference from neighboring cells. In order to overcome these problems, we propose a method based on cooperative model using coalitional games between the concerned eNBs. The goals of this section is to increase the channel rate in the transmission data stage.

5.1 Coalitional Game: System Model

Assuming a mm-wave system, with N cells in the network and $\mathcal{N} = \{1, ..., N\}$ denoting the set of millimeter-wave small cells (MMWSCs) which are connected with each other via a wire backhaul, e.g., fiber, providing reliable links for traffic control. Each MMWSC $i \in \mathcal{N}$ works over the same set of channels, that is the available spectrum, i.e., is the frequency resource shared between all MMWSCs. Therefore the MMWSCs should share these bandwidth allocated in an opportunistic way in order to avoid handoff recurrent, besides diminishing the interference, leading to an improve in the channel rate.

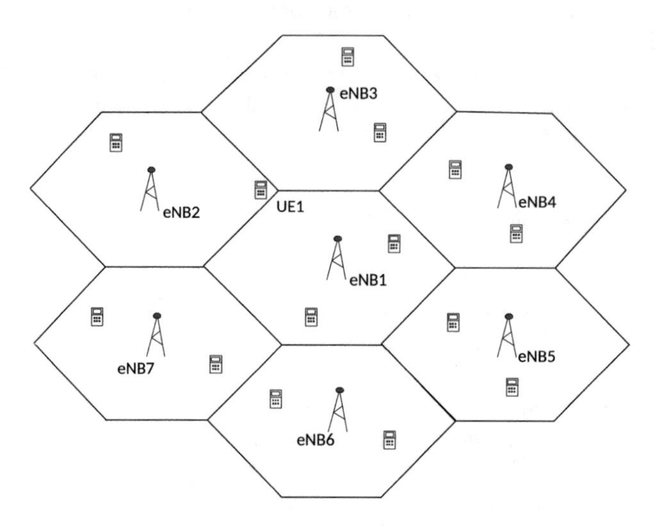

Fig. 4. Multicell network architecture

In the non-cooperative scenario, the access mode is frequency division duplexing (FDD) where each MMWC $i \in \mathcal{N}$ transmits over a set of subchannels Γ, which contains $|\Gamma| = M$ subchannels. The MMWSC i occupies the full time duration of all its $l \in \Gamma$ subchannels, under this non-cooperative scenario the UE1 (Fig. 4) uses the M subchannels available to transmit its data. In this case a UE located near the coverage area border may suffer an important degradation due to interference and coalitions in the subchannels from neigboring MMWSCs. Under this setup and without considering UE mobility, we can rewrite Eq. 15 as:

$$\mathbf{y}_{i,l} = \sqrt{\rho}\mathbf{q}_{i,l}^{H}\mathbf{H}_{i,l}\mathbf{f}_{\hat{i},l}\mathbf{x}_{i,l} + \mathbf{v} + \bar{\mathbf{I}}_{S}, \tag{18}$$

where $\bar{\mathbf{I}}_S = \sum_{j\in\mathcal{N},j\neq i}\sqrt{\rho}\ \mathbf{q}_{j,l}^{H}\mathbf{H}_{j,l}\mathbf{f}_{j,l}\mathbf{x}_{j,l}$, is the interference from neighboring cells during the mm-wave CE process in the eNB. Here is easy to notice how the interference affect the estimation process. By the other hand the rate in downlink transmission of MMWC $i \in \mathcal{N}$ to an UE is given by

$$R_{Multi} = \sum_{l\in\Gamma}\log_2(1 + \frac{\rho_{i,l}\mathbf{q}_{i,l}^{H}\mathbf{H}_{i,l}\mathbf{f}_{i,l}\mathbf{f}_{i,l}^{H}\mathbf{H}_{i,l}^{H}\mathbf{q}_{i,l}}{\sigma_o^2 + \mathbf{I}_S}) \tag{19}$$

where $\rho_{i,l}$ denotes the downlink transmit power by MMWSC i to a UE on subchannel l, $\mathbf{f}_{i,l}$ is the beamforming vector in the MMWSC i pointing to UE on subchannel l, $\mathbf{H}_{i,l}$ is the massive MIMO channel between the MMWSC i and the UE on the l-th subchannel, $\mathbf{q}_{i,l}$ is the combining vector in the UE pointing to MMWSC i on subchannel l and $\mathbf{I}_S$ denotes the power interference suffered by an UE from neighboring MMWSCs, as

$$\mathbf{I}_S = \sum_{j \in \mathcal{N}, j \neq i} \rho_{j,l} \mathbf{q}_{j,l}^H \mathbf{H}_{j,l} \mathbf{f}_{j,l} \mathbf{f}_{j,l}^H \mathbf{H}_{j,l}^H \mathbf{q}_{j,l}, \tag{20}$$

The term $\mathbf{I}_S$ can significantly reduce the rates achieved. Specifically, depending on the signal to noise and interference ratio (SNIR) feedback from the UE, the MMWCs can decide to form cooperative groups called coalitions, in order to overcome the interference between neighboring cells and the handoff management. Under this coalitional game approach, the MMWCs are modeled as players that access the spectrum, avoiding coalitions among them by jointly scheduling their transmissions, we state the following definition for a coalition:

Definition 1. *A coalition $S \subseteq \mathcal{N}$ is a non−empty subset of N in which players inside the set access the spectrum via a coordinated manner.*

Therefore we consider that if a coalition S is formed, the transmissions inside S are managed by a local scheduler as in [20], using Time Division multiple access (TDMA) mode, such that the subchannels are split in time slots allocated for every MMWSC. As a result no more than one MMWSC will access the every channel in each time slot, mitigating the interference inside the coalition S.

Although coordination can help to increase the channel rate by decreasing the interference, it also incurs in a coordination cost. Here we consider this coordination cost in terms of transmit power. Thus, the power spent by a MMWSC i to reach the farthest MMWSC $\hat{j}$ in a coalition S is $\rho_{\hat{j},i}$. Then the power cost needed to form a coalition S is:

$$\rho_S = \sum_{i \in S} \rho_{\hat{j},i} \tag{21}$$

In addition, we define a maximum tolerable power cost ρ_{lim} for every coalition S as in [21]. By considering the coalition cost in this way we take account the spatial distribution of the MMSCs and the coalition size.

5.2 MMWCs Cooperation as a Coalitional Game

The main goal is to deal with interference from neighboring MMWSC in the border coverage area by forming coalitions, i.e., using coalitional game theory [18], we denote as $\mathfrak{B}$ the set of all partitions $G_{\mathcal{N}}$ of $\mathcal{N}$, this problem can be modeled as coalitional game in partition form with transferable utility as [19]:

Definition 2. *A coalitional game in partition form with transferable utility (TU) is defined by the pair $(\mathcal{N}, v)$ where $\mathcal{N}$ is the set of players in the game, and a value function $v(S, G_{\mathcal{N}})$ assigning a real value to each coalition S.*

We also assume that $v(\emptyset) = 0$. Thus, the definition above imposes a dependence on the coalitional structure $\mathcal{N}$ when evaluating the value of $S \subseteq \mathcal{N}$, i.e. to the players $\mathcal{N} \backslash S$ as well. By the other hand the TU property implies that the total utility represented by a real number, (in our case the channel rate) can be divided in any manner between the coalition members.

Clearly, the cooperation model can be stated as a game in partition form where the MMWSC are the players, since the channel rate of a coalition is affected by the interference from others players, thus there is a dependence between the coalition and the players which do not belong to the coalition (coalitional structure). Therefore the utility achieved by the coalition S can be expressed in terms of the channel rate as:

$$\mathcal{U}(\mathcal{S}, G_{\mathcal{N}}) = \sum_{i \in \mathcal{S}} \sum_{l \in \Gamma} \alpha_i^l \ \log_2(1 + \frac{\rho_{i,l} \mathbf{q}_{i,l}^H \mathbf{H}_{i,l} \mathbf{f}_{i,l} \mathbf{f}_{i,l}^H \mathbf{H}_{i,l}^H \mathbf{q}_{i,l}}{\sigma_o^2 + \hat{\mathbf{I}}_S}), \tag{22}$$

where $\alpha_i^l \in [0, 1]$ denotes the fraction of time duration during which MMWSC i transmits on the subchannel l to the UE. In the non-cooperative scenario, i.e., FDD transmission mode, each transmission occupies a whole subchannel, hence $\alpha_i^l = 1$. In addition $\mathbf{I}_{\hat{S}}$ denotes the co-tier interference suffered by the UE served by MMWSC i on subchannel l from players $j \in \mathcal{N} \backslash S$ as follows:

$$\mathbf{I}_{\hat{S}} = \sum_{j \in G_{\mathcal{N}} \backslash S, \ j \neq i} \rho_{j,l} \mathbf{q}_{j,l}^H \mathbf{H}_{j,l} \mathbf{f}_{j,l} \mathbf{f}_{j,l}^H \mathbf{H}_{j,l}^H \mathbf{q}_{j,l}, \tag{23}$$

Therefore thanks to transmission scheduling the interference from players within the coalition S is suppressed, while inter-coalition interference still remain and leads to a game in partition form. Given the power cost and utility function for any coalition $S \in \mathcal{N}$, we can define the value of any coalition, i.e., the total benefit as:

$$v(\mathcal{S}, G_{\mathcal{N}}) = \begin{cases} |S| \ \mathcal{U}(\mathcal{S}, G_{\mathcal{N}}) \ if \ \rho_S \leq \rho_{lim} \\ 0 \qquad\qquad\qquad otherwise, \end{cases} \tag{24}$$

As the utility in (24) represents a sum rate, then the proposed coalitional game has a transferable utility, since the sum rate can be shared among the coalition members by dividing the frequency resource in any manner, while meeting a fairness criterion, consequently our aim is to maximize the sum rate while taking account the constraints in terms of transmit power.

We can define the payoff of a MMWSC $i \in S$ as:

$$x_i(\mathcal{S}, G_{\mathcal{N}}) = \frac{1}{|S|} \left(v(\mathcal{S}, \pi_{\mathcal{N}}) - \sum_{j \in \mathcal{S}} v(\{j\}, G_{\mathcal{N}}) \right) + v(\{i\}, G_{\mathcal{N}}) \tag{25}$$

5.3 Proposed Algorithm

For the stated coalitional game is important to notice that due to the power constraint the grand coalition seldom forms. Therefore cooperation will occurs when the interfering MMWSCs are closely located in a way that $\rho_S \leq \rho_{lim}$, thus coalition with many members is unlikely to happen, in this sense we are not focus on analyze the stability of the grand coalition. Finding an optimal coalitional structure for games in partition form have been studied in, [22] and [23], we will apply the concept of *recursive core* as in [23]. The concept of *recursive core* studies the behaviour of dynamics coalition formation but also considering the interference from neighboring MMWSCs. The detail about how the *recursive core* works is provided in [23].

In the algorithm proposed (Algorithm 1), at the first step an UE is sensing interference which will trigger the coalitional game. Secondly to resolve the coalitional game, i.e., achieving the recursive core, we propose three phases: environment sensing, coalition formation and scheduling transmission. First of all, the network is partitioned in $\mathcal{N}$ single coalitions, this is the non-cooperative case. Then by discovering neighbors stage, the MMWSCs can create a list of existing neighbors in the network, once each MMWSC has a neighboring list, they can start a recursive coalition formation to find a recursive core. Here every MMWSC establishes negotiations with the discovered neighbors, to identify potential partners for cooperation, this information is exchanged by using the wire reliable control channel. Then the cooperation cost for every coalition is calculates as in Eq. 21, and the potential payoff is computed as in Eq. 25 for every member of a coalition. To reach the recursive core, each MMWSC joins to the coalition which provides the highest revenue, i.e., payoff. Then, once the coalition is formed, coalition-level scheduling occurs in each coalition.

6 Numerical Results

In this section, simulations are carried out to evaluate the performance for the proposed channel estimation strategies. We assess the channel estimation error, +the effect of SNR and speed on the method's performance of the proposed methods, additionally the computation time is analyzed.+

The sub-6 Ghz channel works at $f_{6G} = 3$ GHz and mm-wave channel at $f_{mm} = 28$ GHz, bandwidth $B_o = 10$ MHz alike than mm-wave case, the distance between the UE and eNB is set at 50m, the path loss exponent at sub-6 Ghz is equal to 2 while for mm-wave has been set to 3. The angles of arrival and departure for both environments are limited at $[-\frac{\pi}{2}, \frac{\pi}{2})$, the antennas number at sub-6 Ghz and mm-wave will be changing according to every experiment at the UE and eNB, the inter antenna element distance at both cases is half-wavelength. Additionally the signal-to-noise ratio is set as $SNR = \frac{P_o}{\sigma_o^2}$.

To assess the estimation error we express the mm-wave virtual channel $\mathbf{H}_{virt}$ as sparse [16], such that $\mathbf{H}_{virt} = \mathbf{J}_{LN_r}^T \Lambda\ \mathbf{J}_{LN_t}$ where Λ is an $L \times L$ diagonal matrix and the L are the entries different than zero of $\mathbf{H}_{virt}$, the binary matrices $\mathbf{J}_{LN_r}^T, \mathbf{J}_{LN_t}$ are $L \times N_r$ and $L \times N_t$ selection matrices, generated by keeping L rows of $N_r \times N_r$ and $N_t \times N_r$ identity matrices respectively. Therefore we can compute the mean square error as MSE $= E\{\|\mathbf{H}_{virt} - \tilde{\mathbf{H}}_{virt}\|_F^2 / \|\mathbf{H}_{virt}\|_F^2\}$.

Algorithm 1. The Proposed MMWSC cooperation algorithm

Step 1: UE interference sensing

The UE sense the interference UE_{int}, once it overpass a threshold I_{thr}, the UE feedback the information to its attached eNB in order to initiate the cooperation process, thus:

if $UE_{int} \geq I_{thr}$ **then**

Step 2: Coalitional Game Starts

At the beginning when players are not cooperating $G_{\mathcal{N}} = \{1, ..., \mathcal{N}\} = \{S_1, ..., S_{\mathcal{N}}\}$.

Three stages in each round of the algorithm

Stage 1 - Discovering Neighbors:

- Each MMWSC discovers the neighboring coalitions.

Stage 2 - Recursive Coalition Formation:

repeat

- Each MMWSC establishes negotiations with discovered neighboring FAPs, in order to identify potential coalition partners.
- Each MMWSC create a list of the feasible coalitions which ensure $\rho_S \leq \rho_{lim}$
- The payoff for the feasible coalitions is computed and each MMWSC joins to the coalition which ensures the maximum payoff.
- The resulting coalition is included in the recursive core.

until convergence to a stable partition in the recursive core.

Stage 3 - Inner-coalition scheduling:

- The scheduling information is gathered by each MMWSC $i \in S$ from its coalitions members, and transmitted within the coalition S afterwards.

end if

Step 3: High Speed mmWave Communications

- And high data rate transmission starts.

In order to explore the performance of the proposed method, in the first simulation we set the Sub 6-Ghz channel with 16 transmitter antennas and 16 receiver antennas, in the other and we set the mm-wave channel with $N_r = N_t = 64$, $N_r = N_t = 32$ and $N_r = N_t = 16$ antennas, in the Fig. 5 we can see how the spectral efficiency change according to different SNR values for every antenna array, as is expected under the beam codebook design, increasing the number of antennas lead to an increasing in the beam-resolution.

For exhaustive-search this occurs due to the number of $N_r \times N_t$ blocks needed for channel estimation, here the training time occupies most of the channel coherence time leaving no time for data sending, by the other hand under this method thanks to prior spatial information obtained from sub-6 Ghz the number of beam candidates decrease, such that $P \ll N_r$ and $R \ll N_r$ therefore the blocks for training are highly reduced, leaving more time for data transmission.

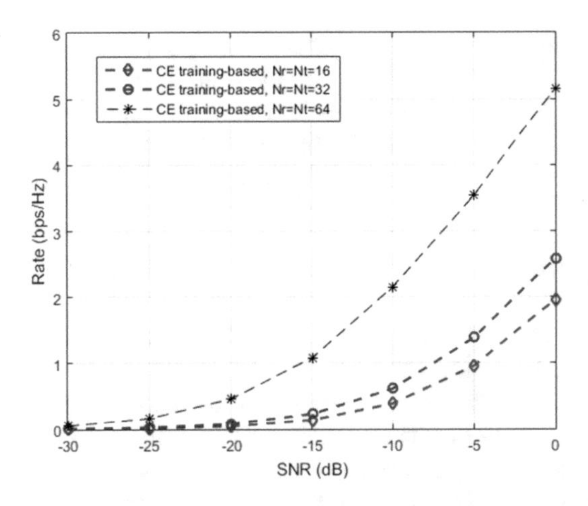

Fig. 5. Rate achieved by CE training-based over different SNR values.

The last experiment takes account the system complexity between the refined CE method and FCE detailed at [17] in term of their computation time, the complexity cost are $O(N_rT + LN_tT + LT)$ and $O(N_rN_tL + N_rL)$ respectively. All simulations are conducted at Matlab R2015a by the Intel Core i5 CPU, in Fig. 6(a) the N_r is set to 16, while N_t is increasing, then in Fig. 6(b) theN_t is set to 16 and N_t is changing. For the sake of fairness we comparative both method

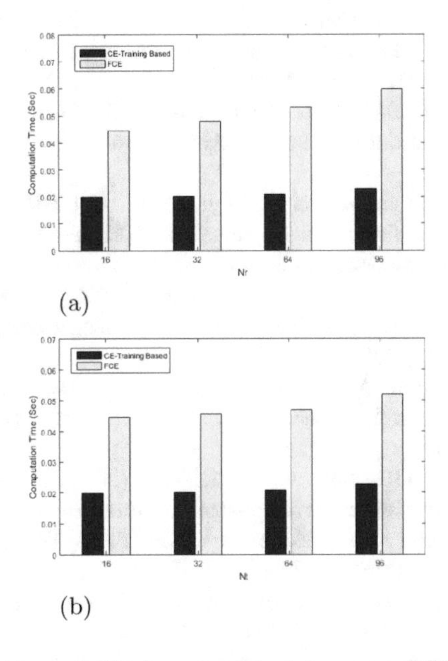

Fig. 6. Computation time of CE based on training and FCE, varying: (a) Nr and (b) Nt.

setting the parameters to obtain a higher accuracy, that is L=18 according to [16]. In the Fig. 6 we observe the computation time for both methods grows linearly when the number of antennas increase, although this proposed method run faster than FCE, this is mainly because the searching space is diminished thanks to prior information obtained from sub-6 Ghz channel.

7 Conclusions

In this work, we proposed a channel estimation method based on coalitional game for a multicell case that improves the throughput and its performance over UE in mobility cases. The prior based on an algorithm that improves intercell interference. We analyze how the coalitional game improves the throughput and its performance over user equipments (UEs) in mobility cases. The proposed algorithm allows sharing the bandwidth allocated to the UE in an opportunistic manner to avoid recurring handover, in addition to reducing interference, which leads to an improvement in the channel speed. In addition, authors expect that the proposed method can be applied to reduce complexity and improve efficiency in terms of probability of non-detection of the system for NOMA (non-orthogonal multiple access systems).

References

1. Wang J, Zhu H, Gomes NJ (2012) Distributed antenna systems for mobile communications in high speed trains. IEEE J Sel Area Commun 30(4):675–683
2. Hassibi B, Hochwald BM (2003) How much training is needed in multiple-antenna wireless links? IEEE Trans Inf Theory 49(4):951–963
3. Heath J, Robert W (2013) What is the role of MIMO in future cellular networks: massive? Presentation delivered at IEEE international conference on communications (ICC)
4. Maschietti F, Gesbert D, Kerret P, Wymeersch H (2013) Robust location-aided beam alignment in millimeter wave massive MIMO. In: GLOBECOM 2017-2017 IEEE global communications conference, pp 1–6
5. Hur S, Kim T, Love DJ, Krogmeier JV, Thomas TA, Ghosh A (2013) Millimeter wave beamforming for wireless backhaul and access in small cell networks. IEEE Trans Commun 61(10):4391–4403
6. Alkhateeb A, El Ayach O, Leus G, Robert RW (2014) Channel estimation and hybrid precoding for millimeter wave cellular systems. IEEE J Sel Top Signal Process 8(5):831–846
7. Alkhateeb A, Ayach OE, Leus G, Heath RW (2013) Hybrid precoding for millimeter wave cellular systems with partial channel knowledge. In: Proceedings of information theory and applications workshop (ITA), February 2013, p 15
8. Ayach OE, Rajagopal S, Surra SA, Pi Z, Heath RW (2013) Spatially sparse precoding in millimeter wave MIMO Systems. In: Proceedings of information theory and applications workshop (ITA), February 2013, p 15
9. Wang Y, Venugopal K, Andreas F, Heath RW Jr (2017) Analysis of urban millimeter wave microcellular networks. IEEE Trans Veh Technol 66(10):8964–8978

10. Va V, Heath RW Jr (2017) Basic relationship between channel coherence time and beamwidth in vehicular channels. IEEE Trans Veh Technol 66(10):8964–8978
11. Garcia N, Wymeersch H, Strom EG, Slock D (2016) Location-aided mm-wave channel estimation for vehicular communication. In: IEEE 17th international workshop on signal processing advances in wireless communications (SPAWC), July 2016, pp 1–5
12. Ali A, Gonzalez-Prelcic N, Heath RW Jr (2016) Estimating millimeter wave channels using out-of-band measurements. In: Proceedings of information theory and applications (ITA) workshop, February 2016, p 15
13. Ali A, Gonzalez-Prelcic N, Heath RW Jr (2018) Millimeter wave beam-selection using out-of-band spatial information. IEEE Trans Wirel Commun 17:1038–1052
14. Sayeed AM (2002) Deconstructing multiantenna fading channels. IEEE Trans Signal Process 50(10):2563–2597
15. Rappaport TS, Sun S, Mayzus R, Zhao H, Azar K, Wang GN, Wong JK, Schulz M, Samini F Gutierrez (2013) Millimeter wave mobile communications for 5G cellular: it will work! IEEE Access 1:335–349
16. Sayeed AM, Raghavan V (2016) A fast channel estimation approach for millimeter-wave massive MIMO systems. In: Proceedings of IEEE global conference on signal and information processing, Arlington, VA, USA
17. Va V, Choi J, Shimizu T, Bansal G, Heath RW Jr. Inverse multipath fingerprinting for millimeter wave V2I beam alignmen. IEEE Access
18. Palacios P, Castro A (2018) Cognitive radio simulator for mobile networks: design and implementation. i-manager's J Commun Eng Syst 7(2):1–9
19. Saad W, Han Z, Debbah M, Hjorungnes A, Basar T (2009) A distributed coalition formation framework for fair user cooperation in wireless networks. IEEE Trans Wirel Commun 8(9):4580–4593
20. Pantisano F, Ghaboosi K, Bennis M, Verdone R (2010) Interference avoidance via resource scheduling in TDD underlay femtocells. In: Proceedings of the IEEE PIMRC workshop on indoor and outdoor femto cells, Istambul, Turkey, September 2010
21. Zhang Z, Song L, Han Z, Saad W (2014) Coalitional games with overlapping coalitions for interference management in small cell networks. IEEE Trans Wirel Commun 13(5):2659–2668
22. Huang C-Y, Sjostrom T (2006) Implementation of the recursive core for partition function form games. J Math Econ 42:771–793
23. Pantisano F, Bennis M, Saad W, Verdone R, Latva-Aho M (2011) Coalition formation games for femtocell interference management: a recursive core approach. In: Proceedings of 2011 IEEE wireless communications and networking conference, Cancun, Quintana Roo, Mexico, March 2011, p 2831
24. Bogomonlaia A, Jackson M (2002) The stability of hedonic coalition structures. Games Econ Behav 38:201–230

Resource Allocation in WDM vs. Flex-Grid Networks: Use Case in CEDIA Optical Backbone Network

Rubén Rumipamba-Zambrano[1](✉), Luis Vargas[2], Claudio Chacón[2], Flavio Rodríguez[2], and Juan Pablo Carvallo[2]

[1] Corporación Nacional de Telecomunicaciones CNT E.P., Quito, Ecuador
ruben.rumipamba@outlook.com
[2] Corporación Ecuatoriana para el Desarrollo de la Investigación y la Academia (CEDIA), Gonzalo Cordero 2-122 y José Fajardo, 010203 Cuenca, Ecuador

Abstract. In this paper, we evaluate in a network planning scenario the performance of typical Wavelength Division Multiplexing (WDM) networks versus networks based on flexible spectrum grid (i.e., Flex-Grid networks). To this end, we propose several lightweight heuristics with *adaptive routing* strategies and a Simulated Annealing (SA)-based metaheuristic in order to minimize the spectrum occupation. Moreover, we use the CEDIA backbone network under realistic traffic scenarios. The results obtained with the most promising heuristic disclose that, for foreseeable traffic conditions, spectrum savings up to 42% can be obtained if elastic spectrum allocation is considered compared to rigid spectrum allocation. Furthermore, according to the results the current spectral resources of CEDIA network will be exhausted by mid-2023 using spectrum allocation based on Mixed-Line-Rate (MLR) WDM, while this estimation will extend until near 2025 if allocation based on Flex-Grid is considered.

Keywords: Network planning · WDM networks · Flex-Grid networks

1 Introduction

The emerging telecommunication services such as the advent of 5G/Internet of Things (IoT) require every day more bandwidth. These stringent bandwidth requirements have to be supported in both access and transport networks, leveraging on fiber-optics communications. Thus, the optical technology and protocols also have to evolve in order to increase network Grade of Service (GoS).

As for transport networks, on which this work will be focused, their enabling technologies have been changing over the last decades. Firstly, Synchronous Digital Hierarchy (SDH) was able to support up to 40 Gb/s (STM-256) by multiplexing different signals of level 1 (STM-1). Later on, Wavelength Division Multiplexing (WDM) technology enhanced the bandwidth capacity by multiplexing several wavelengths in a single fiber link. In this way, WDM allows high data rates of 2.5 G, 10 G, 40 G and recently 100 G by deploying a fixed spectral channel spacing (i.e., a fixed grid) of 50/100/200 GHz [1]. The optical connections, called *lightpaths*, can be

M. Botto-Tobar et al. (Eds.): TICEC 2018, AISC 884, pp. 18–33, 2019.
https://doi.org/10.1007/978-3-030-02828-2_2

established using Single Line Rate (SLR) or Mixed Line Rate (MLR) systems when traffic demands request very heterogeneous bit-rates [2]. More recently, Elastic Optical Networks (EONs) [3] suggest to increment the transmission capacity by using Flexible Grid (Flex-Grid, synonym term of EON). Here, the minimum spectral granularity to be allocated is typically 12.5 GHz (according to the ITU-T recommendation [4]), called a Frequency Slot (FS), and the incoming demands can use a flexible number of them. This allows offering different high transmission bit-rates in the order of hundreds of Gb/s or even Tb/s by forming super-channels (SChs) over multiple FSs. Different bit-rates can be obtained by adapting the modulation format to optical reach or by increasing the number of carriers per SCh. Lately, in order to continue increasing the transmission capacity, Space Division Multiplexing (SDM) have been presented as the next frontier of fiber optics [5]. This technology in combination with Flex-Grid, known as spectral and spatially flexible optical networking [6], is a promising network scenario to overcome the foreseeable *capacity crunch* of the current Single-Mode Fiber (SMF) systems.

One important aspect in optical networking is the resource allocation, in order to accommodate efficiently the connections carrying data traffic. In traditional WDM networks, the well-known Route and Wavelength Assignment (RWA) problem was widely studied [7]. In the case of Flex-Grid networks, different allocation schemes are applicable because there exist several degrees of flexibility such as modulation format, baud-rate, code-rate, lunch-power, guard-bands, etc. The simplest upgrade of RWA problem only taking into account the assignment of flexible spectrum ranges is addressed by Route and Spectrum Assignment (RSA) problem [8]. Currently, the study of RSA-related problems [9, 10] and enabling technologies related to devices as well as to control and management of bandwidth-variable network elements continue attracting the attention of research community [11, 12].

In general, the resource allocation is executed for either static (planning) or dynamic (operation) network scenarios. Network planning scenarios refer to the cases where traffic demands are known in advance, while operation cases when the connections arrive and departure to/from the network over the time in a random fashion. The planning during network operation is also possible and it is known as in-operation planning [13]. Network planning problems can be addressed by solving Integer Linear Programs (ILPs) or Mixed ILPs (MILPs). The advantage of this method is the accurate solution provided, but the computational cost is high. Heuristic approaches, in contrast, offer an alternative to mathematical programming, reducing the computational cost but they offer sub-optimum solutions. In this work, we focus on a network-planning study in order to evaluate future network scenarios under available resources of *"Corporación Ecuatoriana para el Desarrollo de la Investigación y la Academia"* (CEDIA) optical network. The main contribution of this work is to take a realistic scenario of CEDIA network (as a use case) to compare MLR WDM versus Flex-Grid performance under present and future traffic conditions. The rest of the paper is organized as follows. Section 2 presents the strategies for resource allocation in fixed-grid- and flexible-grid-based optical networks divided in two subsections. Subsection 2.1 explains the resource allocation problem, while subsection 2.2 presents different methods to solve it. Numerical results are presented in Sect. 3 in three subsections. Subsection 3.1 describes the network scenario and assumptions; the comparison of different proposed heuristics is presented in Subsect. 3.2, while Subsect. 3.3 evaluates the performance of MLR WDM and Flex-Grid in the CEDIA backbone network with different foreseeable future traffic conditions. Finally, Sect. 4 draws up the main conclusions of this work.

2 Strategies for Resource Allocation in Fixed-Grid- and Flexible-Grid-Based Optical Networks

As stated in the previous section, in order to set-up lightpaths at the transport level is necessary to assign them some resources. Essentially, one lightpath is defined by the physical path (route) plus the spectrum portion over which optical signals are transmitted. Initially, the spectrum portion was denoted as *wavelength* in traditional WDM networks arising the well-known RWA problem. Later on, when the concept of flexible-grid-based networks appeared, the term *wavelength* was generalized as Frequency Slot (FS). According to the ITU-T recommendation [4], the spectrum is discretized in FSs of 12.5 GHz width, while central frequencies of the laser sources can be tuned in multiples of 6.25 GHz. In the context of Flex-Grid networks, the assignment of other tunable parameters (e.g., via Software Defined Network (SDN)-based controller), such as modulation format (M), code-rate (C), baud-rate (B), transponder (T), launch power, etc., have been incorporated to the RSA problem [9, 14–16] as additional subproblems, thus giving rise to other terms such as RMSA, RCMSA, RSTA, etc.

In this section, we aim to describe the resource allocation problem jointly with optimal and suboptimal solutions, which will be applied, as use case, in CEDIA backbone network in next section.

2.1 Problem Statement

Considering that any spectrum portion can be expressed in terms of different FS widths (e.g., 12.5, 25, 37.5, 50 GHz, as assumed in [17]), let us consider a general RTSA problem for both WDM and Flex-Grid networks. In fact, in the practice from the allocation point of view, the unique singularity when allocating spectral resources over Flex-Grid networks is they allow accommodating contiguous FSs. This latter aspect is translated in additional constraint, namely, *spectrum contiguity,* which is going to be explained later in this subsection.

The targeted RTSA problem in WDM and Flex-Grid optical core networks can be formally stated as:

Given:

(1) A network represented as a directed graph $\mathcal{G}(\mathcal{N}, \mathcal{E})$, where $\mathcal{N}$ is the set of optical nodes and $\mathcal{E}$ the set of unidirectional fiber links.
(2) A spectral grid consisting of an ordered set of FSs, denoted as $\mathcal{S}$, available in every fiber link. FSs have a spectral width (in Hz) equal to W.
(3) A set of transponders, denoted as $\mathcal{T}$. For a given $t \in \mathcal{T}$, r_t represents its line-rate in b/s, n_t the number of FSs occupied and l_t the maximum optical reach in km.

(4) A set of offered unidirectional demands to the network, denoted as $\mathcal{D}$. Each $d \in \mathcal{D}$ has associated a source (s_d), a destination (t_d) node in $\mathcal{G}(\mathcal{N}, \mathcal{E})$ and a requested bit-rate (in b/s), denoted as r_d.
(5) We define as $\mathcal{P}_d$ the set of pre-computed candidate physical paths for demand $d \in \mathcal{D}$; l_p is the physical length (in km) and $\mathcal{P}$ is the set of all pre-computed candidate physical paths in the network ($\mathcal{P} = \cup_{d \in \mathcal{D}} \mathcal{P}_d$).

Objective: minimize the spectral requirements allocating the most suitable route, transponder and spectrum for every offered demand in $\mathcal{D}$, subject to the following *constraints*:

(1) *Spectrum continuity*: FSs supporting a lightpath must be the same in all links along the path from s_d to t_d. That is, no spectrum conversion and regeneration (transparent transmission) is considered.
(2) *Spectrum clashing*: a given FS in any fiber link can only be allocated to one lightpath.
(3) *Spectrum contiguity*: This constraint is applied for Flex-Grid networks and mandates that if one lightpath requires more than one FSs, they are assigned using a contiguous subset of FSs. In the case of WDM networks, the RTSA problem is relaxed by not requiring the *spectrum contiguity* constraint. In fact, in WDM networks the lightpaths are suitably fit in one FS (*wavelength*). However, if one lightpath requires more than one FS (e.g., when multiple optical carriers are necessary), the demand is split in several sublightpaths, each one with independent transmission. That is, different route, transponder and spectrum can be assigned to each sublightpath. This concept, in WDM networks is known as *inverse multiplexing* [3]. Meanwhile, for Flex-Grid networks, if multiple optical carriers are necessary they are jointly grouped in a form of SCh, which is routed as a single entity.

Regarding the objective function of the RTSA problem, our goal is to minimize the number of FSs used in any fiber link (hereinafter called as *spectral occupation* and represented as Φ). The number of FSs (n_t) used by one transponder given the guard-band (GB) between channels and the spectral efficiency of the used modulation format (ef_m) in (b/s $\cdot$ Hz) can be computed by Eq. 1.

$$n_t = \left\lceil \frac{r_t/ef_m + GB}{W} \right\rceil \quad (1)$$

In addition, when multiple (n_{oc}) adjacent optical carriers are necessary (i.e., SCh for Flex-Grid networks) the total number of FSs assigned to one demand (n_d) can be computed by Eq. 2, where n_{oc} is equal to r_d/r_t

$$n_d = \left\lceil \frac{(r_t/ef_m) \cdot n_{oc} + GB}{W} \right\rceil \quad (2)$$

The compact switching of SChs and the using of Nyquist-WDM (N-WDM) makes possible to suppress inter-carrier GBs [18], while inter-SCh GBs are always necessary in order properly extract optical signals by waveband filters. Moreover, note that the entire SCh has to be expressed in terms of integer number of FSs, but not necessarily its individual optical carriers (i.e., optical carriers that form a SCh could be accommodated in a grid-less form). Figure 1(a, b) illustrates the spectrum accommodation using fixed and flexible grid, respectively. Let us consider three demands d1, d2, d3 of r_d equal to 200 Gb/s, 40 Gb/s and 10 Gb/s, respectively. For the allocation purposes, we have MLRs with a set of transponders of $r_t \in \{10, 40, 100\}$ Gb/s, where 10 and 40 Gb/s optical signals are modulated with QPSK modulation format and 100 Gb/s with Dual-polarized (DP)-QPSK modulation format. Assuming the selected path for every demand is within the maximum reach, in case of WDM demand d1 would be served by two lightpaths/carriers (in blue, not necessarily contiguous) by using *inverse multiplexing*. That is, d1 would be served by two transponders of 100 Gb/s, while d2 and d3 only need one transponder/carrier (in red and green, respectively) with line-rate equal to r_d. As shown, although the optical signals corresponding to each demand occupies different spectrum ranges, the spectrum allocated per lightpath is fixed and equal to 50 GHz. As d1 has been split into two lightpaths, we have 4 lightpaths of 50 GHz (i.e., total occupied spectrum is equal to 200 GHz). As for the Flex-Grid network scenario, d1 does not need to be split, but rather a SCh can be formed with contiguous optical carriers of 100 Gb/s each. Meanwhile, for d2 and d3 demands, identical to WDM, it has been assigned them a 40 and 10 Gb/s transponders. However, the significant difference is about the spectrum allocated per demand. Unlike fixed-spaced channels, the optical channels occupy different spectrum ranges as they need, according to Eqs. 1 and 2 considering a 10 GHz GB [3]. Under flexible grid allocation, the total spectrum occupied for these 3 demands is 125 GHz. Therefore, in this basic example we foresee that Flex-Grid reduces the spectrum occupation in about 38%.

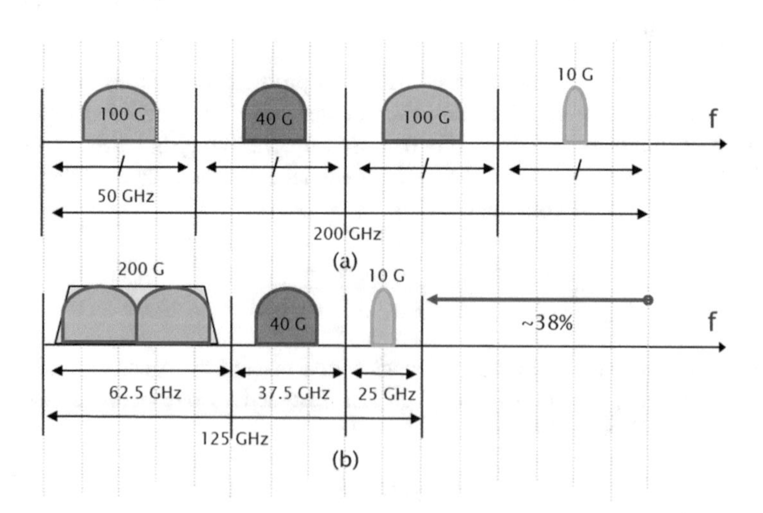

Fig. 1. Spectrum accommodation for (a) MLR WDM vs. (b) Flex-Grid networks.

2.2 Optimal and Suboptimal Solutions

Typical RWA and RSA problems in WDM and EON networks, respectively, aresolved by means of two strategies, namely, Integer Linear Programming (ILP)- and/or metaheuristic-based algorithms [19]. The former ones are mathematical formulations of the problem and provide optimal solutions. An optimal solution follows a maximization or minimization objective. The main drawback of this type of solutions is that time complexity can significantly grow with the size of the problem. In fact, several works in the literature have demonstrated that RWA and RSA problems are Non-deterministic Polynomial (NP)-hard (i.e., the problem cannot be optimally solved in a deterministic time by means of an algorithm) [8, 20]. Meanwhile, metaheuristics are a kind of stochastic algorithms, i.e., algorithms that provides non-deterministic solutions. Generally, as they take as input several candidate parameters, the combination of them results in different possible solutions. The size of one problem determines the feasibility of exploring all possible combinations yielding valid solutions. Indeed, for large-problem instances this is practically impossible in a deterministic time. So, suboptimal solutions are returned, e.g., after reaching a certain number of iterations and/or execution time. The advantage of these algorithms is that they can address complex problems with good accuracy if they are carefully designed.

There are different types of metaheuristics, e.g., the interested reader can review a good book about network planning and optimization, where is summarized several metaheuristics [19]. The idea consists of exploring the solution space finding each time better solutions. If only one region is explored, myopic solutions are provided and these type of heuristics are known as Local Search (LS). Instead, if this LS procedure is repeated several times in different regions of the solution space, the quality of the solutions can be improved. Examples of LS-based metaheuristics are Iterated LS (ILS), Tabu Search (TS), Simulated Annealing (SA) and Greedy Randomized Adaptive Search Procedure (GRASP). Moreover, SA metaheuristic and others like Ant Colony Optimization (ACO) and evolutionary/genetic algorithms are inspired in natural processes.

In our case, we propose a SA-based metaheuristic (Pseudo-code 1) using several greedy heuristics that provides a valid solution (x) to RTSA problem and returns the evaluation of the objective function [i.e., $f(x) = \Phi$], lines 4, 13 and 23. SA metaheuristic has a good trade-off between accuracy and execution time, as demonstrated in [21]. This is inspired in the annealing processes in metallurgy to produce crystals [22]. The input parameters are the network graph $\mathcal{G}(\mathcal{N}, \mathcal{E})$, the set of demands $\mathcal{D}$, the set of candidate paths $\mathcal{P}$, the set of transponders $\mathcal{T}$ and the maximum number of iterations (*maxIter*). In principle, any order of demands in $\mathcal{D}$ is a valid solution (ω) of the metaheuristic since all possible permutations of its elements yields a valid RTSA solution x. However, the challenge of the metaheuristic is to find the best order (not necessarily the global optimum) of elements in $\mathcal{D}$ which improves the objective function Φ. In this manner, the SA metaheuristic starts (lines 2, 3) with an initial solution (ω_o) sorting the demands in $\mathcal{D}$ in descending order, according to their required number of FSs using shortest path (SP) and continues permuting following certain criterions. Thus, the solutions of SA metaheuristic evolve in the neighborhood (ω') of the current solution (ω) not only by accepting improving solutions (like, *hill climbing* movements), but also worse solutions (*uphill* movements) to provide diversification (jump to unexplored regions) within the solution space. A neighboring solution (ω'), as intensification strategy, is defined as a swap movement (line 11) between two randomly chosen demands of the current order (ω) in $\mathcal{D}$. The annealing process starts with an initial temperature $T(0)$ and continues decreasing in each iteration (line 20) with a cooling rate $\alpha \in (0, 1)$. The temperature affects the acceptance probability (P_{acc}) of non-improving solutions. In fact, the P_{acc} depends on the objective function worsening ($\Omega = \Phi_{\omega'} - \Phi_{\omega}$) and the temperature, namely, $P_{acc} = e^{-\Omega/T}$. For example, if we initially decide to accept with probability $P_{acc}(0) = 0.3$ a solution yielding an objective function worsening $\Omega(0)$ in one FS. Then, $e^{-1/T(0)} = 0.3$, and $T(0) = -1/\ln(0.3)$. After the evaluation of each neighboring solution (ω'), if ω' improves the resulting Φ value of the incumbent (best feasible) solution (ω^*), then this is updated ($\omega^* = \omega'$, line 17). This process ends when either *maxIter* parameter or freezing state (e.g., temperature equal to 0) is reached.

Pseudo-code 1: SA metaheuristic

Input: $\mathcal{G}, \mathcal{D}, \mathcal{P}, \mathcal{T}, \Omega(0), P_{acc}(0), maxIter$
Output: Φ_{ω^*}
1: $\omega_o \leftarrow \mathcal{D}$ sorted in descending order by their required n_d over SP
2: $\omega = \omega_0$ **comment:** Current solution
3: $\omega^* = \omega_0$ **comment:** Incumbent solution
4: $(\mathcal{A}, \Phi_\omega) \leftarrow$ RTSA heuristic $(\mathcal{G}, \omega, \mathcal{P}, \mathcal{T})$
5: $\Phi_{\omega^*} = \Phi_\omega$ **comment:** Obj. function of the incumbent solution
6: $T(0) = -\frac{\Omega(0)}{\ln[P_{acc}(0)]}$, $T = T(0)$ **comment:** Initial temperature
7: *iter* = 0
8: **while** *iter*<*maxIter* and *T*>0 **do**
9: $d_x \leftarrow$ Select one demand randomly from $\mathcal{D}$
10: $d_y \leftarrow$ Select one demand randomly from $\mathcal{D}$ different than d_x
11: $\omega' \leftarrow$ **swap** (ω, d_x, d_y)
12: Release all established lightpaths $\mathcal{A}$ in $\mathcal{G}$
13: $(\mathcal{A}, \Phi_{\omega'}) \leftarrow$ RSTA heuristic $(\mathcal{G}, \omega', \mathcal{P}, \mathcal{T})$
14: $\Omega = \Phi_\omega - \Phi_{\omega'}$
15: $prob \leftarrow$ **random** [0,1) **comment:** Random probability
16: **if** $\Omega < 0$ or $prob < e^{-(\Omega/T)}$ **then**
17: $\omega = \omega'$ **comment:** Jump to neighboring solution
18: **if** $\Phi_{\omega'} < \Phi_{\omega^*}$ **then**
19: $\Phi_{\omega^*} = \Phi_{\omega'}$ **comment:** Update incumbent
20: $T = \alpha \cdot T$ **comment:** Decrease temperature
21: *iter* = *iter*+1
22: Release all established lightpaths $\mathcal{A}$ in $\mathcal{G}$
23: $(\mathcal{A}, \Phi_{\omega^*}) \leftarrow$ RTSA heuristic $(\mathcal{G}, \omega^*, \mathcal{P}, \mathcal{T})$
comment: Allocate demands with the incumbent solution
24: return Φ_{ω^*}
25: **End.**

As for the greedy heuristics to solve RTSA problem mentioned in lines 4, 13 and 23 of the previous pseudo-code 1, we propose several strategies adapted from others available in the literature. For both WDM and Flex-Grid networks, we propose two heuristics inspired in the Lowest Indexed Spectrum Allocation (LISA) [20] and Balanced Load Spectrum Allocation (BLSA). Specifically, for WDM networks we consider a Bin-Packing-based Spectrum Allocation (BPSA) heuristic [23], while for Flex-Grid networks the Maximum Reuse Spectrum Allocation (MRSA) heuristic [24]. These heuristics are summarized in pseudo-codes 2, 3 and 4. All heuristics are based on which is known as *adaptive routing*, because the route assignment is based on network state instead of predefined routes as in *fixed-alternate routing* strategies. In particular, as RTSA LISA and BLSA only differ in the routing path selection, we present both heuristics in a single pseudo-code 2, while BPSA and MRSA heuristics are presented in pseudo-code 3 and 4, respectively.

According to pseudo-code 2, for each demand $d \in \mathcal{D}$ we select from $\mathcal{P}_d$ (e.g., $k = 3$ SPs) the Lowest Indexed Starting FS (LISFS) or the Least Congested (LC) path p_j for BLSA and LISA, respectively (line 3). The rest of the code performs the selection of the transponder according to r_d and transmission reach which has to be longer or equal than l_{p_j}. If $r_t < r_d$, demand has to be split in n_{OC} chunks in case of WDM by means of *inverse multiplexing*, while it has to be formed a SCh in case of Flex-Grid (line 5). As a result of the analysis of lines 4 and 5, the required number of FSs (n_t or n_d) are computed in line 6. Then, First-Fit (FF) spectrum assignment is realized (line 7). Finally, the active lightpaths and the objective value Φ is returned (line 9).

Pseudo-code 2: RTSA LISA/BLSA

```
    Input: G, D, P, T
    Output: A, Φ
1:  A ← ∅ comment: Set of established lightpaths
2:  for each d in D do
3:      Select path p_j from P_d according to heuristic criterion (LISFS or LC)
4:      Select transponder t ∈ T according to r_d with reach l_t >= l_pj
5:      Determine if inverse multiplexing or SCh is necessary
6:      Compute n_d, if needed for SCh allocation
7:      (a, Φ) ← accommodate (G, d, p_j, [n_t|n_d])
8:      A ← A ∪ {a}
9:  return (A, Φ)
10: End.
```

Pseudo-code 3 shows the RTSA BPSA algorithm for MLR WDM case. The algorithm is based on the selection of Disjoint Shortest Paths (DSPs) over several copies of the original network graph $\mathcal{G}$. Each copy of the original graph is added to a set called BINS (line 3). For each pending demand (outer loop, line 4) to be served is tested if there exists a SP (e.g., computed by means of the Dijkstra algorithm) in each copy of the original graph in BINS (inner loop, line 6). If so, the transponder and spectrum allocation is realized according to the steps from 4 to 8 of the previous pseudo-code 2. In addition, in order to ensure the allocation over only DSPs, the SP p_i found in the current copy of the graph $\mathcal{G}_i$ is deleted from it. If a feasible SP is not found among all network graphs in BINS, a new copy of the original network graph $\mathcal{G}$ is added to BINS (line 13). Again, the transponder and spectrum allocation is realized over the SP of demand in $\mathcal{G}$ by executing steps from 4 to 8 of the previous pseudo-code 2 (line 16).

Pseudo-code 3: RTSA BPSA

```
    Input: 𝒢, 𝒟, 𝒫, 𝒯
    Output: 𝒜, Φ
1:  𝒜 ← ∅ comment: Set of established lightpaths
2:  𝒢_1 ← 𝒢 comment: Copy of the original graph
3:  BINS = {𝒢_1} comment: Set of copies of the original graph
4:  while any pending demand in 𝒟 do
5:      p_j = ∅
6:      for each 𝒢_i in BINS do
7:          if there exist a SP p_i between s_d and t_d in 𝒢_i then
8:              p_j = p_i
9:              Execute steps from 4 to 8 of pseudo-code 2
10:             Delete p_i from 𝒢_i
11:             break
12:     if p_j = ∅ then
13:         𝒢_{|BINS|+1} ← 𝒢 comment: Create a new copy of the original graph
14:         BINS ← BINS ∪ {𝒢_{|BINS|+1}}
15:         Find a SP between s_d and t_d in 𝒢_{|BINS|+1}
16:         Execute steps from 4 to 8 of pseudo-code 2
17:     end for
18: end while
19: return (𝒜, Φ)
20: End.
```

Finally, RTSA MRSA algorithm suitable for elastic spectrum allocation is shown in pseudo-code 4. This algorithm has a similarity with the previous one RTSA BPSA, in view of that it is also based on DSPs. The fact is that, for Flex-Grid networks, a set of contiguous FSs (not only one such as in WDM networks) can be assigned to lightpaths. Consequently, non-uniform allocation of spectrum ranges makes impractical to assign one copy of the graph (bin) per each. Instead, only one copy of the original graph $\mathcal{G}$ per iteration is used for path selection. According to the pseudo-code 4, two iterative processes are followed to allocate lightpaths over the DSP in the first available and consecutive FSs. In each iteration j of the outer loop (line 2), the first pending demand in $\mathcal{D}$ is served over the shortest path p_j on the original network graph $\mathcal{G}$. After that, in the inner loop (line 7) each pending demand $d \in \mathcal{D}$ is served over the shortest path p_i, which is also disjoint with all already established routing paths $\mathcal{R}$ (line 8) in the current outer loop iteration j. The demands, for which finding a DSP or available spectrum portion in the current iteration j is impossible, have the opportunity to be accommodated in subsequent iterations. As in pseudo-code 2, transponder and spectrum assignment is realized following its steps from 4 to 8.

Pseudo-code 4: RTSA MRSA

Input: $\mathcal{G}, \mathcal{D}, \mathcal{D}_{\text{MiMo\_Candidates}}, \mathcal{P}, \eta, V$
Output: $\mathcal{A}, \Phi$
1: $\mathcal{A} \leftarrow \emptyset$ **comment:** Set of established lightpaths
2: **while** any pending demand in $\mathcal{D}$ **do**
3: $\mathcal{R} \leftarrow \emptyset$
4: Find SP p_j between s_d and t_d in $\mathcal{G}$
5: Execute steps from 4 to 8 of pseudo-code 2
6: $\mathcal{R} \leftarrow \mathcal{R} \cup \{p_j\}$
7: **for each** pending demand $d \in \mathcal{D}$ **do**
8: Find SP p_i and disjoint with all paths in $\mathcal{R}$ between s_d and t_d in $\mathcal{G}$
9: Execute steps from 4 to 8 of pseudo-code 2
10: $\mathcal{R} \leftarrow \mathcal{R} \cup \{p_i\}$
11: return $(\mathcal{A}, \Phi)$
12: **End.**

3 Numerical Results

In this section, we evaluate different allocation strategies by means of proposed heuristics and their subsequent improvement by running the SA metaheuristic in a specific network topology and traffic load conditions. To this end, firstly, we detail the network scenario and assumptions. After that, we concentrate on the evaluation of different proposed heuristics against SA metaheuristic. Finally, we extrapolate the traffic matrix to foreseeable future load conditions of a specific backbone network in order to compare the performance of MLR WDM versus Flex-Grid networks.

3.1 Network Scenario and Assumptions

This work has been carried out in cooperation with a National Research and Education Network (NREN) in Ecuador, called *"Corporación Ecuatoriana para el Desarrollo de la Investigación y la Academia"* (CEDIA). CEDIA promotes the development of the science and innovative projects between several academic and private Ecuadorian institutions. As stated in Sect. 1, the goal of this collaboration is to evaluate future network scenarios under available resources of CEDIA optical network. For this purpose, we consider the topology of Fig. 2 with its main characteristics shown in Table 1. The network connectivity parameter of column 4 is obtained according to the reference [25], using the concept of *natural connectivity.*

We assume a traffic matrix corresponding to services to be deployed in 2020. This is composed of 160 IP traffic demands with an overall offered traffic of $\sim$2.1 Tb/s. The average bit-rate per IP demand is $\sim$13 Gb/s. The IP demands are transmitted over a MLR WDM or Flex-Grid underlying transport networks. The heterogeneity of traffic demands in terms of bit-rate justify using MLR because the higher the line-rates, the lower the cost per bit [2] and the lower the number of sublightpaths. For this purpose,

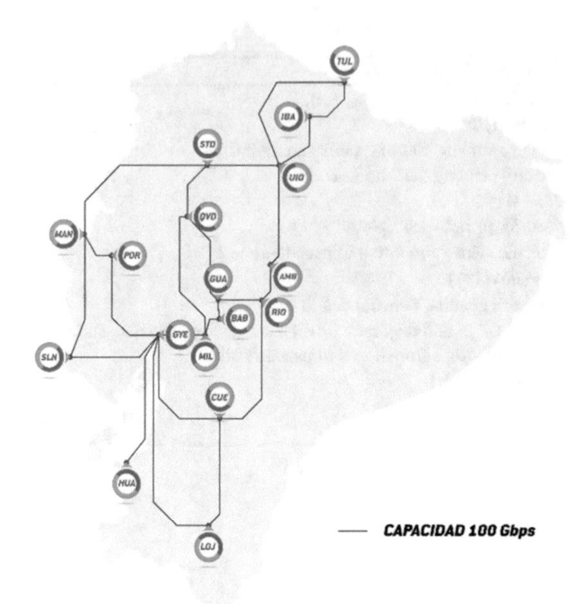

Fig. 2. Reference CEDIA network topology

Table 1. Main characteristics of CEDIA topology

Network ($\|\mathcal{N}\|, \|\mathcal{E}\|$)	Avg. Link Length [km]	Network diameter [km]	Nodal degree Min/Avg/Max	Network connectivity
CEDIA (16, 46)	175	1,160	2/2.81/4	4.41

we assume a set of transponders $\mathcal{T}$ with $r_t \in \{10, 40, 100\}$ Gb/s. The same set of transponders are also considered for Flex-Grid scenario in order to provide a fair comparison and not increase the network cost. In addition, the optical signals are modulated with QPSK and DP-QPSK modulation formats for 10, 40 and 100 Gb/s line-rates, respectively. The maximum optical reach of these transponders are 1000, 2000, 2500 km according to reference [17]. For MLR WDM case, IP demands can be served by different sublightpaths, where the sum of the corresponding line-rates of assigned transponders has to be higher or equal to the requested bit-rate (r_d). In the case of Flex-Grid, each IP demand is served by a single lightpath. If r_d is higher than the line-rate of the assigned transponder, then adjacent optical carriers using N-WDM are accommodated in form of SCh. Finally, according to the current CEDIA infrastructure, the available spectrum capacity per fiber link is 25 channels of 50 GHz, i.e., 1.25 THz. For MLR WDM, we consider FSs of 50 GHz, while for Flex-Grid the minimum unit is a channel of 12.5 GHz width. In order to compute the required number of FSs per (sub)lightpath Eqs. 1 and 2 are considered taking into account a 10 GHz GB [3]. Note that for MLR WDM the spectrum occupation per carrier, regardless the line-rate of the assigned transponder, is equal to 1 FS of 50 GHz.

3.2 Heuristics vs. SA Metaheuristic Evaluation

In this subsection, we present the evaluation of different proposed heuristics LISA, BLSA, MRSA, BPSA and the objective function improvement by means of SA metaheuristic. For this purpose, we use the previous introduced CEDIA topology and traffic matrix. The results are summarized in Tables 2 and 3 for WDM and Flex-Grid networks, respectively. The objective function for the initial solution (Φ_{ω_0}), i.e., without running SA metaheuristic, is shown in column 2, while the objective function for the incumbent solution after running SA metaheuristic (Φ_{ω^*}) is shown in column 4 of each table. Furthermore, the execution time for each type of solution is shown in column 3 and 5, respectively.

Table 2. Objective function of heuristics vs. SA metaheuristic for MLR WDM

Heuristic	$\mathbf{\Phi}_{\omega_0}$	Exec. Time (s)	$\mathbf{\Phi}_{\omega^*}$	Exec. Time (s)	% $\mathbf{\Phi}$ improvement
LISA	10	0.022	10	1.397	0%
BLSA	11	0.035	10	4.807	9%
BPSA	11	0.051	10	7.382	9%

Table 3. Objective function of heuristics vs. SA metaheuristic for Flex-Grid

Heuristic	$\mathbf{\Phi}_{\omega_0}$	Exec. Time (s)	$\mathbf{\Phi}_{\omega^*}$	Exec. Time (s)	% $\mathbf{\Phi}$ improvement
LISA	56	0.027	27	3.448	51%
BLSA	32	0.041	27	11.952	13%
MRSA	36	0.057	27	15.098	23%

It is worth mentioning that the SA metaheuristic has several parameters to be tuned. For example, the initial temperature, in turn defined by the initial acceptance probability [$P_{acc}(0)$] and the initial objective worsening value [$\Omega(0)$], the cooling rate (α), the number of maximum iterations (*maxIter*) and the group size to do the swap (perturbation) process. After executing several tests, we have probed that the latter parameter has to be proportional to the size of traffic matrix in order to shorten the number of iterations (therefore, the execution time) needed to find an incumbent solution. Thus, according to the size of the traffic matrix we have found that best group size to do the swap operation is equal to one. Meanwhile, *maxIter* parameter has been set to 5000 and 10000 for MLR WDM and Flex-Grid scenario, respectively, because higher values do not improve the objective function. Moreover, we have also found that the performance of SA metaheuristic is not sensible for the rest of the parameters due to the traffic load conditions and the available spectrum capacity explained in the previous subsection. Conversely, the higher the spectrum capacity, the higher the sensibility to mentioned parameters, as demonstrated in [21]. In fact, for these experiments, if we set the temperature to a very low value (i.e., equivalent to avoid accepting non-improving solutions) the results are the same to the ones shown in Table 2. In that case, the metaheuristic would be equivalent to a pure *hill climbing* one.

The results show that all heuristics converges to the same objective value after running the SA metaheuristic for MLR WDM as well as for Flex-Grid network scenario. Note that the results of Table 2 are expressed in number of FSs of 50 GHz width, while the ones of Table 3 in terms of 12.5 GHz FSs. Therefore, the spectrum occupation (in the highest congested link) to allocate all traffic demands over MLR WDM is 500 GHz (10 $\times$ 50) against 337.5 GHz (27 $\times$ 12.5). That is, Flex-Grid optimizes the spectrum usage by $\sim$33%. It is worth mentioning that BPSA and MRSA heuristics explore DSPs while demands are set-up, what means that $k > 3$ SPs (conversely to what happens in LISA and BLSA heuristics) can be considered as long as the maximum reach is not surpassed. This latter aspect allows improving the objective function in topologies with a good connectivity and a moderate network diameter, as demonstrated in [21]. However, in the case of CEDIA topology due to low connectivity, the results of BPSA/MRSA are equal to the ones of LISA and BLSA heuristics.

Moreover, looking at the results, we can observe that execution time of all heuristics is very low (lower than 1 s.); however, among them LISA heuristic has the lowest execution time, which can yield more benefits when a metaheuristic is used. For instance, after 5000 or 10000 iterations of SA metaheuristic, the execution time of BLSA and BPSA/MRSA heuristics are $\sim$3.5$\times$ and $\sim$5$\times$ of LISA, respectively. That is, since BPSA/MRSA heuristics explore DSPs in each iteration the execution times can be significantly higher than the lightweight LISA heuristic. Finally, the objective function improvement is more significant (up to 51%) for SA metaheuristic using LISA heuristic in Flex-Grid scenario. That said, by means of metaheuristics a poor initial solution of one heuristic can reach equal or better solutions than others.

3.3 Performance Evaluation of MLR WDM vs. Flex-Grid Networks in CEDIA Backbone Network

In this subsection, we aim to compare the performance of MLR WDM versus Flex-Grid network in CEDIA network scaling the traffic matrix in order to forecast when the available spectral resources will be exhausted. To this end, we assume different scaling factors (f), taking as a reference, values available in the literature. In [5] is stated that traffic of transport networks is increased by 10$\times$ every 4 years (hereinafter only called *10$\times$ scaling factor*). In addition to this estimation, we also consider a conservative 5$\times$ increasing every 4 years (hereinafter only called *5$\times$ scaling factor*). Then, the scaling factor f per year can be computed according to the function $f = a^{x/4}$, where a takes the value 5 or 10 and $x \in \mathbb{Z}^+$; $x = 0$ corresponds to year 2020, 1 to 2021, and so on. For this set of simulations, we have chosen the LISA heuristic since it provides the lowest execution times with good accuracy after running the SA metaheuristic, as demonstrated in the previous subsection. Figure 3 shows 7-year forecast spectrum occupation (in GHz) of the highest congested link (i.e., objective function Φ of the incumbent solution ω^*) since 2020 for MLR WDM as well as for Flex-Grid network scenario. The upper limit of the vertical axis corresponds to available spectrum capacity per link (1.2 THz). The curves in blue represents the spectrum occupation in a MLR WDM network scenario, while the curves in red correspond to the spectrum occupation of Flex-Grid

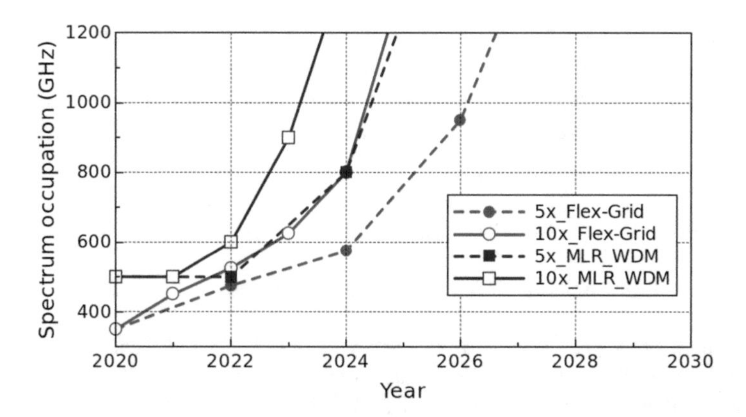

Fig. 3. 7-year forecast spectrum occupation

network scenarios. Solid lines are for 10× scaling factor, while dotted ones are for 5× scaling factor.

According to the results, the exponential increment of spectral occupation reveals that in the worst case, the current available spectral resources of CEDIA optical network will be exhausted by mid-2023 considering a 50 GHz fixed grid and 10× scaling factor. Meanwhile, if a more conservative traffic increment is considered (5× scaling factor) the foreseeable year to meet the upper limit of the available spectral resources could be extended until near 2025. Regarding the Flex-Grid network scenario with 12.5 GHz FSs, for 10× scaling factor the foreseeable spectral exhaustion would occur near 2025, while near 2027 for 5× scaling factor.

Comparing WDM vs. Flex-Grid network performance, we can observe that the spectral occupation of Flex-Grid network scenario can be reduced by ~42% (1200 vs. 700 GHz) with respect to spectral occupation of MLR WDM one for both 10× and 5× scaling factors. For a target 0.75 spectral utilization (900/1200 GHz), by computing the traffic-scaling factors *f*, we can determine that, for this working area and experiment, the Flex-Grid scenario is able to support up to two times more traffic than the MLR WDM scenario.

4 Conclusions

In this work, we have assessed the performance of resource allocation based on fixed versus flexible grid. Due to the heterogeneity of CEDIA traffic matrix, we use transponders with different bit-rate, i.e., MLRs. For the evaluation, we propose several lightweight heuristic based on different *adaptive routing* strategies and a SA-based metaheuristic to improve the objective function, namely, spectrum occupation. Among different heuristics, the one called LISA provides the best execution times with good accuracy. According to the obtained results, the spectrum savings of Flex-Grid scenario can be up to 42% against WDM scenario. This allows extending the foreseeable exhaustion of current available spectral resources of CEDIA optical network until 2025 rather than mid-2023 for WDM case considering a 10× scaling factor every 4 years, while in 2027 vs. 2025, respectively, if a more conservative 5× scaling factor is considered. Future works can be focused on both physical and virtual topology design considering infrastructure conditions of CEDIA network operator.

References

1. Ellbrock G, Tiejun JX (2010) The road to 100G deployment. IEEE Commun Mag 48(3): S14–S18
2. Christodoulopoulos K, Manousakis K, Varvarigos EM (2011) Reach adapting algorithms for mixed line rate WDM transport networks. J Light Technol 29(21):3350–3363
3. Gerstel O, Jinno M, Lord A, Ben Yoo SJ (2012) Elastic optical networking: a new dawn for the optical layer? IEEE Commun Mag 50(2):12–20
4. International Telecommunication Union - ITU-T (2012) G.694.1 (02/2012), Spectral grids for WDM applications: DWDM frequency grid, Ser. G.694.1, pp 1–16
5. Richardson DJ, Fini JM, Nelson LE (2013) Space division multiplexing in optical fibres. Nat Photonics 7:354–362
6. Klonidis D, Cugini F, Gerstel O, Jinno M, Lopez V, Palkopoulou E, Sekiya M, Siracusa D, Thouénon G, Betoule C (2015) Spectrally and spatially flexible optical network planning and operations. IEEE Commun Mag 53(2):69–78
7. Rouskas GN (2003) Routing and wavelength assignment in optical WDM networks. In: Wiley encyclopedia of telecommunications, Hoboken
8. Klinkowski M, Walkowiak K (2011) Routing and spectrum assignment in spectrum sliced elastic optical path network. IEEE Commun Lett 15(8):884–886
9. Rottondi C, Boffi P, Martelli P, Tornatore M (2017) Routing, modulation format, baud rate and spectrum allocation in optical metro rings with flexible grid and few-mode transmission. J Light Technol 35(1):61–70
10. Lu W, Jin X, Zhu Z (2017) Game theoretical flexible service provisioning in IP over elastic optical networks. In: 16th international conference on optical communications and networks (ICOCN), Wuzhen, pp 1–3 (2017)
11. Sambo N, Giorgetti A, Cugini F, Castoldi P (2017) Sliceable transponders: pre-programmed OAM, control, and management. J Light Technol 36(7):1403–1410
12. De Dios GO, et al (2016) Experimental demonstration of multivendor and multidomain EON with data and control interoperability over a Pan-European test bed. J Light Technol 34 (7):1610–1617
13. Velasco L, Castro A, King D, Gerstel O, Casellas R, López V (2014) In-operation network planning. IEEE Commun Mag 52(1):52–60
14. Dallaglio M, Giorgetti A, Sambo N, Velasco L, Castoldi P (2015) Routing, spectrum, and transponder assignment in elastic optical networks. J Light Technol 33(22):4648–4658
15. Sambo N, Meloni G, Cugini F, DErrico A, Poti L, Iovanna P, Castoldi P (2015) Routing Code and Spectrum Assignment (RCSA) in elastic optical networks. J Light Technol 33 (24):5114–5121
16. Ives DJ, Bayvel P, Savory SJ (2015) Routing, modulation, spectrum and launch power assignment to maximize the traffic throughput of a nonlinear optical mesh network. Photonic Netw Commun 29(3):244–256
17. Pedrola O, Castro A, Velasco L, Ruiz M, Fernández-Palacios JP, Careglio D (2012) CAPEX study for a multilayer IP/MPLS-over-flexgrid optical network. J Opt Commun Netw 4 (8):639–649

18. Palkopoulou E, Bosco G, Carena A, Klonidis D, Poggiolini P, Tomkos I (2013) Nyquist-WDM-based flexible optical networks: exploring physical layer design parameters. J Light Technol 31(14):2332–2339
19. Pavón-Mariño, P.: Optimization of computer networks : modeling and algorithms: a hands-on approach. 1st. edn. Wiley (2016)
20. Christodoulopoulos K, Tomkos I, Varvarigos E (2011) Elastic bandwidth allocation in flexible OFDM-based optical networks. J Light Technol 29(9):1354–1366
21. Rumipamba-Zambrano R, Perelló J, Spadaro S (2018) Route, modulation format, MIMO and spectrum assignment in Flex-Grid/MCF transparent optical core networks. J Light Technol 36(16):3534–3546
22. Kirkpatrick S, Gelatt CD, Vecchi MP (1983) Optimization by simulated annealing. Science 220(4598):671–680
23. Skorin-Kapov N (2007) Routing and wavelength assignment in optical networks using bin packing based algorithms. Eur J Oper Res 177(2):1167–1179
24. Wang Y, Cao X, Pan Y (2011) A study of the routing and spectrum allocation in spectrum-sliced Elastic Optical Path networks. In: Proceedings of IEEE INFOCOM, Shangai, pp 1503–1511
25. Wu J, Barahona M, Tan Y-J, Deng H-Z (2012) Robustness of random graphs based on graph spectra. Interdiscip J Nonlinear Sci Interdiscip J Nonlinear Sci. 221(10):1–7

NFC-Based Payment System Using Smartphones for Public Transport Service

Diego Veloz-Cherrez(✉) and Jaime Suárez

Escuela Superior Politécnica de Chimborazo, Riobamba 060155, Ecuador
diego.veloz@espoch.edu.ec, jgabriel.531@gmail.com

Abstract. This paper presents an implementation of software applications and an infrastructure for payments using NFC compatible Android smartphones. This prototype is suited for, but not limited to, payments in a public transport system. The importance of security in communication plays a significant role, especially when sensitive information and personal data are sent. Therefore, the authors analyze vulnerabilities during throughout communication in order to prevent and detect information leakage points. Each app has specific roles and handles encryption to guarantee security during data transmission. On the other hand, Apache server runs in a secure mode that executes some measures as Unicode coding, limits upload memory, decrease waiting time for connection to prevent attacks. With this in mind, the system has been evaluated to measure its performance and gauge how secure it is. Man in the middle and brute force attacks were run as a result of previous research works that analyzed among different threats in NFC communication and concluded these vulnerabilities could affect its security environment. This paper describes a methodology to implement this system efficiently, covering all security requirements throughout communication between users and the platform to guarantee data protection and gives the results of test vulnerabilities that prove the system resistance.

Keywords: NFC · Android · Public transport · Payment system · Smartphones

1 Introduction

Near Field Communication (NFC) is a technology that was standardized in 2003. It is based on Radio Frequency IDentification (RFID) operating at the frequency of 13.56 MHz and its short range (0.1 m) makes it efficient for electronic and mobile payments. However, both technologies differ from each other because of their operation modes and their architectures. NFC works on the ISO-14443 (RFID), ISO-18092 (NFCP-1), ISO-21481 (NFCP-2) [1] standard.

In recent years, wireless technologies development has been very broad, allowing the data exchange to be done in an easier way. Similarly, NFC technology has had fast growth and now is present in most smartphones, ID cards and passports in some countries and electrical appliances due to its flexible architecture and continuous development.

Nowadays, financial companies have applications to perform financial transactions using NFC, being the first companies that made use of this technology. In Europe many

M. Botto-Tobar et al. (Eds.): TICEC 2018, AISC 884, pp. 34–44, 2019.
https://doi.org/10.1007/978-3-030-02828-2_3

countries are already developing applications that use NFC to make payments for public transport services. The first city to do so was Valencia in Spain with the Mobilis NFC project which was developed by mobile operator Orange, later joining Movistar and Vodafone. This application was intended for public transport infrastructure like buses, metro, tram and bicycle rental [2].

By 2015, one out of four smartphone users in the world makes different types of payments or access to public transport with NFC, and it is estimated this type of transactions will be common by 2020 [3], which means developers are called to improve data security for these transactions since sensitive data is sent and they are highly desired by computer attackers.

Most applications for payments with NFC are developed in the architecture of smart card emulation, since it is not necessary to write labels or carry different objects to the smartphone [4]. Smart card emulation on most smartphones with Android operating system (OS) can be implemented working in smart card emulation mode with secure element or working in emulation of a host-based smart card mode (HCE) [5].

Smart card emulation mode, using secure element, has many advantages in terms of security, but its complexity, when implemented, has made HCE mode the most accepted. Nowadays, there are secure elements embedded in NFC devices but usually these are controlled by the device manufacturer or by a trusted service manager (TSM); so they put barriers to independent developers for developing payment applications that use this hardware [6].

HCE is an alternative for developing mobile payment applications; this mode provides greater control to developers and less dependence on third parties. On the other hand, its security is poor compared to secure element, so that the security offered by HCE mode must be implemented by the application developer, creating secure communication channels [7].

This project describes the design and implementation methods of a payment platform for public transport service with NFC-based smartphones. In addition, to analyze the vulnerability of the systems, man in the middle and brute force attacks were performed in order to identify what future actions need to be enforced in terms of data security.

1.1 NFC Vulnerabilities and Electronic Payment Platforms

NFC has inherently security vulnerabilities due to its operating modes; so that, NFC security depends on which mode the application works.

According to Haselsteiner and Breitfus [9], NFC vulnerabilities are eavesdropping, data corruption, data modification, insertion of data and man in the middle. They conclude that these attacks are controllable by adding a secure channel. Furthermore, it is essential to use 3DES or AES ciphers to establish this secure channel.

Additional conclusions of these authors above [9]. The man-in-the-middle attacks are very difficult to perform in an NFC communication because of the distance in which the smartphone reader is located, but Lee et al. [8] support this idea and pronounce that this attack is feasible in other points of communication because of an attacker could be masqueraded as another user and get easily data when they are making transactions and go unnoticed. Moreover, the ISO/IEC 14443 standard does not handle any security protocol, so in a man-in-the-middle attack, a NFC reader can be masqueraded with an unauthorized one and get all the data that is sent through and therefore the communication is vulnerable.

On the other hand, in a study carried out on 7 Android mobile applications which work in 5 different countries, are used by millions of users and carry out financial transactions, their information security were analyzed in the processes of data entry, registration and payment transactions. The results were that 6 out of 7 applications have critical security problems like non-use of secure channels and the lack of encryption in sensitive data such as names, telephones, account numbers among others [10]. For these reasons, it is necessary to analyze the vulnerabilities throughout the transmission of information, not only taking into account the weaknesses of a particular network point or technology but the whole communication system. Therefore, it is very important that electronic payment platforms and those, in which sensitive information is sent, implement security measures.

2 Methodology

The prototype integrates different parts, which establish a complete and secure transmission of users information, taking into account the results of previous studies on NFC security, taking care of data throughout the communication platform. Figure 1 shows the general structure of the system that displays its physical components and its communication topology. However, the logical structure will be explained later.

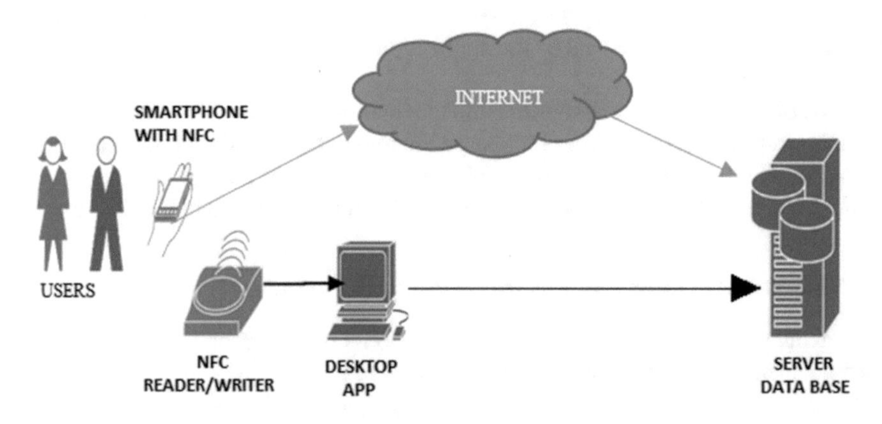

Fig. 1. System architecture

The components of the architecture are (from left to right):

Users. People, who use the platform to make payments, records and queries of balances and trips.

Smartphone. It is used to perform actions of payments, queries and records. This device must be NFC compatible. Likewise, a mobile application will be performed in host-based smart card emulation (HCE) mode and will also have access to the Internet if users want to make balance and travel inquiries.

NFC Reader. It receives data sent from the smartphone, processes it and sends it to the desktop application. This will always be in a passive state waiting for entry communication.

Desktop Application (Manager Application). This application is responsible for data processing and is the intermediary between the payment system and the user's phone. It would be installed in the infrastructure of the transportation service provider company and has three fundamental modules: payment processing application which handles payments through NFC; registration/queries application, which will register users and process queries through NFC; and the administrator/operator module, where the managers and operators execute administrative, maintenance and control actions. All these modules will encrypt each data field that is sent to the database.

Server and Database. Manage all connections and queries, as well as store user data and transactions. It allows handling all the information of the system.

The system is based on a prototype that controls different types of applications that allow the processing of data through NFC technology. The decision to use NFC technology is given by the benefits it offers but above all by the integration and easy management with smartphones. This technology offers a simple and spontaneous interaction between users and the payment platform as well as minimizes the operations that users should achieve to convey information using a different communication technology. The requirements for its implementation can be divided into hardware and software needs.

2.1 Hardware

The hardware used for the implementation of the prototype was as follows:

- An NFC ACR122U reader/writer, which can operate in active and/or passive communication mode
- A processor for the installation of desktop application. It will emulate the terminals of the transport provider in the corresponding stations.
- A CPU in which will be established as the server with the database
- Smartphones compatible with NFC. For this project, three different brands of NFC-based smartphones were used to check the communication and compatibility of NFC: Sony Z5 compact, Samsung Galaxy S7 and Sony Xperia XA.
- A router which communicates users' smartphones directly with the server to make queries through the Internet. In a real scenario, Internet access can be replaced by data transfer through cellular network.

2.2 Software

According to the statistical portal "Statista" [17], mobile OS market share, in terms of sales, was led by Android with 61 percent running on the smartphones, while iOS had 35 percent of the market and only 4 percent for the rest, including Windows phone between them. It is clear that mobile developers choose Android and iOS to develop their applications.

It must be specified that the mobile application was developed in JAVA language compatible only for smartphones with NFC and Android operating systems.

The software used to make the prototype was:

- Netbeans IDE 8.2 used to develop the desktop application in JAVA language.
- Android Studio 3.0 used to develop the mobile application in JAVA language.
- XAMPP used to tune the Apache server and create the MySQL database.

The mobile application allows users to make payments quickly and safely on a public transport service platform. For secure data communication to exist, each node of the network must be analyzed and define the possible vulnerabilities or requirements to guarantee protection for information. In this way, the mobile application with HCE mode was developed, which allows software to emulate the behavior of a smart card and allows the implementation of secure transactions.

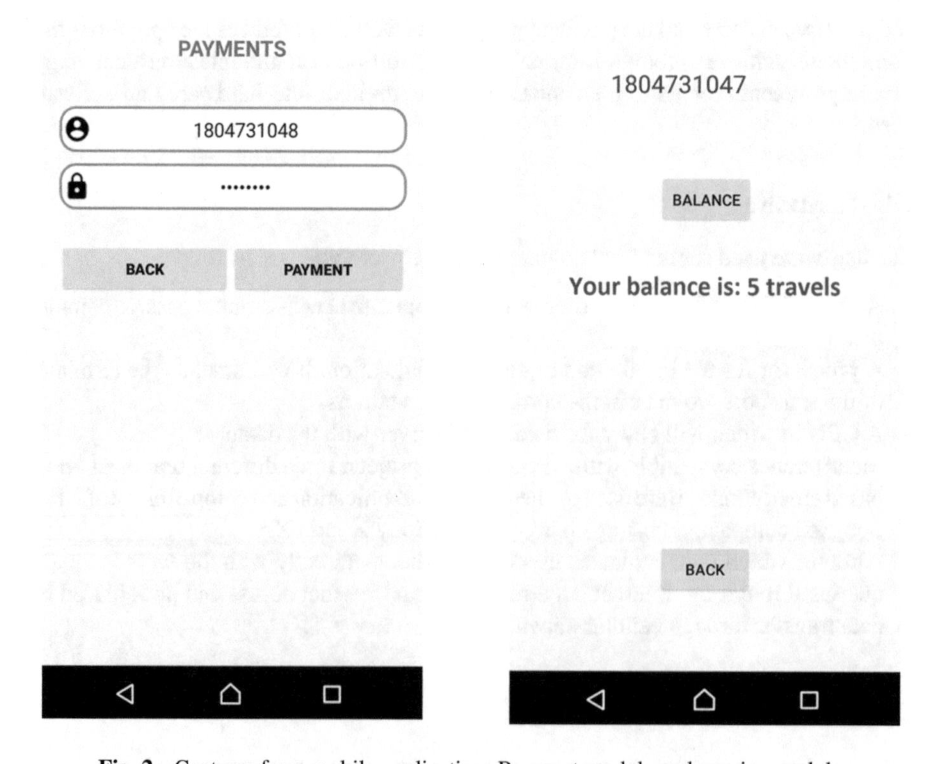

Fig. 2. Captures from mobile application: Payment module and queries module

Then, this application allows the user, through secure transactions, to communicate with the platform to make user records, payments, as well as balances inquiries and transaction history but the latter through Internet. In Fig. 2, two of the main actions of the mobile application are shown: left figure shows the module for payments using users ID number and their passwords. Right figure shows an image of the module of balance query. As it can be seen, the application is user friendly and the interaction is easy with basics operations.

On the other hand, the desktop application in the station is divided into two: one module is for user registration that also has the ability to perform balance inquiries as well as travel history query through NFC technology. The other module is intended to process payments automatically, just bringing near to the NFC reader. Figure 3 shows both modules of desktop application.

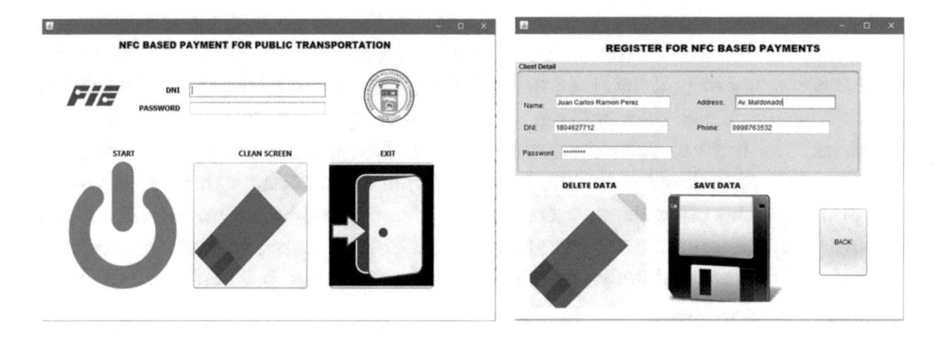

Fig. 3. Captures from mobile application: Payment module and register module

Both the desktop application and the mobile application encrypt the information that is transmitted in each transaction. As a consequence, it is ensured that information cannot be disclosed to unauthorized entities of the communication. On the other hand, Fig. 4 shows the coding in the applications to implement encryption as part of the communication. The figure shows the code that allows using the SHA-1 algorithm, a unidirectional code that prevents decrypting the information. Additionally, each desktop application contains a unique identification that is used to communicate with the main application (desktop application). In this way, it is avoided that information from the system can be extracted through masked false readers as part of the system.

```
MessageDigest md = MessageDigest.getInstance("SHA-1");
byte[] digestOfPassword = md.digest(secretKey.getBytes("utf-8"));
byte[] keyBytes = Arrays.copyOf(digestOfPassword, 24);
```

Fig. 4. Application code to implement SHA-1 algorithm

3 Tests and Results

3.1 Man in the Middle Attacks

For the testing phase, man-in-the-middle attacks were performed over the whole system network, with two protocol analyzers (sniffers) that are Ettercap and Wireshark running on the Wifislax Linux distribution.

Man-in-the-middle attack was performed since with this method attackers can be placed in the middle of the communication between the server and the desktop applications, allowing them to have access to users and system data. With this in mind, it is important to analyze what happens if this kind of attack would be executed in the payment platform and what kind of information the attacker would obtain if the attack were successful.

A man-in-the-middle attack was performed in the network segment between the desktop application (station terminals) and the NFC reader. Ettercap was used to perform an ARP poisoning to find the IP addresses of the terminals containing the desktop and server applications, and then perform the protocol analysis (sniffing), at the same time. Wireshark was used to scan the information that is passing through the attacker's computer. Figure 5 shows a capture from Wireshark where data can be seen with no encryption. On the other hand, users data were encrypted while transmissions were carrying out and communication was established. Consequently, any attacker could not get plain information and understand it, as it is shown in Fig. 6. Therefore, data confidentiality is protected.

```
> Frame 96: 91 bytes on wire (728 bits), 91 bytes captured (728 bits)
> USB URB
  Leftover Capture Data: 804100000000da0081002a313132323333343434312a4a75...
```

```
0000  1b 00 00 c8 56 60 87 a5  ff ff 00 00 00 00 09 00   ····V`·· ········
0010  01 02 00 02 00 82 03 40  00 00 00 80 41 00 00 00   ·······@ ····A···
0020  00 da 00 81 00 2a 31 31  32 32 33 33 34 34 34 31   ·····*11 22334441
0030  2a 4a 75 6c 69 6f 2a 41  6e 64 72 65 73 2a 4d 6f   *Julio*A ndres*Mo
0040  72 61 2a 43 65 72 6f 6e  2a 31 32 33 34 34 31 2a   ra*Ceron *123441*
0050  33 38 33 38 33 39 32 37  33 37 2a                  38383927 37*
```

Fig. 5. Data captured by Wireshark with no encryption

```
˅ Frame 61: 91 bytes on wire (728 bits), 91 bytes captured (728 bits)
    Encapsulation type: USB packets with USBPcap header (152)
    Arrival Time: May 31, 2018 15:45:54.940864000 Hora est. Pacífico, Sudamérica
    [Time shift for this packet: 0.000000000 seconds]
    Epoch Time: 1527799554.940864000 seconds
    [Time delta from previous captured frame: 0.046880000 seconds]
    [Time delta from previous displayed frame: 0.046880000 seconds]
    [Time since reference or first frame: 1.756470000 seconds]
    Frame Number: 61
    Frame Length: 91 bytes (728 bits)
    Capture Length: 91 bytes (728 bits)
    [Frame is marked: False]

0000  1b 00 80 f7 aa 5f 87 a5  ff ff 00 00 00 00 09 00   ....._.. ........
0010  01 02 00 02 00 82 03 40  00 00 00 80 39 00 00 00   .......@ ....9...
0020  00 94 00 81 00 2a 35 61  61 66 32 61 65 66 63 65   .....*5a af2aefce
0030  36 37 30 33 30 38 35 34  36 37 32 66 63 37 31 39   67030854 672fc719
0040  35 64 64 64 65 64 66 30  30 32 65 36 39 63 2a 54   5dddedf0 02e69c*T
0050  70 50 47 63 70 47 6a 63  6a 67 3d                 pPGcpGjc jg=
```

Fig. 6. Data captured by Wireshark while a payment transaction was carried out

Similarly, a man-in-the-middle attack was performed over the network segment between the desktop application and the server. In the same way, using Ettercap and Wireshark, users data were captured but payload frames were encrypted as it can be seen in Fig. 7. As a result, communication between software applications and server are secure and protects data to be disclosure by an untrustworthy entity.

```
GROUP 1 : 192.168.0.104 E8:03:9A:DD:3D:28
GROUP 1 : 192.168.0.100 6C:F0:49:77:FF:37

GROUP 2 : 192.168.0.1 F8:E9:03:B3:91:80
GROUP 2 : 192.168.0.104 E8:03:9A:DD:3D:28
Unified sniffing already started...
MYSQL : 192.168.0.100:3306 -> USER:jaimesuarez Seed:736a3a2134245663702b7e5f40624a2f475a6c34 Encrypted:7061676f5f6e666300000045000000109171b700
192.168.0.100*3306*jaimesuarez:$mysqlna$736a3a2134245663702b7e5f40624a2f475a6c34*7061676f5f6e666300000045000000109171b700
Unified sniffing was stopped.
Host 192.168.0.100 added to TARGET1
Host 192.168.0.104 added to TARGET2
Starting Unified sniffing...

MYSQL : 192.168.0.100:3306 -> USER:jaimesuarez Seed:746d45524b602122273d3d5b307e4b3d2c663c3b Encrypted:7061676f5f6e666300000045000000109171b700
192.168.0.100*3306*jaimesuarez:$mysqlna$746d45524b602122273d3d5b307e4b3d2c663c3b*7061676f5f6e666300000045000000109171b700
MYSQL : 192.168.0.100:3306 -> USER:jaimesuarez Seed:7c4a3c2f7170606c7e58797a7a725077285b755a Encrypted:7061676f5f6e66630000002d000000100d80b338
192.168.0.100*3306*jaimesuarez:$mysqlna$7c4a3c2f7170606c7e58797a7a725077285b755a*7061676f5f6e66630000002d000000100d80b338
MYSQL : 192.168.0.100:3306 -> USER:jaimesuarez Seed:44456c737a2d6e642d502d2f43415e352d226146 Encrypted:7061676f5f6e666300000045000000109171b700
192.168.0.100*3306*jaimesuarez:$mysqlna$44456c737a2d6e642d502d2f43415e352d226146*7061676f5f6e666300000045000000109171b700
MYSQL : 192.168.0.100:3306 -> USER:jaimesuarez Seed:2b3a233d4b424a2b397b22327474693157453165 Encrypted:7061676f5f6e66630000002d000000980d80b338
192.168.0.100*3306*jaimesuarez:$mysqlna$2b3a233d4b424a2b397b22327474693157453165*7061676f5f6e66630000002d000000980d80b338
MYSQL : 192.168.0.100:3306 -> USER:jaimesuarez Seed:5a73583b5e482d45555d515e645c223274736625 Encrypted:7061676f5f6e666300000045000000109171b700
192.168.0.100*3306*jaimesuarez:$mysqlna$5a73583b5e482d45555d515e645c223274736625*7061676f5f6e666300000045000000109171b700
```

Fig. 7. Data captured by Ettercap while registering information in the database

Each user parameter (Name, last name, ID number, date of birth, address), entered in the registration process, is encrypted in 3DES from the mobile application. This cypher algorithm shows robustness to man-in-the-middle attacks [11]. On the other hand, user passwords were encrypted with SHA-1. This method is very effective to protect data especially passwords by delivering a solution to many of the problems found in database storage since this method of encryption is one way [13]. More specifically, attackers can reach information stored in databases in order to use against the system itself or obtain privileges escalation to access more resources, especially when it comes to passwords. For this reason, it is essential this type of information remains encrypted with a robust algorithm that prevents the attacker from performing reverse engineering and getting passwords in plain text.

3.2 Brute Force Attacks

Brute force attack is one of the most performed attacks in cryptographic systems [12]. Likewise many other transactional platforms, this system carries out several authentications during the communication with the server using username and password; as a consequence, this type of attack can be used to decrypt data. It is very important to execute the brute force attack on the platform because if the attacker manages to obtain a valid user and password, new vulnerabilities could be exploited in the system [14].

For these reasons, it was necessary to measure the time that would take an attacker in order to perform a successful brute force attack but also taking into account the processing capacity of his resources. To achieve this purpose, three processing capacities were considered to calculate the time:

- A standard computer with 4000 million operations per second [15].
- A processor cluster with an average of 340000 million operations per second and a supercomputer with 1000 trillion operations per second [16].

As it can be seen in Table 1, attackers would need a lot of time to perform an attack by brute force. Nevertheless, if a computer, with greater processing capacity, is used, the time to succeed would decrease drastically.

Table 1. Time it takes an attacker to decrypt each field with brute force attack

Network segment	Standard computer [s]	Clúster [s]	Supercomputer [s]
NFC reader – Desktop app (registration)	6.15×10^9	7.06×10^7	24.6
Desktop app (registration) – Server	6.15×10^9	7.06×10^7	24.6
NFC reader – Desktop app (payment)	6.47×10^8	7.44×10^6	2.59
Desktop app (payment) – Server	1.92×10^9	2.21×10^7	7.7

It must be clarified that times for clusters and supercomputers were estimated from the number of operations explained above.

4 Conclusions

NFC-based payment system can be implemented assuring security during data transmission. However, it should be reinforced with encryption techniques, since the data is also transmitted in the public network. In fact, NFC technology has a lot of potential for the development of payment applications since HCE mode allows easier and less complicated application development than implementation with secure element but performing additional security features. The implemented system has much potential in the immediate future, since most smartphones in recent years have NFC technology enabling more people to use these applications for the payment of public transport services and offering a simpler and dynamic interaction with smartphones. Unlike similar communications as Bluetooth, NFC uses a lightweight protocol that allows transferring small amounts of information; thus, implementing a heavyweight encryption code would be not recommended in terms of processing.

Regarding to security communication, although there are stronger encryption codes, 3DES and AES algorithms that provide a robustness encryption enough for the amount of bits that are part of a NFC frame. Additionally, the inherent NFC short range and configuring servers, in other to avoid and prevent possible attacks, complement an efficient and effective information security to guarantee data protection and data confidentiality.

References

1. Minihold R (2011) Near Field Communication (NFC) technology and measurements white paper. In: Rohde&Schwarz
2. 20minutos. https://www.20minutos.es/noticia/2565576/0/valencia-primera-ciudad-europa/tecnologia-nfc/transporte-publico. Accessed 12 June 2018
3. Nombela JJ (2013) Pagar con el movil NFC, p 19
4. Instituto Nacional de Tecnologías de la Comunicación (INTECO) (2013) La tecnología NFC: Aplicaciones y gestión de seguridad, pp 1–21. http://observatorio.inteco.es
5. Android Developer. https://developer.android.com/guide/topics/connectivity/nfc/hce. Accessed 12 June 2018
6. Roland M (2012) Software card emulation in NFC-enabled mobile phones: great advantage or security nightmare? In: Fourth international workshop on security and privacy in spontaneous interaction and mobile phone use
7. Alliance SC (2014) A smart card alliance mobile & NFC council white paper host card emulation (HCE) 101, August 2014. http://www.smartcardalliance.org/wp-content/uploads/HCE-101-WP-FINAL-081114-clean.pdf
8. Lee Y, Kim E, Jung M (2013) A NFC based authentication method for defence of the man in the middle attack. In: 3rd international conference on computer, p 5. http://psrcentre.org/images/extraimages/113113.pdf
9. Haselsteiner E, Breitfus K (2006) Security in Near Field Communication (NFC) strengths and weaknesses. In: Semiconductors, vol 11, no 71, p 71. ISSN 00010782. https://doi.org/10.1145/358438.349303. http://books.google.com/books?hl=en&lr=&id=iHwsuHUFq0EC&oi=fnd&pg=PA71&dq=Security+in+Near+Field+Communication+(+NFC+)+Strengths+and+Weaknesses&ots=we-RvNBror&sig=-ZTxz9uRt0pPWkOCQczpa5yNy9Y
10. Reaves B, Scaife N, Bates A, Traynor P, Butler KRB (2015) Mo(bile) Money, Mo(bile) problems: analysis of branchless banking applications in the developing world. In: 24th USENIX security symposium (USENIX security 15), pp 17–32. ISSN 2471-2566. https://doi.org/10.1145/3092368. https://www.usenix.org/conference/usenixsecurity15/technical-sessions/presentation/reaves
11. Anaya KB, Ordoñez ML, Donado SA, Freddy L, Sanabria M (2012) Criptografía simétrica. Análisis del algoritmo criptográfico TDES y sus vulnerabilidades, pp 3–5
12. Sandoval Acosta SRS (2014) La criptografía. repositorio.unapiquitos.edu.pe/handle/UNAP/4622
13. Johnson D (2012) Password hashing. campus.murraystate.edu/academic/faculty/wlyle/540/2012/DJohnson.docx

14. Guataquira NM (2017) Seguridad para iot, una solución para la gestión de eventos de seguridad en arquitecturas de internet de las cosas. https://repositorio.escuelaing.edu.co/bitstream/001/693/6/Moreno%20Guataquira%2c%20Nicolas%20-%202017.pdf
15. Labaca R (2014) La matemática de las claves: ¿numérica o alfanumérica?. Consulta 5 junio 2018. https://www.welivesecurity.com/la-es/2014/06/13/matematica-claves-numerica-alfanumerica/
16. BBC Mundo (2015) La supercomputadora más poderosa del mundo con la que EE.UU. quiere superar a China. Consulta 5 junio 2018. http://www.bbc.com/mundo/noticias/2015/07/150731_tecnologia_eeuu_supercomputador_mas_poderoso_autorizo_obama_lv
17. Statista "The Statistics portal". https://www.statista.com/statistics/716053/most-popular-smartphone-operating-systems-in-us/. Accessed 24 Aug 2018

An Open Source Synchronous and Asynchronous Approach for Database Replication

Marcos Orellana Cordero(✉), Gerardo Orellana Cordero, and Esteban Crespo Martinez

Universidad del Azuay, Av. 24 de Mayo 7-77, Cuenca, Ecuador
{marore,gorellana,ecrespo}@uazuay.edu.ec

Abstract. Information is one of the most important assets of an organization. Therefore, database management systems need to ensure that this information is adequately safeguarded. Moreover, non-functional characteristics such as availability, performance, and security become critical during the use and management of a database. Therefore, to assure such capabilities, database management systems implement high availability and fault-tolerance systems. However, most times these solutions have high costs with privative licenses and specific implementation requirements. This paper presents a method to address server replication and failover by using open source utilities available in the Linux operating system. We accomplish such goals through the implementation of synchronous and asynchronous techniques which use the activity logs provided by the database management system. Finally, in order to demonstrate the feasibility of this proposal, it has been carried out a proof of concept implementation with Oracle databases deployed on a Linux operating system, in which the high-availability issue is solved successfully through dual back up with synchronous and asynchronous methods.

Keywords: High-availability · Database replication · Failover · Fault tolerant Distributed system

1 Introduction

A database is a group of software and hardware components coordinated by a Database Manager System (DBMS) [1]. Nowadays, databases are considered to be an essential element in the daily activities of public and private enterprises, assuring characteristics such as data security, integrity, and availability [2]. However, these characteristics are not enough for large-scale systems, where high availability, fault-tolerance, and disaster recovery are imperative [3]. Fault-tolerance is the capability of a system to work correctly in the case of faults and excessive charge [4]. To maintain high availability in a database system, this characteristic becomes essential. Keeping a high availability, however, is not a trivial task. Different problems such as network accessibility, software, and hardware failure, malware, and others, can alter the continuity of the service. These problems can affect the performance to the point of not being able to meet the users' needs. In a narrower sense, [2] defines high availability as a computer system that can

M. Botto-Tobar et al. (Eds.): TICEC 2018, AISC 884, pp. 45–56, 2019.
https://doi.org/10.1007/978-3-030-02828-2_4

implement fast failure recovery by the use of a set of software components. The unavailability of a database system is frequently related to faults in hardware or software resources. These faults may occur when a physical resource is not available or such resource does not have enough capabilities to keep the database operating [2], as well as when catastrophic events happen. Replication is seen as the mechanism to provide a fault-tolerance system. Therefore, a number of replication schemes have been constructed by the databases' vendors and third parties. Nonetheless, these solutions involve additional costs and specific expensive hardware components. There are two well-known models for database replication. In [5], the authors define synchronous and asynchronous replication. A database working with synchronous replication allows the primary and secondary nodes to be synchronized in real time, where every single transaction is demanded to be in every node before is committed. However, this introduces latency across the network. In asynchronous replication, the transactions are not replicated in real time, thus, when a failure happens in the primary node, there is the risk of losing transactions.

The synchronization between a primary and a single secondary node is called dual-machine backup [2]. Moreover, there are different approaches to accomplish dual-machine backup described in [2] (a) sharing storage (RAID), (b) full redundancy (dual servers and dual storage), and (c) data replication. In this work, we implement a high availability environment through a dual-machine backup based on data replication using Linux programs and current and historical logs of an Oracle database. The purpose of this work focuses on the use of non-proprietary techniques to obtain a symmetrical replication on an Oracle database.

2 Related Work

High availability and fault tolerance systems have existed for decades [6]. Initially, servers included components' redundancy to prevent systems' faults, however, faults are not originated only by hardware failures but also other issues can trigger them. Some of these issues are: (i) the communication and networks, (ii) software bugs and crashes, (iii) malicious code, and (iv) operating systems failures [7].

Distributed systems and high availability databases developed in the last decade have motivated the creation of several technologies and methods to achieve high availability. Furthermore, some open source solutions have been created for non-proprietary databases. Some databases implement systems called active-standby with synchronous and asynchronous propagation [8]. An example of these systems can be seen in [9], where they implement a proprietary replication system of type master-slave on MySQL, which achieves high availability and reliability by synchronizing data on real-time between the servers automatically. Other non-proprietary solutions solve the replication by the generation of an incremental security copy of the database's logs. These logs are fundamental for the recovery processes on the secondary nodes [10]. Some other solutions propose the use of asynchronous copies not tied to the DBMS, as it is the case with RemusDB [8]. Our solution differs from the mentioned because it offers the advantage of a synchronous and asynchronous replication without dependency of the DBMS by assigning the replication responsibilities to open source tools and utilities of the operating system. In consequence, we obtain a calibration feature on the grade of synchronism (synchronous, semi-synchronous, asynchronous) and an improvement on the performance.

Solutions such as [11], provides a taxonomy of the data replication across multiple data centers with partitioned data where each site has a replica of every partition. Other such as [12], provide a solution focus on different contexts such as communication systems, transaction manager, and the concurrency and replica control. In [13], a survey of the available open source and commercial tools for data replication are presented, in the context of open-source tools, they introduced the following (1) Postgres-R: a read-one write-all approach for Postgres databases, (2) Slony-I: a master to multiple slaves replication system which allows cascade propagation, (3) Escada Replication Server: a synchronous master/slave replication and zero data-loss over the WAN (4) database Replicator, uses the principle of merge replication with a semi-automatic conflict resolution in which the DBA has minimal involvement, (5) Pgpool-II, a tool that acts as middleware between the clients and a Postgres server, it takes care of the replication, the fail-over in case of issues with the primary node and load balancing when there are too many transactions in the queue. As it can be seen in the survey done by [13], there are several open source solutions with different approaches to cover the replication problem. However, most of these solutions are specific to an open-source database engine and do not provide a replication mechanism for a proprietary database such as Oracle Database, thus, we researched methods to replicate an Oracle database with only open source and built-in utilities of the Linux OS.

Within the domain of Oracle Databases, there is the data guard solution described in [13–15] which provides a real-time protection which assures zero data loss and maintains a hot copy of the database in the secondary nodes which can resume to a fully functional state of the services in case of a critical event. In addition, it provides a manual or automatic fail-over, therefore a failure would be transparent to the end-users. However, the data guard solution involves additional costs in licensing to the private or public organization using an Oracle database. As a consequence of this, we provide a solution which achieves data replication with synchronous and asynchronous methods which use only open source software.

3 The Replication Basics

This section describes the required steps and structures to build a high availability environment between a primary server also known as production node, and a secondary server, known as a standby node. Moreover, Oracle Data Guard solution introduces the term "active standby node" [15]. Such term explains a standby node which is active for querying, which is applicable to our solution. For the sake of clarity, this study refers to the primary server as the *primary node* and the active standby node as a *secondary node*.

Oracle DBMS generates parallel traces of the database activities, which means that for every update on the database structure or data, a portion of the working memory is registered in the permanent storage as a log record. We aim to transfer these logs from the primary to the secondary node to set up a usable recovery environment, providing the bases for a high availability environment. We define that a full integrity of the database is reached when the log records are totally transferred from the primary to the secondary node. Then, the recovery processes is a separate step which is done only when a failure occurs. If the synchronization between nodes is done in real time, the integrity of the database is kept. This means that replication of the logs must correspond to the same point in time.

Databases have a variety of mechanisms to avoid losing their data and structure after a breakdown. However, this concept does not apply to a high availability mechanism, but to a recuperation of its instance when the services are restarted. A high availability mechanism implies the presence of at least two nodes; when the primary node gets a hardware or software failure, it is replaced by the secondary node, which has the same data of the first one at the time a breakdown happened. The first node is in charge of the CRUD services to the users in a production environment, and the second node contains the synchronized data of the first node.

This work is not the traditional replication case, where a single directory of a file system contains all the data files, such case can be seen in [16], where a MySQL database is replicated. Two type of logs can be identified in an Oracle database.

Historical redo logs, we refer to these as *archived redo log files*.
Online redo logs, we refer to these as *redo log files*.

Figure 1, illustrates the deferred storage mechanism implemented by an Oracle database when there is a change in the working memory, the deferred storage mechanism replicates this data to the redo log buffer. There are several redo log files which alternate from one to another every time they are full or a timer demands the change.

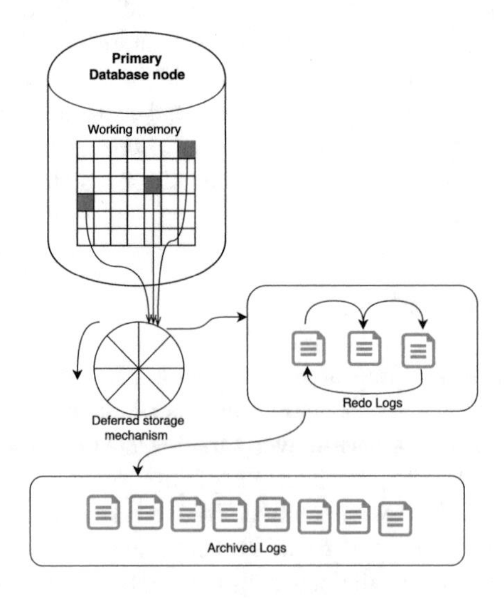

Fig. 1. Redo log and archived log mechanism

The DBMS mechanism performs the activities of online storage; this means that for every transaction which alters or modifies the database, the information is copied to a redo log file. The use of redo logs for replication is optimal given that updates in the database cannot be generated faster than the redo log can be written to disk [5]. This process is done because the hard disk operations are always slower than memory. Therefore, in order to keep the database as efficient as possible, hard disk IOs are reduced with the mentioned process. In this work, the replication is done with an Oracle database, which possesses features of online redo log, archived redo log. A detailed description of the redo log files can be seen below:

1. *Online log records*: The online redo log is a portion of memory destined to storage the users' activities and the data dictionary; this contains the objects' structures and the data of the running applications. Databases work with metadata which is also registered in the redo log files when a change occurs. Therefore, all the activities are registered by the online redo log, and once it is registered, the means to provide contingency operations are in place [17]. In order to ensure the high availability of the database, it is necessary to synchronize this log in real time with the secondary node, this involves synchronizing every change from the first to the secondary node.

Figure 2, illustrates the process of redo log files creation. The redo log buffer has to redo log groups which have one or several redo log files [18]. Logs are organized in a circular mode, and when one of them is full, the process continues with the next one until it reaches the last one, when this occurs, the process goes back to the first one and begins to overwrite the content of the logs in the same order. It must be taken into consideration that only one redo log can have an active status. To ensure the correct local replication, the method used for the redo log file is synchronous. The DBMS has two running processes in memory, a first one known as log writer (LGWR) which moves the updated blocks from the memory to the hard disk, and a second one called database writer (DBWR) which moves the updated blocks to the data files; this process is known as deferred storage, which can be seen in Fig. 1.

2. *Historical redo log record*: In the case of the redo log files, the log is generated by the activity of the database every time that the file is overwritten, which means that any change registered in a log, is not available subsequently its overwritten. To avoid this problem, we store the archived log files in a different place than the redo logs, this helps us to keep a historical record of the database and therefore, enable us to make a complete reconstruction at any point in time. Previous the generation of the redo logs, it is necessary to execute a backup of the database; this is considered as a starting point for the achieved log systems. It is important to note that the last transactions are always saved in the redo log files. Figure 2, shows the sequence of activities in a synchronous mode.

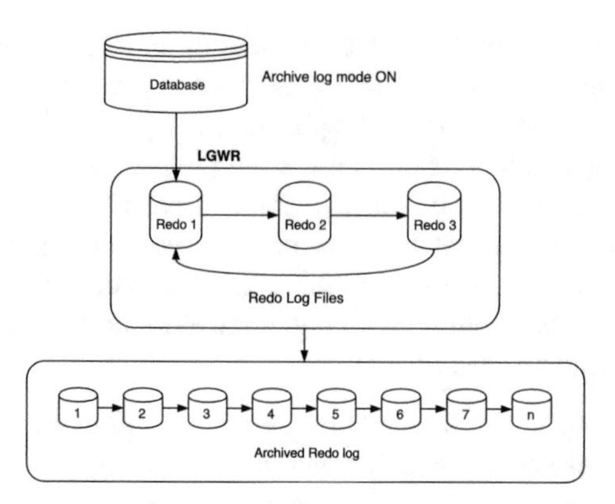

Fig. 2. Synchronous replication schema - online redo log files

In the paragraphs above, we have described a part of the Oracle database architecture needed to understand the further process. Therefore, from this point forward we describe the necessary steps to set up a replication environment. This process starts from the backup of the primary and its recovery in the secondary node, both with an active archived log configuration to the synchronous and asynchronous replication of the database through open source software, these steps are described below:

A. *Backup*: The replication to a secondary node must start with a backup of the database at a specific point in time; this backup is used as a starting point for the activities in the secondary node. An initial backup is set up after the activation of the database in archived log mode, by doing it in this way, errors area avoided in the archived logs on the secondary node, this keeps a coherence in both nodes. We can say that, if the database starts its backup in a point in time p1 and finishes in a point p2, during the time p2–p1 the database does not stop, thus, the changes p1' and p1'' are not registered in the backup. Figure 3 shows this behavior. Since the database has been initiated in archive log, all the changes that happen in the database during the time p2–p1 and after it is archived on a storage section, this ensures the coherence of the database at the time the redo log has been registered.

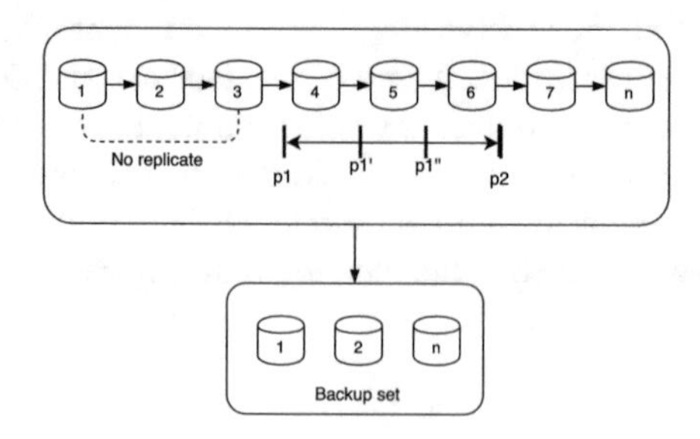

Fig. 3. Asynchronous replication schema - archived redo log files

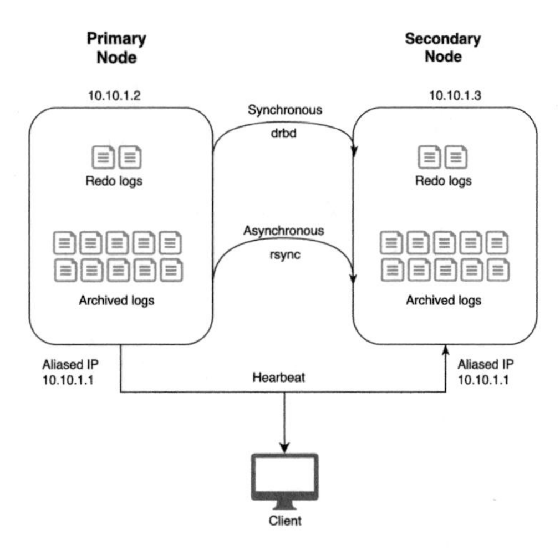

Fig. 4. An approach to database replication

B. *Activation of the log storage mode*: The database does not come with the storage of log records active in a default state. Therefore, it was necessary to configure the storage directory and this required the following steps: (i) shut down the database, (ii) start the administration mode, (iii) activate the storage of the log record (archived log), and (iv) start the database again.
C. *Replication structure*: A characteristic of Oracle database is that, once a redo log file is full, it passes its control to the next one; thus, the construction of a script to detect the generation of a newly archived redo log file and start an asynchronous copy process of it to the secondary node is written. In the case of the online redo log, a synchronous copy to the secondary node is needed in order to maintain coherence at the current point in time. Therefore, low-level replication techniques are used through the Distributed Replicated Block Device (DRBD). The previously described replication processes in the replication environment are:
(a) Synchronous replication of the online redo log
(b) Asynchronous replication of the archived redo log

Figure 5, shows the division of the described replication segments. This is: replication of the archived redo log in asynchronous mode with OS utilities for the file system, and another one which performs the replication of the redo log online in a synchronous mode through the DRBD in a hard disk block level.

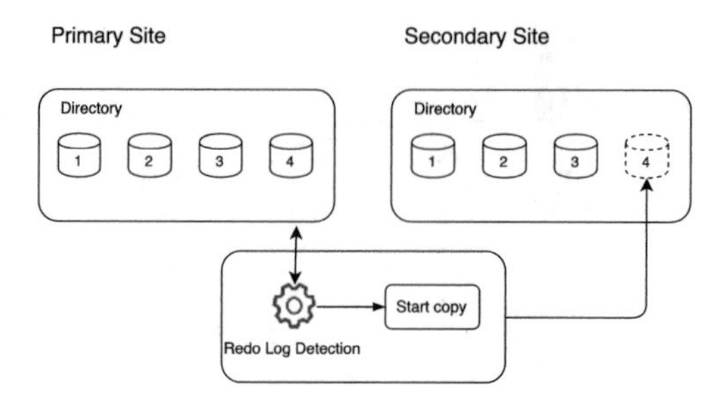

Fig. 5. Replicate of archived redo log files by RSync (Remote Sync)

4 Proof of Concept

The previous section described the basic concepts for our replication process. Thus, this section describes in detail the tools and processes to achieve a proof of concept implementation of a replication platform. Figure 4, introduces the main open source tools and processes used for synchronous replication, asynchronous replication, and fail-over transition. The details described below are required to settle a replication environment:

A. *Initial configurations*: It was stated in Sect. 3, that there are initial configurations to allow the replication. We describe the main ones in the following lines:
   - We defined in the database configuration the directory location of the archived log.
   - We set up the size of the redo log files to a value which suits the servers and network capabilities.
   - We set up the automatic storage of the redo log files to be able to replicate them and use in the secondary node.
B. *Synchronization service archived redo logs*: We created an asynchronous service to copy the archived redo log from the primary to the secondary node, this process implements a file modification detector, trust certificates to allow continuous replication and data synchronization process.
   - The update detector on the archived log directory is done by using the *inotifywait* OS utility.
   - We used *rsync* utility to synchronize the changes from the primary to the secondary node. Rsync implements the "rsync algorithm" [19], which computes the differences between two files and updates the second to achieve equality with efficiency.
   - We implemented trust certificates with the Linux *ssh* utility to avoid constant validation of credentials from the nodes.
C. *Synchronous copy via DRBD*: We implemented the synchronous copy of the redo log files using the DRBD tool; this software makes a low-level copy of a determined disk partition. The database was configured from the beginning with a standard directory for the redo log files. Figure 6, illustrates the low-level replication through the DRBD utility. The following are the necessary configurations for the synchronous replication:

- We set the blocks size. We used *mke2fs* to format the file system.
- We configured DRBD with the following parameters:
  - Resource name: The resource that identifies the replication parameters.
  - Protocol A: Replication level which compares security against performance.
  - Syncer: Bandwidth to transfer the files.
  - Device: ID of the replication device.
  - Disk: ID of the device which acts as source for the replication process.
  - Address: IP Addresses and ports for every node involved.

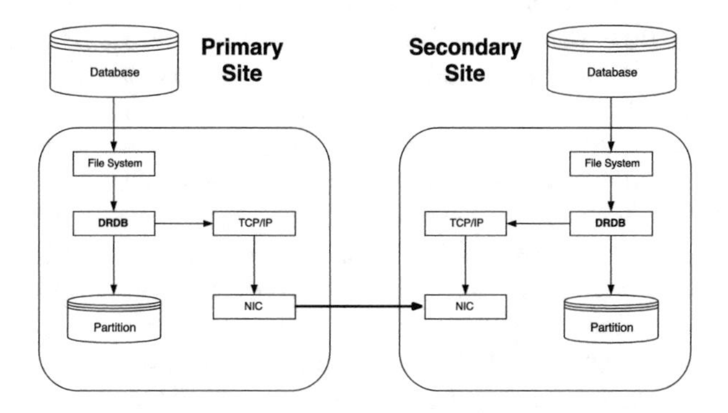

Fig. 6. Replication of online redo log files with the DRBD utility

D. *Recovery logs*: At this point, we have a running replication environment where the synchronous and asynchronous replication services are up and running. Therefore, we could proceed to test the recovery process, this involves three steps: recovery of the backup set, recovery of the archived log files and recovery of the online redo log files.

1. The database was recovered through the backup set; we use the utility Recover Manager (RMAN) for these purposes.
2. Once we have recovered the database to the point of the backup, we recovered the historical records of modifications saved in the archived redo log files.
3. We enabled the directory with the online redo log files which were synchronously replicated to the secondary node. The RMAN utility helps with this process.

The replication was fulfilled with these steps. However, when the primary node was recovered, the secondary node could not receive the archive redo log files because of a conflict between the recovered blocks given by the online redo log files and the archived redo log files. Thus, we recreated a complete recovery of the backup set and the archived redo log files, excluding the online redo log files. We were able to recover the archived redo log files a number of times without any problem.

5 Results

We analyzed the feasibility of the data files replication case under synchronous and asynchronous methods. We established that for the asynchronous replication, as a requirement, copies of the backup point must exist until the last transferred archive log. The database automatically verifies the necessary archived log files to get coherence at this point in time.

The backup of the database can also include a backup of the archived log files; however, after running tests, it was not possible to recover all the archived log files. Therefore, we performed the backup in one step and the copy of the archive log files, in this way we ensure the necessary files for keeping the coherence of the database in the backup point of time. At this point, we verified the suitability of the recovery by starting up the database and, immediately after, we executed SQL (Structured Query Language) queries to several high transactional tables of the database; the result showed the exact data at the expected point in time. However, when it was necessary to copy the redo log files after the backup of the primary node, the database of the secondary node already had new redo log records, and consequently, the recovery was not successful.

The problem with the coherence after the recovery was solved by starting the database in a read-only mode. This enables just query activities with modification of records, in consequence, we could recover the archived log files any number of times. It is important to note that the generation of the archived log files must not overcome 20 min according to the vendor's handbook.

We performed a coherence test by simulating a breakdown in the communication channel while the system was copying an archived redo log file; this caused the loss of approximately the last 20 min of activity. Although the archived redo log file was not copied, it was possible to make a complete recovery through the online redo log files.

We verified the synchronous copy and corroborate the presence of 3 online redo log files; one of these had an active state while the other two represented the last 40 (2 archived redo log files of 20 min each) minutes before the switch to the active one. Thus, we could recover all the transactions until the last known point in time. Even though we accomplish a successful recovery of transactions, we also lost the synchronization of the archive redo log files copied after the recovery of the communication.

6 Conclusion and Future Work

This study introduces a replication environment with synchronous and asynchronous design, reaching the desired high availability and fault tolerance services. This method only needs the log-files generated by the database along with open source tools to accomplish replication from a primary to a secondary node. We took advantage of the database architecture which generates current and historical log-files to build two replication environments: one based on a recurrent copy of archived redo log files with an update detection mechanism, and another based on a synchronous low level copy through the DRBD software for the online redo log files. Moreover, this method can be easily applied to databases which have a similar structure to Oracle databases. We fulfilled our replication goals, accomplishing in all the cases the recovery of the database to the last point in time, however, we did not accomplish the replication of the database once the primary node is recovered. The tests which recover the blocks from the online redo log files cannot activate the incremental recovery because of lost synchronization.

This work shows that a high availability and fault tolerant services are feasible to implement for an Oracle database with only open source tools which are included in a Linux operating system. Our experiments show a high-level recovery in the presence of failures. Furthermore, this method shows promising results, but bidirectional replication after a failover should be further improved. The implementation of new techniques is necessary to get bidirectional synchronization when the primary node recovers from previous failures. We conclude that this type of replication differs from similar products because it uses open source tools and existing utilities of the operating system with a proprietary database such as Oracle. Therefore, the implementation costs are low, offering also improvements in the automation.

The presented design solves the problem of replication in order to maintain high availability. However, some further work can be done in optimization of some processes. We consider the following as the main points of improvement: The present design presents a unidirectional replication constraint when the nodes perform a fail-over. Thus, the backup from the node acting as primary, to the new secondary node has to be done again. In this case, the design could perform better by decreasing the amount of manual work loaded to the database administrator. The method is designed for a dual-backup. However, when more than two nodes are necessary, there is the need for a decision algorithm to determine the node that will become primary in case of failure.

Acknowledgment. This research was supported by the vice-rectorate of investigations of the Universidad del Azuay. We thank our colleagues from Laboratorio de Investigación y Desarrollo en Informática (LIDI) de la Universidad del Azuay who provided insight and expertise that greatly assisted this research.

References

1. Mukherjee N, Chavan S, Colgan M, Gleeson M, He X, Holloway A, Kamp J, Kulkarni K, Lahiri T, Loaiza J, Macnaughton N, Mullick A, Muthulingam S, Raja V, Rungta R (2016) Fault-tolerant real-time analytics with distributed oracle database in-memory. In: 2016 IEEE 32nd international conference on data engineering (ICDE), pp 1298–1309
2. Wu X, Wang K, Su Z, Liu Y (2012) Research and implementation of the high-availability spatial database based on oracle. In: 2012 international conference on computer science and service system, pp 1713–1716
3. Yu K, Gao Y, Zhang P, Qiu M (2015) Design and architecture of dell acceleration appliances for database (DAAD): a practical approach with high availability guaranteed. In: 2015 IEEE 17th international conference on high performance computing and communications, 2015 IEEE 7th international symposium on cyberspace safety and security, and 2015 IEEE 12th international conference on embedded software and systems, pp 430–435
4. Amirishetty AK, Li Y, Yurek T, Girkar M, Chan W, Ivey G, Panteleenko V, Wong K (2017) Improving predictable shared-disk clusters performance for database clouds. In: 2017 IEEE 33rd international conference on data engineering (ICDE), pp 237–242
5. Wong L, Arora NS, Gao L, Hoang T, Wu J (2009) Oracle streams: a high performance implementation for near real time asynchronous replication. In: IEEE 25th international conference on data engineering, 2009. ICDE 2009. IEEE, pp 1363–1374

6. Somani AK, Vaidya NH (1997) Understanding fault tolerance and reliability. Computer 30(4):45–50
7. Kreutz D, Ramos F, Verissimo P (2013) Towards secure and dependable software defined networks. In: Proceedings of the second ACM SIGCOMM workshop on hot topics in software defined networking. ACM, pp 55–60
8. Proskurin S, McMeekin D, Karduck AP (2012) Smart camp building scalable and highly available it-infrastructures. In: 2012 6th IEEE international conference on digital ecosystems and technologies (DEST), pp 1–6
9. Jaiswal C, Kumar V (2015) DBHAaaS: Database high availability as a service. In: 2015 11th international conference on signal-image technology internet-based systems (SITIS), pp 725–732
10. Ping Y, Hong-Wei H, Nan Z (2014) Design and implementation of a MySQL database backup and recovery system. In: Proceedings of the 11th world congress on intelligent control and automation, pp 5410–5415
11. Agrawal D, El Abbadi A, Salem K (2015) A taxonomy of partitioned replicated cloud-based database systems. IEEE Data Eng. Bull. 38(1):4–9
12. Kemme B, Alonso G (1998) A suite of database replication protocols based on group communication primitives. In: Proceedings. 18th international conference on distributed computing systems 1998. IEEE, pp 156–163
13. Moiz SA, Sailaja P, Venkataswamy G, Pal SN (2011) Database replication: a survey of open source and commercial tools. Database **13**(6)
14. Ray A (2002) Oracle data guard: ensuring disaster recovery for the enterprise. An Oracle white paper
15. Meeks J (2009) An oracle technical white paper-oracle data guard with oracle database 11g release 2. Oracle
16. Chaurasiya V, Dhyani P, Munot S (2007) Linux highly available (HA) fault-tolerant servers. In: 10th international conference on information technology (ICIT 2007). IEEE, pp 223–226
17. Litchfield D (2007) Oracle forensics part 1: Dissecting the redo logs. NGSSoftware Insight Security Research (NISR), Next Generation Security Software Ltd, Sutton
18. Candea G, Cutler J, Fox A, Doshi R, Garg P, Gowda R (2002) Reducing recovery time in a small recursively restartable system. In: Proceedings international conference on dependable systems and networks, pp 605–614
19. Tridgell A, Mackerras P et al (1996) The RSync Algorithm

Forensics Analysis on Mobile Devices: A Systematic Mapping Study

Jessica Camacho[1], Karina Campos[1], Priscila Cedillo[1,2(✉)], Bryan Coronel[1], and Alexandra Bermeo[2]

[1] Faculty of Engineering, University of Cuenca, Cuenca, Ecuador
{jessica.camachoc,karina.campos,priscila.cedillo, bryan.coronel}@ucuenca.edu.ec

[2] Computer Science Department, University of Cuenca, Cuenca, Ecuador
alexandra.bermeo@ucuenca.edu.ec

Abstract. Nowadays, mobile devices have evolved vertiginously due to their massive adoption by users, who have several devices with different purposes. These devices contain greater capacity/functionality to manage information, with the embedded characteristics they become an important digital evidence container. In recent years, considerable research has been conducted on various types of digital electronic evidence, acquisition schemes and methods of extracting evidence from mobile devices. In this paper, a systematic mapping of the Forensics Analysis on Mobile Device is presented; this research has been conducted following the guidelines of Kitchenham's methodology. The aim of this study is to provide a background of relevant activities that are considered by investigators to handle with potentially useful digital evidence from mobile devices. A total of 36 primary studies were selected and categorized to extract information regarding the aforementioned classification. The results presented in this contribution provide a detailed study about current analysis in research forensics field by the use of mobile devices.

Keywords: Forensics · Digital evidence · Devices · Mobile

1 Introduction

Nowadays, mobile devices are being used massively, becoming one of the best inventions that have ever existed, mainly because of their functionality, contents, and versatility. Smartphones are mini computers that provide the functionality of conventional telephones, wireless Internet access, and, recently, many booming applications. They also provide sources of information in real time, exchange of data and information on a daily basis. Besides, they represent an interesting source of proof for crime research due to the content that can be found stored on one of those devices (e.g., bank transactions, social interaction). Fraudsters and other cyber criminals can use different services provided by platforms with false identities, in order to hide their malicious intentions behind profiles that seem to be reliable.

M. Botto-Tobar et al. (Eds.): TICEC 2018, AISC 884, pp. 57–72, 2019.
https://doi.org/10.1007/978-3-030-02828-2_5

The digital forensics analysis has been defined as the use of scientifically derived and proven methods for the preservation, collection, validation, identification, analysis, interpretation and presentation of digital evidence [S01]. The challenge of preserving and managing the evidence existing in mobile devices has motivated the creation of methods and solutions to manage the evidence in a properly manner.

As far as it is known, no evidence-based studies (e.g., systematic mapping studies, systematic literature reviews) have been reported recently about the considerations with which a forensics tool treat the electronic digital evidence of mobile devices. A systematic mapping study is a way to categorize and summarize the existing information around a research question in an unbiased manner [1]. A last general revision of the literature date back to 2013, proposed by Barmpatsalou et al. [S01] and Tajuddin and Manaf [S02], those papers report a systematic literature review where they are identifying, evaluating and interpreting all research relevant to a particular research question, topic area, or phenomenon of interest. Later on, in 2017 a study oriented to devices with Android operative system was addressed by Scrivens and Lin [S03]; and, in 2015 there is a secondary study for the IOS operative system in the iPhone 5S version which was performed by Mushcab and Gladyshev [S04]; nevertheless, there is not forensic analysis information for the current versions of IOS and Android devices. Moreover, in said studies, the approaches are specific to a situation (e.g., mobile applications, e-mail, text messages). Finally, Garfinkel [2] presents a study with more than 15 years of experience in computer forensics, digital forensics research, and discussions with forensics community.

Therefore, the present paper presents a systematic mapping study that addresses the different methodologies and tools for the acquisition of digital evidence when it is gathered from mobile devices. In the forensics investigation of a mobile device, there are considered three main requirements [S03]: (i) location of data storage, (ii) data extraction, and (iii) data analysis. That said, there are forensic methodologies for different cases of information extraction in which it is important to know what type of evidence is going to be manipulated; otherwise, the mishandling of a method or tool could potentially damage the investigation. Each extraction method has its pros and cons, for such reason this study addresses several primary contributions in this field.

The present work is organized as follows: Sect. 2 presents the related work that addresses secondary studies in this field, Sect. 3 presents an analysis of different methods and tools by means of a secondary study by defining a protocol, employed and validated to conduct the systematic mapping study, then, Sect. 4 describes the results obtained. Section 5 discusses the threats to the validity of the results and finally, the conclusions and next steps related to forensics in mobile devices are presented.

2 Related Work

A number of surveys and reviews aimed at analyzing digital evidence in mobile devices have been reported in recent years [S01]–[S04].

Kitchenham [3] proposed a guideline for systematic reviews where there are three phases: planning, conducting and reporting the review. A systematic review of the literature is a secondary study where individual studies called primary studies are evaluated, identified and interpreted. In particular, Kitchenham et al. [4], state that a systematic review is methodologically rigorous of the results of the investigation. The goal is not just to add all the existing evidence to a research question; it also supports the development of evidence-based guidelines for professionals.

Sánchez et al. [5] present a systematic mapping of the technologies developed with the use of Intelligent User Interfaces (IUIs), this research is based on Kitchenham's methodology for systematic reviews and its involvement with software engineering techniques. Also, Cedillo et al. [6] present a study aimed to intelligent users' interfaces, which also uses tools for a systematic mapping study in order to categorize and summarize the information in recent years.

Szvetits and Zdun [7] perform a Systematic Literature Review (SLR) with three phases. It includes: initial search, filter with defined selection criteria, and a final classification. Also, this study analyzes objectives, techniques, types, and architectures when using models at runtime. While, Petersen et al. [8] describe the differences between systematic review and systematic mapping studies. These authors characterize and summarize ten systematic reviews, where it was found that the methods differ in goals, scope and depth of the studies.

Petersen et al. [8] try to improve systematic mapping guidelines, the current practice of systematic mapping studies in software engineering is analyzed and it propose updates to systematic mapping guidelines. Alherbawi et al. [9] show a systematic literature review (SLR) with the following topics: realistic data sets, validation under fragmented data storage, and semantic validation to reduce false positive rates. While, Alharbi et al. [10] present a SLR concerning digital forensics investigation processes. This study states that SLR results are reproducible and there is a lesser possibility of missing an important reference.

Robinson and Clem in [11] present a SLR for examining and organizing findings from available studies with respect to benefits of service-learning identified within forensic proposal. The difference of the present study with all the previously named studies lay on the research questions. In this paper, the research question is related specifically with evidence digital on mobile devices.

Although several related surveys and reviews have been reported, they present two main limitations:

(a) There is a need of a more systematic process in order to summarize the existing knowledge in the forensics area.
(b) There is a need of surveys or reviews, specifically focused on solutions, to preserve, collect, validate, identify, analyze, interpret and preserve digital evidence.

3 Research Method

The digital forensic science of mobile devices has become an essential field of study when it comes to crimes; however, according to Scrivens and Lin [S03] it still is relatively new. For a competent evaluation of the research topic, a review of the current state is made through the analysis of primary studies. A systematic review involves several stages and activities [3, 4]: (i) planning the review, (ii) carrying out the review, and (iii) notification of the review. This research method has gained popularity in the last few years and it has been adopted in several other studies related to computer sciences and web engineering field [3].

3.1 Planning the Review

After the need for the mapping is identified, the systematic mapping is divided into six important steps that will influence the research [3, 4]: (i) Establish the research question and the sub-questions. (ii) Define the search strategy. (iii) Select primary studies. (iv) Evaluation of quality. (v) Definition of the data extraction strategy. (vi) Selection of synthesis methods. The basis for all research is a general question of the subject which will be called the research question, it is very important to state this question since all the unknowns for this review are divided into secondary questions, which are defined below.

Research question: "Which are the most popular tools and methods to extract, iden*tify, collect, preserve and manage the evidence of a mobile device?"*

Research sub-questions. (a) RQ1: *What kind of digital evidence can be found in mobile devices for a forensic analysis?* (b) RQ2: *Where can the digital evidence be found in the mobile device?* (c) RQ3: *What tools can be used to automate the collection and analysis of digital evidence?* (d) RQ4: *How are the solutions being evaluated?*

Data sources and search strategy: Papers related to this area have been selected from books, important journals, conferences and workshops in the forensics area, they are shown in Table 1. In order to perform the automatic search, the selected sources of information include: ACM Digital Library, IEEEXplore, SpringerLink, ScienceDirect. It has been selected a set of keywords, which allow the retrieving of the related papers. The search string defined is: *"(FORENSIC) AND (HAND OR SMART OR MOBILE OR DEVICE) AND (DIGITAL) AND (EVIDENCE)".*

Table 1. Conferences, workshops, books and journals.

Conferences and workshops
International Conference on Digital Forensics and Cyber Crime
Systematic Approaches to Digital Forensic Engineering
International Workshop on Systematic Approaches to Digital Forensic Engineering
International Conference on Cybercrime Forensics Education and Training
Books and Journals
International Journal of Electronic Security and Digital Forensics
IEEE Transactions on Information Forensics and Security

Search period. The period was selected starting with the appearance of wireless networks, internet, and mobile devices which evolved to adapt to all these new changes. The date of 2007 was chosen because it was then when the first application for mobile phones was downloaded [S01], hence, only studies carried out after that date have been considered.

Selection of primary studies. Each identified study was evaluated by the researchers to decide whether or not it should be included. The discrepancies were solved by consensus. The studies that met the following conditions were included:

- Studies presenting methods to collect and process digital evidence from mobile devices.
- Studies presenting methods to safeguard digital evidence in a mobile device.
- Studies presenting tools that allow to automate computer forensic processes.
- The exclusion criteria are as follows.
- Introductory papers for special issues, books and workshops.
- Duplicate reports of the same study in different sources.
- Short papers with less than five pages.
- Papers not written in English.

3.2 Quality Assessment

In addition to general inclusion/exclusion criteria, it is considered critical to assess the "quality" of the primary studies [3, 4], likewise Cedillo et al. [4], where a three-point Likert-scale questionnaire was used to provide a quality assessment of the selected studies. The questionnaire contains the following aspects: (a) forensics information related to the gathering, management, and preservation of the evidence found in mobile devices; (b) the journal or conference in which the paper was published (e.g., journal, proceedings, core ranking); and (c) if the study has been cited by another author (Google Scholar). The score for each closed-question is the arithmetic mean of all the individual scores from each reviewer. The sum of the three closed-question score of each study provides a final score which is not used to exclude papers from the systematic mapping study, but is rather used to detect representative studies.

3.3 Conducting the Review

The search to identify primary studies in the selected libraries was conducted on October 23rd, 2017. The application of the review protocol determined that 36 primary studies were selected, after the application of the inclusion criteria.

4 Results

The data extraction strategy was defined by breaking down each research question into more specific criterion in which a set of possible options was established. A summary of the results of the mapping is presented in Table 2. The included papers cited in the following section are concerning to RQ1, here the type of evidence that has the most relevant presence in the studies is related to the records and applications that state that more research has been developed on these items.

Table 2. Results of systematic mapping

Code	Research sub-questions	Possible answers	# Studies	% Percentage
RQ1: What kind of digital evidence can be found in mobile devices for a forensic analysis?				
EC1	Origin of Evidence	Browser	5	6,02%
		Application	18	21,69%
		Network	6	16.66%
		Register	20	55.55%
		Multimedia	13	36.11%
		System process	16	44.44%
RQ2: Where can the digital evidence be found in the mobile device?				
EC2	Local Artifacts	Temporary files	26	72,22%
		Browsing history	6	16,67%
		(e.g., caches, cookies, others)	27	32,53%
EC3	End Device	Smartphone	30	83,33%
		Another device	13	36,11%
RQ3: What tools can be used to automate the collection and analysis of digital evidence?				
EC4	Dependencies	Software	30	83,33%
		Methods	5	13,89%
RQ4: How are the solutions being evaluated?				
EC5	Evaluation Methods	Case of study	2	6%
		Controlled Experimentation	18	50%
		Proof of concepts	15	42%

Referring to question RQ2, the studies are mostly focused on mobile devices, and data to be extracted is in temporary files and logs. While the question RQ3 depends on the type of software that is used to extract the evidence, among the 30 studied papers, several tools are mentioned, which depend on the type of data to be analyzed.

Finally, in question RQ4, the methodologies given in the different items have a large percentage of controlled experiments and proof of concepts, which can be performed in order to demonstrate the feasibility of the primary studies.

5 Discussion of the Results

In this section, the most relevant topics found in the primary studies are presented: (i) Origin of evidence, (ii) Evaluation Methodologies, (iii) Tools and software for extracting digital evidence, (iv) Platforms, and (v) Application. The data shown on Table 3 presents the criteria used to perform the analysis, and the papers that were used.

Table 3. Papers selected for each extraction criteria

Data	Articles
Evaluation methodologies	[S03], [S05]–[S24]
Origin of evidence	[S02]–[S05], [S13], [S18]–[S22] [S24]–[S34]
Application	[S01]–[S05], [S11], [S14], [S18]–[S21], [S23], [S28]–[S33], [S35]
Tools and software	[S02], [S03], [S05], [S08], [S12], [S14], [S16], [S18], [S19], [S22] [S25], [S33], [S36]
Platforms	[S01]–[S05], [S11], [S14], [S18]–[S21], [S23], [S28]–[S32], [S35]

5.1 Origin of Evidence

In the area of forensics investigation there are legal aspects that are not fulfilled, these aspects entail the improper use of applications, fraud, theft, dissemination of copyrighted materials, etc. Taylor et al. [S27] have analyze the processes for obtaining digital evidence (i.e., computer forensic investigation procedures, digital evidence acquisition from mobile telephone applications, legal aspects). Moreover, this study explains that if personal data is obtained in an investigation, this information should be accessible only for the research team. Besides, all data sent, downloaded, and saved leaves a trace once it is deleted, this is called footprint [S28]. It means that there are always evidence about the actions performed within the smartphone. "Locard's Exchange Principle" states that "every contact leaves a trace", this principle is also relevant to forensic investigations [S37].

Moreover, there are different types of digital evidence (e.g., images, videos, media messages, files, metadata, logs, e-mails.), it depends on the researchers and the information that is needed to be extracted. Still, several studies [S01], [S03], [S05], [S19]–[S21], [S23], [S28]–[S30] [S32]–[S34] are focused on instant messaging applications. They extract information from different sources (e.g., deleted messages, images, videos, live multimedia, calls). This information could be either, in the volatile memory (RAM) or in the non-volatile memory (ROM) [S13], [S16]. It can also be extracted from the SIM card (Subscriber Identity Module), which must be analyzed immediately after the internal memory, while the device is still on [S25]. Later, Mylonas et al. [S24] divide the evidence according to its source, such as: messaging data, device data, SIM card data, usage history, application data, sensor data, user input data. Another study by Mylonas et al., [S26] presents security considerations like avoid permissions, in which users are delegated to grant administrator privileges when an application is going to be installed. Other considerations are the avoidance of visual notifications to protect sensors, such as the camera and GPS (Global Position System), through visual notifications. Finally, the Android security model, which does not distinguish between creative error messages by third-party applications and those created by the operating system, this flaw helps the malware to trick smartphones.

Mutawa et al. [S18] Alyahya and Kausar [S19], and Mushcab and Gladyshev [S04], all examine and analyze the folders of the devices that contain information of social networks apps (e.g., cache, databases). This evidence could be photos, videos, status, chat messages or logs. As it is known, there are several applications for the different existing branches. Ntantogian et al. [S22] states that the applications that are considered are: mobile banking, e-shopping/financial, password manager, finally encryption and data hiding. Such scenarios are established by creating users on an Android device as data in motion and based on the provided functionality of each application. To obtain the data of the volatile memory of the telephone, the privacy will be evaluated, whether the authentication of the credentials can be discovered.

5.2 Evaluation Methodologies

Scrivens and Lin [S03] propose a methodology based on three steps: (i) data storage location, (ii) data extraction and (iii) data analysis. Firstly, it is necessary to know the location of data, which is going to be extracted; also, it is necessary to have the appropriate permissions in order to prioritize the integrity of the data. Later, to do the data extraction it is necessary to use open source recovery images (e.g., Chip-Off, Forensic Software Suites, and Backup Applications). Varma et al. [S16] use non-encrypted images that are evaluated with the LIFTR method which has also three steps: (i) acquisition, (ii) filtering, and (iii) data recovery with a reprocessing of data. Similarly, Mutawa et al. [S18] presents three stages: (i) scenarios, (ii) logical acquisition, and (iii) analysis. It should be noted that all of those methodologies are evaluated by means of controlled experiments; the use of them helps to corroborate the results in a systematic way. A basic model for the process of a forensic investigation was described by the authors of [S08]. This model has eight steps, which are: (i) identification, (ii) preparation, (iii) preservation, (iv) collection, (v) examination, (vi) analysis, (vii) presentation and (viii) reporting. In the same way, Kubi et al., [S12] present a similar methodology with six activities that are: (i) collection, (ii) identification, (iii) acquisition, (iv) preservation, (v) examination and (vi) report. Then, there is a methodology named Post-mortem, developed by Rueda and Rico [S17], because the devices are not in perfect condition and some are even destroyed. It consists on the nine traditional phases, which are: (i) identification, (ii) preparation, (iii) preservation, (iv) acquisition, (v) examination, (vi) analysis, (vii) report, (viii) presentation, and (ix) review.

Finally, Satrya et al. [S32] describes a methodology with four steps: (i) identifying, (ii) preserving, (iii) analyzing, and (iv) presenting. Anglano et al. [S21] study a methodology that does not follow specific steps (see flowchart of Fig. 1 of the referenced document), rather it is a flowchart that allows identifying, locating, analyzing and offering acceptable results. In synthesis it is a “source code analysis” which analyzes and determines the format of the data stored in folders. Similarly, Mylonas et al. [S24] present a scheme consisting of 6 blocks: (i) investigation engagement, (ii) evidence type selection, (iii) evidence collection, (iv) evidence transmission, (v) evidence storage, and vi) investigation completion. Another methodology, proposed by Amato et al. [S09] has the following phases: (i) data collection, ontological representation, (iii) reasoning, (iv) rule evaluation, and (v) query. This method is based on semantic representation, integration, and correlation. An ontological representation helps forensic tools and improve analytical skills to correlate the evidence more easily. In contrast, Cohen in [S10] makes a comparison between two models: the first was taken from a study conducted by the authors characterized by legal requirement, this model is rather mathematical since variables are taken to obtain a general result. The second model is an alternative to the first one, the difference lays in the fact that the events are hypothetical, but they are also based on a legal context. Omeleze and Venter [S13] in 2013 propose a generic model in the process of standardization called Harmonized Digital Forensics Investigation (HDFI). This model integrates other models studied by the authors and has parallel actions. Rahaditya et al., [S15] focus on the First Digital Forensics Research Workshop (DFRWS) which proposes a six-phase research process.

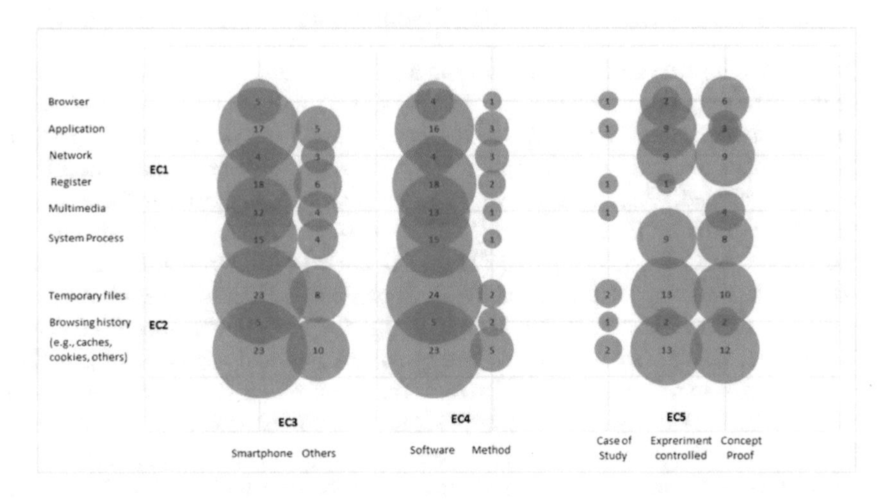

Fig. 1. Origin of evidence, local artifacts, end device, dependencies and evaluation method.

Tso et al. [S05] uses the methodology of the comparative backup process which is created for IOS operating systems and uses the iTunes tool, where basically compares the files of the device with the backup. The method presented by Husain and Sridhar in [S23] start with the creation of test data, followed by data acquisition from iPhone, and finalizing with the analysis of information. Moreover, one of the most complete methodologies for IOS has six steps according to [S14]: (i) preparation, (ii) create information, (iii) documentation and hashes, (iv) capture of packages, (v) location, and (vi) comparison of new data with the original device. Most of the papers deal with the acquisition of information but O'Shaughnessy and Keane [S06], and Sharma et al. [S07] have a traditional preservation phase that involves the event site protection and the digital evidence. It includes: (i) the gathering of data from the volatile memory and from the log files, and (ii) the isolation of the device from the network. However, in data deployed on cloud platforms, the physical preservation is possible only if the devices can be accessed. Therefore, there are no other type of data preservation for this kind of platforms, since the data is stored in virtual images.

Finally, there are forensic analyzes in real cases. Hajdarevic and Dzaltur [S31] analyze information obtained from e-mail, where the primary steps are the collection and analysis of data from the network applied to a company. When referring to hand smart devices, it is not just cell phones, iPad or tablets, but they could be any other device with an Internet connection. Hence, there are several additional devices with Internet connection (e.g., Internet of Things devices) that can provide useful information in a forensic investigation. Harbawi and Varol [S11] discuss a theoretical model of acquisition where the last-on-scene algorithm with seven steps is proposed (i.e., inspect seized, digital evidence retrieval, inspect irregularities, acquire digital evidence and produce backup copies, seizure for device, restricted, report). Finally, [S11] states that the forensic area is still young and there is not extensive research in this field.

5.3 Tools and Software for Extracting Digital Evidence from a Mobile Device

Scrivens and Lin [S03] analyze software and hardware methods. In the hardware field, some of the most popular methods are the Chip-Off, JTAG (Join Test Action Group), Forensic Software Suites, ADB (Android Debug Bridge) [S38]. Also, there are tools that are enabled, such as the one presented by Varma et al. [S16], where the LIFTR method is developed, which uses recovery engines such as Bulk Extractor. Other important tools presented by Yates and Chi [S25], Shortall and Azhar [S33], Tajuddin and Manaf [S02] are: FTK [S39], Mobile Phone Examiner [S40], Oxygen Forensic Suite [S41], EnCase Neutrino [S42], UFED [S43]. In particular, Ntantogian et al. [S22] use open source forensics tools, which is very useful for the examination of credential authentication. Other studies, such the one by Jahankhani and Azam [S36] and Kubi et al. [S12] perform reviews about all the tools that are useful for forensic investigation in mobile devices. Alyahya and Kausar [S19] show a comparison between two tools, AXIOM and Autopsy, this experiment is done in a Snapchat-specific application. The results of this study show that the application with the best tool is AXIOM, since it recovers Snapchat artifacts with high percentages. Yadav et al. [S08] present a comparison between six commercial and open source forensic tools: Encase, DFF, FTK, TSK, Helix, and Liveview.

Tso et al., [S05] and Mutawa et al. [S18] analyze the data through a proprietary Apple Company application. iTunes is not only an audio and video player but also has the synchronization function which can make backup copies. Another point to consider according to Mutawa et al. [S18] is the deactivation of the automatic synchronization in order to conserve the integrity of the information since it prevents the interchange of information between the device and the computer. Meanwhile, the tools Oxygen Forensics, Katana Forensics and Elcomsoft are used to extract evidence from Android based operating systems [S14].

5.4 Platforms

Nowadays, the most popular operating systems are Android and IOS, each with a specific architecture. The iOS architecture is based on layers, where the top-level layers interact as intermediaries between the hardware and applications. Applications communicate through system interfaces, and that mechanism provides the development of applications that work on devices with different capabilities of hardware. In contrast, Android, which is a Linux based operating system, uses several partitions to organize the file-folders on the device, each one with its own functionality.

Other studies [S03], [S16], [S19], [S20], [S22], [S25] apply the analysis on Android devices. Firstly, it is important to know the path in which data are stored, the type of privileges granted and which method and tool can be used for the extraction of such data. Android devices use an isolated space in which the user's applications and data are located and another one where users have their own storage location [S13].

Anglano [S20] does not use a device, but a virtualization platform called YouWave. This platform emulates an Android device. Here, a virtual machine (VirtualBox) is created, where applications are analyzed with an emulator. Similarly, Anglano et al. [S21] present a virtualization of Android devices called Android Virtual Devices or AVD, which acts as a real physical device. In IOS devices investigators have to consider possible options between those that limit the visualization (e.g., Apple File Communication) and those in which the access for certain files of the device are located in the media and are not part of the system [S05], [S18], [S23], [S25].

Yates and Chi [S25] analyze other operating systems such as Blackberry, which have devices with OS or Windows Mobile, in which the logical acquisition is not possible; therefore, the next step is the physical acquisition.

Also, there are a lot of applications for different operative systems (e.g., Android, IOS, Linux, and Windows Mobile.), some of which are used for calls and video conferences. These applications are useful when a forensic investigation needs to be performed [S06]. Then, studies such as [S01], [S02], [S03], [S05], [S19]–[S21], [S23], [S28], [S29], [S30], [S32], [S33] perform a forensic analysis in instant messaging applications, this allow users to constantly be in touch with other people in order to exchange information (e.g., phone numbers, texts, video, images, dates). Besides, users are constantly exchanging information such as contacts, texts, video/voice, images, dates with other users. However, there are cases when there are not relevant data for the forensics inspection and investigator to examine impractical information [S05], [S18]. Another type of instant message is Snapchat, in recent years this application has become more popular since it is possible to upload multimedia files or messages, which are available online for a period of 24 h. This new technology has been a challenge when the data generated is required [S19]. In 2015, Walnycky et al. [S28] gave an extensive analysis of 20 social-messaging applications. These tests were conducted in a controlled environment; here, the network traffic was analyzed from the device when messages were sent, whether they were text, video or images.

Chen and Mao [S30] perform a forensic analysis in emails. In order to execute the experimental test, they used the MailMaster and QQMail tools and focus on the volatile memory of the Android devices. On the other hand, Ovens and Morison [S14] perform an analysis for IOS, these types of devices have the Mail application, and it is researched how the transfer of Apple's email and contacts between a client and the cloud data works. Shortall and Azhar [S33] present a study where they examine the metadata of the attachments included in emails.

Mushcab and Gladyshev [S04] analyze Instagram and path social networking applications; where, the results were not favorable since backup copies do not reveal fundamental information. Tso et al. [S05] deal with Apple iTunes, which is mainly an application for computers; however, it is a key tool for extracting evidence. It manages directly on the computer via the synchronization setting management. When connected to the iTunes, the summary information of the device such as device name, version, and serial number, contacts, etc. is downloaded in the computer. Anglano [S20] shows that the messages and contacts can be reconstructed chronologically. They have been exchanged by users but instead of using a handheld device, it is replaced by an emulator of the Android operating system installed on the computer.

Now, this document also considers storage applications deployed on the cloud: For example, Grispos et al. [S35] study the applications of Dropbox and Box. As a result, it can be seen files on the cloud, and users do not perform maintenance in the cache since it is not deleted. Moreover, Ovens and Morison [S14] also give an approach to cloud services where they found that there are not consistent cryptographic hashes.

In Fig. 1, it is possible to summarize and analyze the intersection of some extraction criteria, which help to understand the main relations between the concerns and characteristics of forensics for mobile devices. A bubble chart is shown (see Fig. 1), according to the analysis about the selected articles it was found that, in the category Origin of evidence there are 17 applications on smartphones, while in the category of local artifacts there are 23 others artifacts (e.g., cache, cookies, logs, volatile memory, non-volatile memory) on smartphone and 10 were developed on the others devices (iPad, tablet, iPod).

On the other hand, according to the dependencies topic: in the category of origin of evidence, the greater development is oriented to the software depending on the platform (e.g., Android, IOS, Linux, Windows Mobile); also, there is only one study dedicated to applications; and there are 23 local artifacts for software.

Finally, referring to the evaluation methods: the majority of studies are experiments controlled in the category of local artifacts and few studies are case study for both origins of evidence and local artifacts.

6 Conclusions and Future Work

This paper presents a systematic mapping study on the forensics analysis on mobile devices, centered on four research questions. Here, 36 papers were selected after the application of the inclusion and exclusion criteria. The results show that there is a shortage of origin of evidence in browser because the investigation focus in analysis of applications (i.e., instant messaging, social networks, e-mail, voice, and video).

Moreover, there is a lack of dependencies on platforms that are not Android. Although, there are not standardized methodologies for forensic analysis on mobile devices; there are guidelines that can help to carry out a correct performance of the forensics process. Also, the protection of the evidence against elements that may affect it has not been mentioned in any study (e.g., surrounding it in a Faraday cage). This technique avoids possible remote accesses that can modify or eliminate all or part of the information contained, compromising the evidence. Also, in the investigation, the topic of binnacle with records of all the activities that the user made on the device was not addressed. This would be a good tool for forensic analysis.

Finally, it is necessary to consider that a good forensics analysis can help discovering important information that could be useful for the judge to take an appropriate decision.

The forensic analysis on a mobile device is a discipline which presents a fast growth and is within an area that is under continuous development. This paper can be a means to understand (e.g., type of evidence, tools, methodologies) this fast-evolving area. It is also expected to further research the development of solutions in the protection of evidence, but also to the standardization aspect.

While the review shows that there is a large number of studies related to the forensics area on mobile devices where different tools are used for the extraction of a particular element. It also shows that there are no studies comparing all the evidence extracted from a mobile device, thus obtaining a record of all the activities that the user performs on said device. As future work is planned to implement a tool that obtains a record of activities, this would be achieved with the use of several forensic tools for the extraction of digital evidence, all this information is combined in a single report for the review of the researcher.

ANNEX: Papers Selected

[S01] K. Barmpatsalou, D. Damopoulos, G. Kambourakis, and V. Katos, "A critical review of 7 years of Mobile Device Forensics," *Digit. Investig.*, vol. 10, no. 4, pp. 323–349, 2013.

[S02] T. B. Tajuddin and A. A. Manaf, "Forensic investigation and analysis on digital evidence discovery through physical acquisition on smartphone," *2015 World Congr. Internet Secur. WorldCIS 2015*, pp. 132–138, 2015.

[S03] N. Scrivens and X. Lin, "Android digital forensics," *Proc. ACM Turing 50th Celebr. Conf. - China - ACM TUR-C'17*, pp. 1–10, 2017.

[S04] R. Al Mushcab and P. Gladyshev, "iPhone 5 s Mobile Device," *Int. Work. Secur. Forensics Commun. Syst. 2015*, pp. 146–151, 2015.

[S05] Y. Tso, S.-J. Wang, C.-T. Huang, and W. Wang, "iPhone social networking for evidence investigations using iTunes forensics," *Proc. 6th Int. Conf. Ubiquitous Inf. Manag. Commun. - ICUIMC'12*, p. 1, 2012.

[S06] S. O'Shaughnessy and A. Keane, "Impact of Cloud Computing on Digital Forensic Investigations," *Adv. Digit. Forensics IX*, pp. 291–303, 2013.

[S07] P. Sharma, D. Arora, and T. Sakthivel, "Information and Communication Technology for Intelligent Systems (ICTIS 2017) - Volume 1," vol. 83, no. Mcc, 2018.

[S08] S. Yadav, K. Ahmad, and J. Shekhar, "Analysis of Digital Forensic Tools and Investigation Process," *High Perform. Archit. Grid ...*, pp. 435–441, 2011.

[S09] F. Amato, G. Cozzolino, A. Mazzeo, and N. Mazzocca, "Correlation of digital evidences in forensic investigation through semantic technologies," *Proc. - 31st IEEE Int. Conf. Adv. Inf. Netw. Appl. Work. WAINA 2017*, pp. 668–673, 2017.

[S10] F. Cohen, "Two models of digital forensic examination," *4th Int. Work. Syst. Approaches to Digit. Forensic Eng. SADFE 2009*, pp. 42–53, 2009.

[S11] M. Harbawi and A. Varol, "An improved digital evidence acquisition model for the Internet of Things forensic I: A theoretical framework," *2017 5th Int. Symp. Digit. Forensic Secur. ISDFS 2017*, 2017.

[S12] A. Kubi, S. Saleem, and O. Popov, "Evaluation of some tools for extracting e-evidence from mobile devices," *Int. Conf. Appl. Inf. Commun. Technol. AICT*, no. 10, 2011.
[S13] S. Omeleze and H. S. Venter, "Testing the harmonised digital forensic investigation process model-using an Android mobile phone," *Inf. Secur. South Africa, ISSA 2013*.
[S14] K. M. Ovens and G. Morison, "Identification and analysis of email and contacts artefacts on iOS and OSX," *Int. Conf. Availab., Reliab. Secur. ARES 2016*, pp. 321–327, 2016.
[S15] I. P. Agus, "Prototyping SMS Forensic Tool Application Based On Digital Forensic Research Workshop 2001 (DFRWS) Investigation Model," 2016.
[S16] S. Varma, R. J. Walls, B. Lynn, and B. N. Levine, "Efficient Smart Phone Forensics Based on Relevance Feedback," *Proc. 4th ACM Work. Secur. Priv. Smartphones Mob. Devices - SPSM'14*, pp. 81–91, 2014.
[S17] R. J. S. Rueda and B. Dewar Wilmer Rico, "Defining of a practical model for digital forensic analysis on Android device using a methodology post-mortem," *2016 8th Euro Am. Conf. Telemat. Inf. Syst. EATIS 2016*, pp. 0–4, 2016.
[S18] N. Al Mutawa, I. Baggili, and A. Marrington, "Forensic analysis of social networking applications on mobile devices," *Digit. Investig.*, vol. 9, no. SUPPL., pp. S24–S33, 2012.
[S19] T. Alyahya and F. Kausar, "Snapchat Analysis to Discover Digital Forensic Artifacts on Android Smartphone," *Procedia Comput. Sci.*, vol. 109, pp. 1035–1040, 2017.
[S20] C. Anglano, "Forensic analysis of whats app messenger on Android smartphones," *Digit. Investig.*, vol. 11, no. 3, pp. 201–213, 2014.
[S21] C. Anglano, M. Canonico, and M. Guazzone, "Forensic analysis of Telegram Messenger on Android smartphones," *Digit. Investig.*, vol. 23, pp. 31–49, 2017.
[S22] C. Ntantogian, D. Apostolopoulos, G. Marinakis, and C. Xenakis, "Evaluating the privacy of Android mobile applications under forensic analysis," *Comput. Secur.*, vol. 42, pp. 66–76, 2014.
[S23] M. I. Husain and R. Sridhar, "iForensics: Forensic analysis of instant messaging on smart phones," *Lect. Notes Inst. Comput. Sci. Soc. Telecommun. Eng.*, vol. 31, pp. 9–18, 2010.
[S24] A. Mylonas, V. Meletiadis, B. Tsoumas, L. Mitrou, and D. Gritzalis, "Smartphone forensics: A proactive investigation scheme for evidence acquisition," *IFIP Adv. Inf. Commun. Technol.*, vol. 376 AICT, no. September 2011, pp. 249–260, 2012.
[S25] M. Yates, "Practical investigations of digital forensics tools for mobile devices," *2010 Inf. Secur. Curric. Dev. Conf.*, pp. 156–162, 2010.
[S26] A. Mylonas, V. Meletiadis, L. Mitrou, and D. Gritzalis, "Smartphone sensor data as digital evidence," *Comput. Secur.*, vol. 38, no. 2012, pp. 51–75, 2013.
[S27] M. Taylor, G. Hughes, J. Haggerty, D. Gresty, and P. Almond, "Digital evidence from mobile telephone applications," *Comput. Law Secur.*, vol. 28, no. 3, pp. 335–339, 2012.

[S28] D. Walnycky, I. Baggili, A. Marrington, J. Moore, and F. Breitinger, "Network and device forensic analysis of Android social-messaging applications," *Digit. Investig.*, vol. 14, no. S1, pp. S77–S84, 2015.
[S29] H. C. Chu, C. H. Lo, and H. C. Chao, "The disclosure of an Android smartphone's digital footprint respecting the Instant Messaging utilizing Skype and MSN," *Electron. Commer. Res.*, vol. 13, no. 3, pp. 399–410, 2013.
[S30] L. Chen and Y. Mao, "Forensic analysis of email on android volatile memory," *10th IEEE Int. Conf. Big Data Sci. Eng.*, pp. 945–951, 2016.
[S31] K. Hajdarevic and V. Dzaltur, "An approach to digital evidence collection for successful forensic application: An investigation of blackmail case," *2015 38th Int. Conv. Inf. Commun. Technol. Electron. Microelectron. MIPRO 2015*, pp. 1387–1392, 2015.
[S32] G. B. Satrya, P. T. Daely, and M. A. Nugroho, "Digital forensic analysis of Telegram Messenger on Android devices," *2016 Int. Conf. Inf. Com. Technol. Syst.*, pp. 1–7, 2016.
[S33] A. Shortall and M. A. Bin Azhar, "Forensic Acquisitions of WhatsApp Data on Popular Mobile Platforms," *6th Int. Conf. Emerg. Secur. Technol. EST 2015*, pp. 13–17, 2016.
[S34] H. K. S. Tse, K. P. Chow, and M. Y. K. Kwan, "The next generation for the forensic extraction of electronic evidence from mobile telephones," *Int. Work. Syst. Approaches Digit. Forensics Eng., SADFE*, 2014.
[S35] G. Grispos, W. B. Glisson, and T. Storer, "Using smartphones as a proxy for forensic evidence contained in cloud storage services," *Proc. Annu. Hawaii Int. Conf. Syst. Sci.*, pp. 4910–4919, 2013.
[S36] A. Blyth, I. Sutherland, H. Jahankhani, and A. Azam, "Review of Forensic Tools for Smartphones," *Ec2Nd 2006*, pp. 69–84, 2007.
[S37] E. W. a Huebner, D. Bem, and O. Bem, "Computer forensics: past, present and future," *Inf. Secur. Tech. Rep.*, vol. 8, no. 2, pp. 32–36, 2003.
[S38] "Android Developers." https://developer.android.com/. [Accessed: 3-Aug-2018].
[S39] "Forensic Toolkit." https://accessdata.com/products-services/forensic-toolkit-ftk. [Accessed: 3-Aug-2018].
[S40] "Product Downloads." https://accessdata.com/product-download/mobile-phone-examiner-plus-mpe-5.5.4. [Accessed: 3-Aug-2018].
[S41] "Oxygen Forensics - Mobile forensics solutions: software and hardware." [Online]. Available: https://www.oxygen-forensic.com/en/. [Accessed: 3-Aug-2018].
[S42] "EnCase Forensic Software-Top Digital Forensics & Investigations Solution." https://www.guidancesoftware.com. [Accessed: 3-Aug-218].
[S43] "UFED Ultimate/PA-.", https://www.cellebrite.com/es/products/ufed-ultimate-es/. [Accessed: 23-Aug-2018].

References

1. Petersen K, Feldt R, Mujtaba S, Mattsson M (2008) Systematic mapping studies in software engineering. EASE 8:68–77
2. Garfinkel SL (2010) Digital forensics research: the next 10 years. Digital Invest. 7:S64–S73
3. Kitchenham B (2004) Procedures for performing systematic reviews. Keele, UK, Keele University, vol 33, pp 1–26 (2004)
4. Kitchenham B, Pearl Brereton O, Budgen D, Turner M, Bailey J, Linkman S (2009) Systematic literature reviews in software engineering – a systematic literature review. Inf Softw Technol 51(1):7–15
5. Sánchez C, Cedillo P, Bermeo A (2007) A systematic mapping study for intelligent user interfaces. In: IEEE international conference on information systems and computer science (INCISCOS)
6. Cedillo P, Fernandez A, Insfran E, Abrahão S (2013) Quality of web mashups: a systematic mapping study. In: Sheng QZ, Kjeldskov J (eds.) Current Trends in Web Engineering, vol 8295, pp 66–78. Springer, Cham
7. Szvetits M, Zdun U (2016) Systematic literature review of the objectives, techniques, kinds, and architectures of models at runtime. Softw Syst Model 15(1):31–69
8. Petersen K, Vakkalanka S, Kuzniarz L (2015) Guidelines for conducting systematic mapping studies in software engineering: an update. Inf Softw Technol 64:1–18
9. Alherbawi N, Shukur Z, Sulaiman R (2013) Systematic literature review on data carving in digital forensic. Procedia Technol 11:86–92
10. Alharbi S, Weber-Jahnke J, Traore I (2011) The proactive and reactive digital forensics investigation process: A systematic literature review. In: International Conference on Information Security and Assurance, pp 87–100 (2011)
11. Robinson TM, Clemens C (2014) Service-learning and forensics: a systematic literature review. Forensic 99(2):35–49

Software Engineering

Analytic Hierarchy Process of Selection in Version Control Systems: Applied to Software Development

Javier Vargas(✉), Franklin Mayorga, David Guevara, and Edison Álvarez

Facultad de Ingeniería en Sistemas Electrónica e Industrial,
Universidad Técnica de Ambato, Ambato, Ecuador
{js.vargas,fmayorga,dguevara,ealvarez}@uta.edu.ec

Abstract. This article presents a systematic study of selection in version control systems. The requirements established in the software development lifecycle, are coupled to the criteria of the developers in different alternatives of version control. Each developer establishes their ability to work collaborative for the integration of software development with new versions, which are presented in different times, segments and works. The selection of software tools for the different stages of the software development process is a common activity, which is often done ad-hoc or specialized. For this, it is analyzed a hierarchical analytical process represented graphically for decision making in a systematic mapping study, based on the criteria of the developers.

Keywords: Systematic study · Version control systems · Software development
Analytic hierarchy process

1 Introduction

The software development lifecycle examines the growth process of a project. The concept of life cycle is presented by several models such as: Sequential model, V-model, Spiral model, among others, that normalize the systematization process in a project [1]. Each of these models involves a systematic order in stages of a software development; these stages are presented in a time frame procedural of preparation [2, 3]. The implementation stage does not study the behavior of the development process by versions, but seeks to adjust to the new versions that can be adapted [4].

These new versions present the change or update (fetch or pull), union or aggregation (merge or add) of source code [5]. For this, the concept of version control systems (VCS) [6] is included, centrally as a case study in the selection of a VCS, aimed at software development.

It is important to guarantee the best option of a VCS adaptable to the requirements, this is done by analyzing indicators proposed by the developers how: time (version history), segment (backup and recovery) and work (centralized system and collaborative work) [7], these selection indicators are implemented using a hierarchical analytical process model (AHP) [8].

M. Botto-Tobar et al. (Eds.): TICEC 2018, AISC 884, pp. 75–85, 2019.
https://doi.org/10.1007/978-3-030-02828-2_6

Figure 1 shows that the selection within the AHP model is divided into three main approaches: the goal (goal-A), the criterion (criterion-B) and the alternative (alternative-C), each of these approaches is a selection procedure that can be subjected to hierarchical subdivisions [9, 10] for decision making.

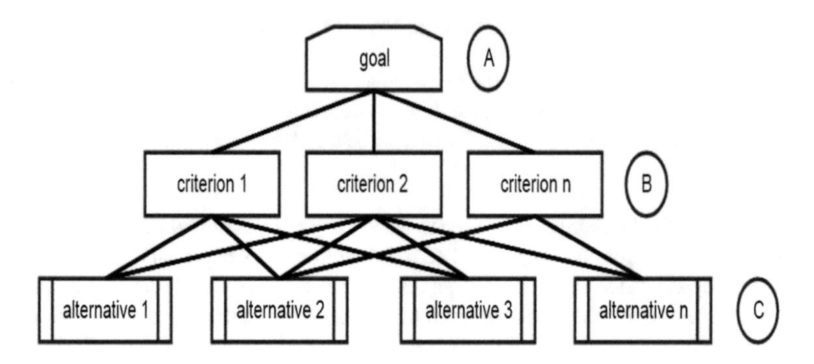

Fig. 1. AHP model.

The three most commonly used VCS alternatives for selection are: Git [11], Mercurial [12] and Apache Subversion (SVN) [13], with the review of articles under the perspective of systematic mapping (SMS) [14]. Determined to help researchers and practitioners to have a complete and valid table of prior art on collaborative work, and to identify potential gaps in current and future research [15].

As an empirical study, tools selection problem is investigated for stages of a software development. For the decision making process, four criteria and three alternatives are revealed. This paper aims to identify the crucial selection criteria for the version control systems. The general selection is first undertaken in order to identify the alternatives and to select those projects with relevant qualification and experience within the research.

This study was born from the need to implement a platform for presenting the results of a research project. Each project in relation of dependence with its participants and the tools used for its execution, seek to solve the problem of how to work collaboratively and manage all versions of its progress. The results within the platform obtained are tabulated from bibliographic research in different areas, which is determined as a process of continuous growth of requirements. Based on these requirements, several alternatives of VCS are presented, but without knowing which is convenient for a research project where several participants are involved, among them external and local, as well as a platform of self-management of bibliographic search.

The rest of the paper is organized as follows. Section 2 describes the literature review. Section 3 presents the AHP Model. Next, in Sect. 5 the results are detailed, and in Sect. 5 Conclusions and Future work are presented.

2 Context and State of the Art

In order to evaluate three VCS (alternative-C) presented in Table 1, four indicators or criteria (criterion-B) delimited by the software developer are initially taken.

Table 1. Version control alternatives.

VCS	alternative-C	
	Description	Objective
Git	Manage open source repositories that can be parameterized to changes between versions	Incorporate source code reviews by pairs, as part of the flow of a developer and hence the life cycle in software development [16]
Mercurial	Contains bifurcation capabilities and multiplatform source code integration	Support alterations between most common operating systems without effort [17]
SVN	It overrides changes between large source versions	Save in a flexible way the changes produced in the repository [18]

The selection criteria in principle are four, by means of the most representative set of a VCS. Each criterion seeks the sustainability of software development, for which the "matching" rule is created [19].

Each article that has matching is taken as a systematic mapping study [20], based on the following type of query (TITLE-ABS-KEY (alternative) AND TITLE-ABS-KEY (criterion)): TITLE, ABS for summary and KEY for key words which fits the section more. In addition, it is necessary to take into account the hierarchical subdivisions that can have this consultation between AND - OR.

The valuation is represented on a scale of 1 to 5:

- value of 1 for single reference (mentioned in development);
- value of 2 for cross reference (probable alternative of development);
- value of 3 for adapted reference (coupled within the Project development);
- value of 4 for applied reference (presented as part of the development process);
- and value of 5 for a valid reference (feasible for development).

For it to take a value 5 on the scale should be the VCS behavior expected by the developer. In Table 2, the values of the evaluated criteria are presented according to the alternatives in VCS referenced, based on the SMS.

The validity of the systematic study of mapping within the articles for each VCS, present reports of the empirical evidence that exists in each of them. For Git and the article of greatest matching is cited in [21], where each development process fits correctly with the project and development team. For Mercurial and article [17] the matching of cross-reference, it presents that the more information, the workflow of the software versions, fit the development process correctly. As well as in SVN in article [25], software development works best with processes of joining or aggregating source code.

Table 2. Inclusion criteria.

criterion-B	alternative-C AND criterion-B				
	alternative-C			SMS	Reference no
	Git	Mercurial	SVN		
Version history	5	5	5	Valid	Git: [21–23] Mercurial: [6, 24] SVN: [22, 25]
Backup and Recovery	2	3	3	Adapted	Git: [21, 26] Mercurial: [9, 27] SVN: [27]
Centralized System	2	2	1	Cross	Git: [17, 28] Mercurial: [17] SVN: [6]
Collaborative work	5	2	4	Applied	Git: [21, 29] Mercurial: [11, 30] SVN: [30]

In the AHP model, the non-representative values are not taken into account, since they do not comply with the hierarchical order of selection of the inclusion criteria.

3 AHP Model

To represent the AHP model, the results obtained in Table 2 are hierarchically selected to meet the goal (goal-A) of a version control system applied to software development. Figure 3 shows the resulting diagram of matching's between the criteria (criterion-B) and alternatives (alternative-C).

The continuous line segment represents a high value (5–4), a dashed line for a mean value (3–2), while for the low value (1), the valuation is not taken into account in the diagram.

For the analysis of the systematic study of mapping the indicators establish that: for the version history, it should have a record of all changes to the files. For the collaborative work, it is necessary to manipulate the flows of responsibility and tools of collaborative work by roles. For backup and recovery, the information must be stored in an agreement repository. While for the centralized system, it must be tolerant to distributed version control.

The values of the scale are of common summation, giving as an interpretation that Git has four matching's with values of 5 + 5 + 2 + 2 that gives us a result of 14. For Mercurial also has four matching's with values of 5 + 2 + 3 + 2 which gives us a result of 12. And finally we have SVN with three matching's with values of 4 + 4 + 3 giving us a result of 11.

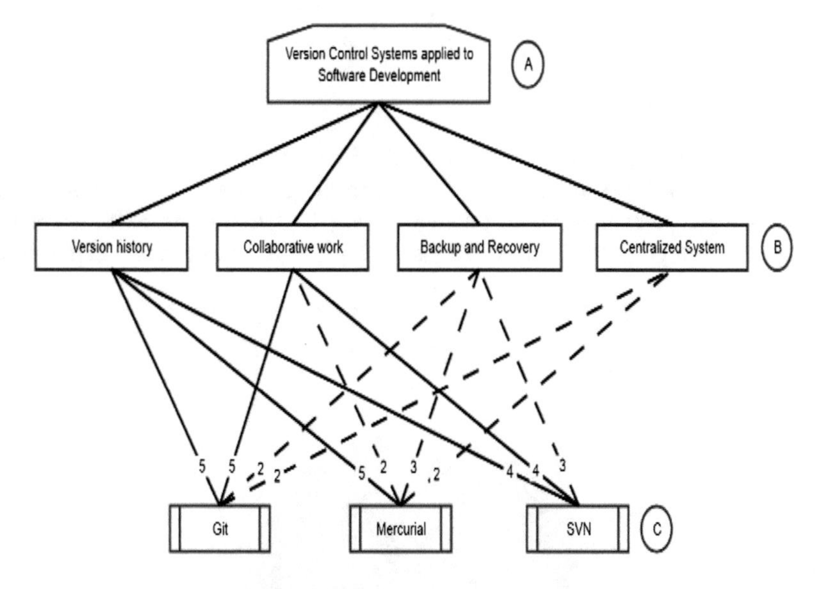

Fig. 2. AHP with reference matching.

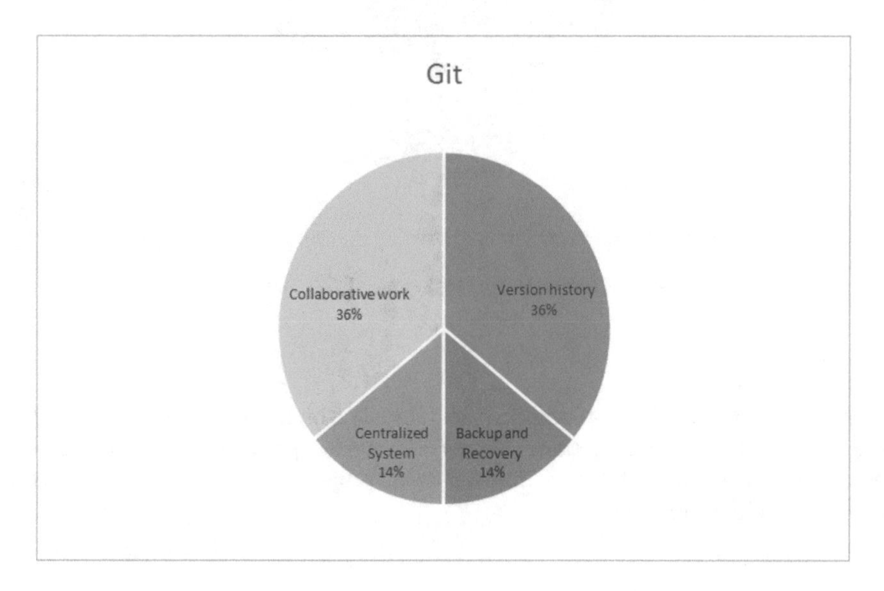

Fig. 3. Git representation, TITLE-ABS-KEY.

4 Results

4.1 Git

It includes four selection criteria (see Table 2) with summation value of 14, carried out by SMS. In Fig. 4 it can be seen that the criteria of version history and collaborative work are of equal percentage value, thus supporting the determinants of the adaptability of software development and the collaborative work with the directors and developers.

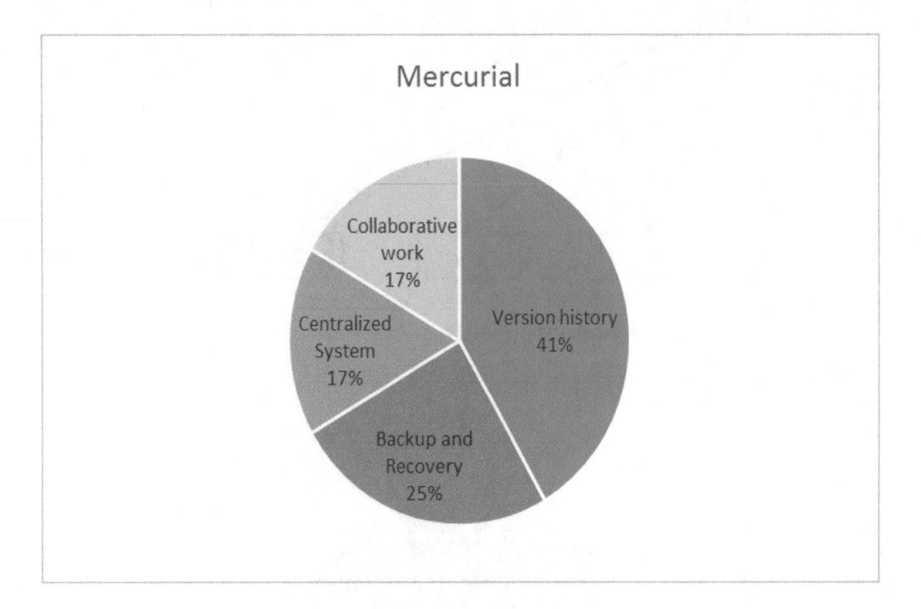

Fig. 4. Mercurial representation, TITLE-ABS-KEY

4.2 Mercurial

It encompasses four selection criteria (see Table 2) with a summation value of 12. In Fig. 5 it can be seen that the version history criterion is a percentage evaluation of 41% over 100% that comes to be the fulfillment of the project goal, thus giving a pattern of uniformity of development.

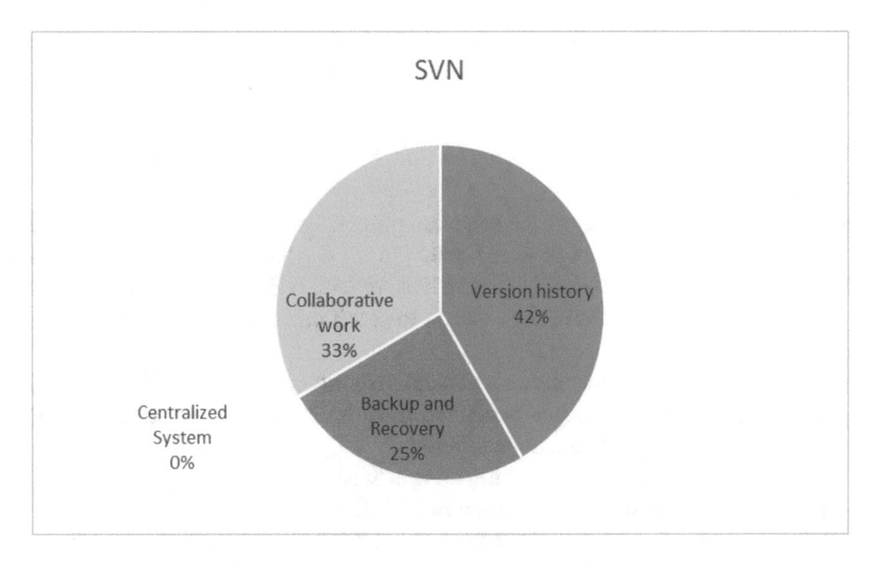

Fig. 5. SVN representation, TITLE-ABS-KEY.

4.3 SVN

For this VCS there are three selection criteria (see Table 2) with summation value of 11. In Fig. 6 it can be seen that the criteria of version history, backup and recovery and collaborative work, its frame of reference is similar percentage assessment, determining

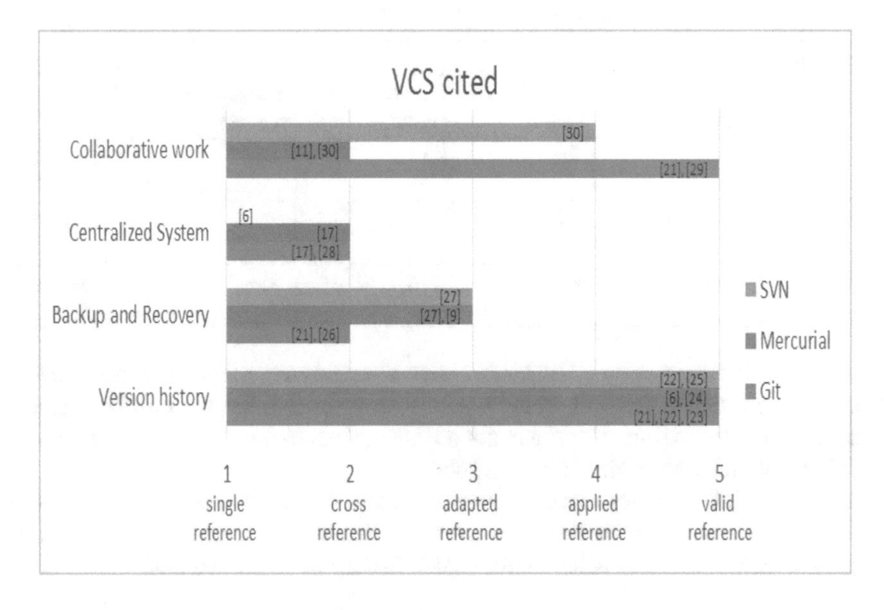

Fig. 6. Systematic mapping study.

that the VCS is half of the software development. Note that it does not fully support all criteria, while Git and Mercurial contemplate the centralized system.

5 Discussion of Results

In an environment of certainty of the selection process of a VCS, it allows the option to add quantitative data in reference to the search of the SMS bibliography. Each of these data is attached to the criteria of the alternatives of each VSC, this selection process of the C-alternative and the B-criterion, allows to reach utilization values in each search and matching obtained.

For example, the matching shows a general mapping in Fig. 6, that Git is feasible or adaptable as VCS. On the other hand, we have SVN, meets the selection criteria including the valuation of 1 for single reference.

This data is either your own selection criteria or a convenience criterion, which does not show a hierarchical order of selection of a VCS.

On the other hand, the selection process using AHP Model seeks the goal based on the analysis of each criterion with respect to the alternatives present. Each thread between the alterative ones has a weighting weight that is established in Sect. 2, this allows to have a scale so that the lines or threads can reach the goal of an adjustment VCS in software development research projects. Each selection process does not seek to rank from best to worst, which is intended to adjust for specific types of research which should be the right VCS to implement in the project.

For the final analysis of AHP the structuring presents a systematic order of selection, in addition to the number of references present in the research (see Table 3).

Table 3. Parallelisms of the distribution of AHP model.

Alternative	Criterion %				References (match)
	Collaborative work	Version history	Centralized system	Backup and recovery	
Git	36	36	14	14	14 (applied reference)
Mercurial	17	41	17	25	12 (adapted reference)
SVN	33	42	0	25	11 (adapted reference)

The selection process by AHP, hierarchizes a systematic order of selection for each alternative, presenting that Git, being one of the most used, has the disadvantages of being a centralized system, but as a major advantage that allows the collaborative work among the participants of the project as such.

For Mercurial, it divides its weighting values between collaborative work and having a centralized system, which allows it to determine that there is an orderly project execution work, as well as being able to have a good history of project versions.

On the other hand, SVN is a VCS has a large percentage of use of the historical version history of the running project, as the main disadvantage is that it cannot be a centralized system.

The criteria may vary according to the procedure taken, for example the backup and recovery can be segmented into more selection criteria. Each thread can be subdivided in each selection process as long as it meets all the alternatives. For the references the selection of Git has a total of 14 on average of the scale proposed by the developer, this determines that the search within those articles is more significant and the use of the VCS is a priority in the present research. Mercurial establishes that it is adapted reference, which determines that there is a wide field of work in versions and each one of them within the articles keeps a history of each proposed project. For SVN that is in the same Mercurial values, there is a difference in the selection criteria to be a centralized system, because they do not present research groups in the articles analyzed, which SVN does not consider as a strong criterion for the hierarchical selection process of AHP.

6 Conclusions and Future Work

The development of a software project is dependent on the collaborative work between directors and developers. Therefore selecting a version control system depends on the indicators that arise from the software development requirements, determined by the development time.

A VCS has its own characteristics, which was taken as a systematic mapping study within different references of the selected articles. The one of greater value of scale within the consultation was the history of versions with a backup and recovery, being the main thing in search of a VCS. One of the VCS that fulfilled the criterion of evaluation within the mentioned articles or references was Git. This VCS determined that the software development lifecycle, in the AHP model the evaluation objectives are adjusted for decision making, equally in its high valuation in the references.

A study of the basic hierarchical analytical process for the VCS section can help complex decision making in an orderly way so that the correct decision can be described. As future work a multi-criteria analysis in diffuse environments for the response times of VCS.

Acknowledgments. To the Ecuadorian Corporation for the Development of Research and Academia CEDIA, for research funding, development and innovation through CEPRA projects, in particular to the CEPRA-XI-2017 project; *New Foods*.

References

1. Hijazi H, Khdour T, Alarabeyyat A (2012) A review of risk management in different software development methodologies. Int J Comput Appl Technol 45:8–12
2. Booch G (1986) Object-oriented development. IEEE Trans Softw Eng 12:211–221
3. Robillard MP, Murphy GC (2007) Representing concerns in source code. ACM Trans Softw Eng Methodol 16:3

4. Boehm B, Clark B, Horowitz E, Westland C, Madachy R, Selby R (1995) Cost models for future software life cycle processes: COCOMO 2.0. Ann Soft Eng 1:57–94
5. Negishi Y, Murata H, Moriyama T (2009) A proposal of operation history management system for source-to-source optimization of HPC programs. In: Proceedings of the 7th workshop on parallel and distributed systems: testing, analysis, and debugging, PADTAD 2009
6. Rama Rao N, Chandra Sekharaiah K (2016) A methodological review based version control system with evolutionary research for software processes. In: ACM international conference proceeding series
7. Robbes R, Lanza M (2007) A change-based approach to software evolution. Electron Notes Theor Comput Sci 166:93–109
8. Al-Harbi KMA (2001) Application of the AHP in project management. Int J Proj Manag 19:19–27
9. Zhu X, Dale AP (2001) JavaAHP: a web-based decision analysis tool for natural resource and environmental management. Environ Model Softw 16:251–262
10. Thanki S, Govindan K, Thakkar J (2016) An investigation on lean-green implementation practices in Indian SMEs using analytical hierarchy process (AHP) approach. J Clean Prod 135:284–298
11. German DM, Adams B, Hassan AE (2015) A dataset of the activity of the git super-repository of Linux in 2012. In: IEEE international working conference on mining software repositories, pp 470–473
12. O'Sullivan B (2009) Making sense of revision-control systems. Commun ACM 52:57–62
13. Ngo L, Apon A (2007) Shibboleth as a tool for authorized access control to the Subversion repository system. J Softw 2:78–86
14. Costa C, Murta L (2013) Version control in Distributed Software Development: a systematic mapping study. In: Proceedings of IEEE 8th international conference on global software engineering, ICGSE 2013, pp 90–99
15. Fiordelli M, Diviani N, Schulz PJ (2013) Mapping mHealth research: a decade of evolution. J Med Internet Res 15:e95
16. Kalyan A, Chiam M, Sun J, Manoharan S (2017) A collaborative code review platform for GitHub. In: Proceedings of the IEEE international conference on engineering of complex computer systems, ICECCS, pp 191–196
17. Rodriguez-Bustos C, Aponte J (2012) How distributed version control systems impact open source software projects. In: Proceedings of the 9th IEEE working conference on mining software repositories. IEEE Press, Piscataway, pp 36–39
18. Schwind M, Schenk A, Schneider M (2010) A tool for the analysis of social networks in collaborative software development. In: Proceedings of the annual Hawaii international conference on system sciences
19. Stoklasa J, Talášek T, Talašová J (2016) AHP and weak consistency in the evaluation of works of art - a case study of a large problem. Int J Bus Innov Res 11:60–75
20. Chou T-C, Cheng S-C (2006) Design and implementation of a semantic image classification and retrieval of organizational memory information systems using analytical hierarchy process. Omega 34:125–134
21. Cosentino V, Izquierdo JLC, Cabot J (2015) Assessing the bus factor of Git repositories. In: Proceedings of 2015 IEEE 22nd international conference on software analysis, evolution, and reengineering, SANER 2015, pp 499–503
22. Greene GJ, Esterhuizen M, Fischer B (2016) Visualizing and exploring software version control repositories using interactive tag clouds over formal concept lattices. Inf Softw Technol 87:223–241

23. Just S, Herzig K, Czerwonka J, Murphy B (2016) Switching to Git: the good, the bad, and the ugly. In: Proceedings of international symposium on software reliability engineering, ISSRE, pp 400–411
24. Rocco D, Lloyd W (2011) Distributed version control in the classroom. In: SIGCSE 2011 - proceedings of the 42nd ACM technical symposium on computer science education, pp 637–641
25. Brindescu C, Codoban M, Shmarkatiuk S, Dig D (2014) How do centralized and distributed version control systems impact software changes? In: Proceedings of international conference on software engineering, pp 322–333
26. Davis RC (2015) Git and GitHub for librarians. Behav Soc Sci Libr 34:158–164
27. Uquillas Gómez V, Ducasse S, D'Hondt T (2015) Visually characterizing source code changes. Sci Comput Program 98:376–393
28. De Alwis B, Sillito J (2009) Why are software projects moving from centralized to decentralized version control systems? In: Proceedings of the 2009 ICSE workshop on cooperative and human aspects on software engineering, CHASE 2009, pp 36–39
29. Biazzini M, Monperrus M, Baudry B (2014) On analyzing the topology of commit histories in decentralized version control systems. In: Proceedings of 30th international conference on software maintenance and evolution, ICSME 2014, pp 261–270
30. Thao C, Munson EV (2011) Version-aware XML documents. In: DocEng 2011 - proceedings of the 2011 ACM symposium on document engineering, pp 97–100

Reliability and Validity of Postural Evaluations with Kinect v2 Sensor Ergonomic Evaluation System

Christian Mariño[1](✉), Rafael Santana[1], Javier Vargas[1], Luis Morales[1], and Lorena Cisneros[2]

[1] Facultad de Ingeniería en Sistemas Electrónica e Industrial, Universidad Técnica de Ambato, Ambato, Ecuador
{christianjmarino,rsantana8412,js.vargas, luisamorales}@uta.edu.ec

[2] Faculty of Business Administration, Bay Path University, Massachusetts, USA
lcisneros@baypath.edu

Abstract. Postural evaluations are becoming a priority in a world where occupational diseases derivatives from ergonomic situations are progressively common with serious consequences. This document proposes a semi-automatic ergonomic evaluation system with Kinect V2, which use the RULA method. It makes possible the discovery of conditions with postural risk and reduces errors in the estimating the evaluators' measurements due to lack of experience, expertise and handling of measurement instruments. Reducing the ergonomic evaluation time, compared to the direct postural evaluation method at a low cost. In a controlled environment, 30 participants with homogeneous characteristics are evaluated their postural work, using a direct method (goniometer) and a non-invasive (Kinect V2) method, to then compared the statistically results using the coefficient of correlation named Cohen's Kappa and vagueness type A. The final RULA scores issued by the direct method of postural evaluation and the proposed system are correlated. The uncertainty results established in the angular measurements on the arm, forearm, wrist, trunk and neck were respectively (±0.36°, ±0.22°, ±0.59°, ±0.27° and ±0.28°). A Cohen's Kappa correlation coefficient of 0.953 was obtained, which means an almost perfect correlation between the evaluations of both systems.

Keywords: Ergonomic evaluation · Kinect V2 · Non-invasive · Uncertainty

1 Introduction

Being ergonomics a multidisciplinary science that embraces principles of biology, psychology, anatomy and physiology in order to adapt the work environment to the capabilities and postural limitations of workers [1, 2], its objective is to prevent the emergence of inflammatory and/or degenerative injuries to muscle, tendons, joints, tissues and nerves known as musculoskeletal disorders (MSDs) caused or aggravated by the work environment [3].

M. Botto-Tobar et al. (Eds.): TICEC 2018, AISC 884, pp. 86–99, 2019.
https://doi.org/10.1007/978-3-030-02828-2_7

The MSDs derived from work are the most common health problems in Europe [4], among the most common risk factors causing them are physical and biomechanical factors, organizational and psychosocial factors, and individual factors [5].

To evaluate the MSDs and the various components by which they occur, there are several methods of ergonomic evaluation of indirect type (based on observation), direct, self-reports [6] and semiautomatic [7]. Indirect assessment methods use data collected through photos, videos and surveys by a professional who observes the activity to be evaluated [8], but they have the disadvantage of variability in the data, it depends on the experience and expertise of the evaluator (this is subjectivity) [9], on the other hand, has advantages economic and requires simple tools [10].

Direct evaluation methods are those that provide the most accurate data [11, 12], but are not frequently used because it requires equipment such as exoskeletons and electromagnetic sensor systems that are expensive and not applicable in the workplace because those are invasive [13].

Self-reports collect data from workers' journals, interviews, questionnaires and web questionnaires. These methods are relatively inexpensive, applicable to many populations and easy to use. The problem is that workers' perceptions of risk exposure are vague and unreliable [14, 15].

Semi-automatic evaluation methods use 3D sensors and high-tech software are increasingly being developed for these methods and there are more options on the market [16, 17]. The advantages of these systems are accuracy of measurement, reduced evaluation time, non-invasive [18], multiple applicability, simple user interface and practical reporting [19].

There are several methods of indirect postural evaluation. For upper extremities assessment, Rapid Upper Limb Assessment (RULA) is commonly used, a method where the postures of the body segments are observed and graded, increasing the score as the postures are more deviated from the neutral position of the person, allowing the evaluation of the postures adopted, the repetitiveness of movements and the force applied in the static activity [20].

In practice, joint angles are estimated through video or captured images of people, so the implementation of low-cost 3D sensors results in the semi-automation of observation methods [7, 16, 21].

When validating new semi-automatic postural evaluation systems, several studies used goniometry as an established, proven and accurate method. In addition, statistical methods are used in which the means of validated methods are related to the system to be evaluated. The statistical methods used are the coefficient correlation called Cohen's Kappa and the coefficient correlation called intraclass [7, 22].

This study is focused on validating the data delivered by the semi-automatic system called Ergonomic Evaluation System with the use of the Kinect V2 sensor (SEEK V2) [23], which is applied in the postural evaluation of the manual cutting activity in materials for shoe manufacturing. The ergonomic postural method programmed in the proposed system is RULA.

Within the industry at national level footwear manufacturing is the fastest growing, for this reason this research is conducted in this sector based on an existing agreement, so the results will be applied later to the same industry.

The concordance of the results issued by the SEEK V2 system evaluations and the direct evaluations applied with the RULA method are studied. The latter measures were carried out with a direct evaluation. Assumptions are defined, H0: SEEK V2 RULA final scores do not match the final scores made with direct measurement. H1: SEEK V2 RULA final scores are consistent with the final scores made with direct measurement. In addition, the measurement uncertainty of the SEEK V2 is calculated taking as a standard the angular measurements taken with the direct method.

2 Materials and Methods

2.1 Type of Research

The current study is experimental in nature as it is carried out under controlled conditions of sensor data capture and measurement with the direct method; on a previously selected population. A correlation study is also performed between both methods of postural evaluation which calculated an uncertainty type A in the measurements.

2.2 Subjects of Study

The type of sampling is convenient because a homogeneous group was sought for the validation of the proposed system, otherwise there could be biases in the statistical analysis.

The participants studied are 30 university students, of which 26 are male and 4 are female, 3 are left-handed and 27 are right-handed. Under the following demographic characteristics, they are all mestizo, in an age range of 20 to 32 years, in a height range of 1.51 to 1.81 meters and with an average body mass index of 23.91 kg/m^2. All participants gave their written consent to participate in this study. We excluded participants with musculoskeletal problems so as not to generate biases in statistical analysis.

2.3 Environmental Conditions

The laboratory experiment was conducted on a single day in the morning in a large room so that all participants would be subjected to the same temperature and light conditions during the evaluation exercise.

2.4 Study Environment

The laboratory study consisted of a folding cutting table that should be in the center of the room isolated from unnecessary objects, in an area of 2 m^2. A cutting mold located in the center of the table was used. The evaluators along with the direct measurement tools and logs are kept away while the participant's data is being captured by the sensor within a set time.

To evaluate a left-handed or right-handed participant, the Kinect must be positioned diagonally to their dominant hand due to the limited capture area of the sensor. For this study the Kinect has been positioned in the same places previously marked on the floor of the room for data capture of left-handed or right-handed participants (see Fig. 1).

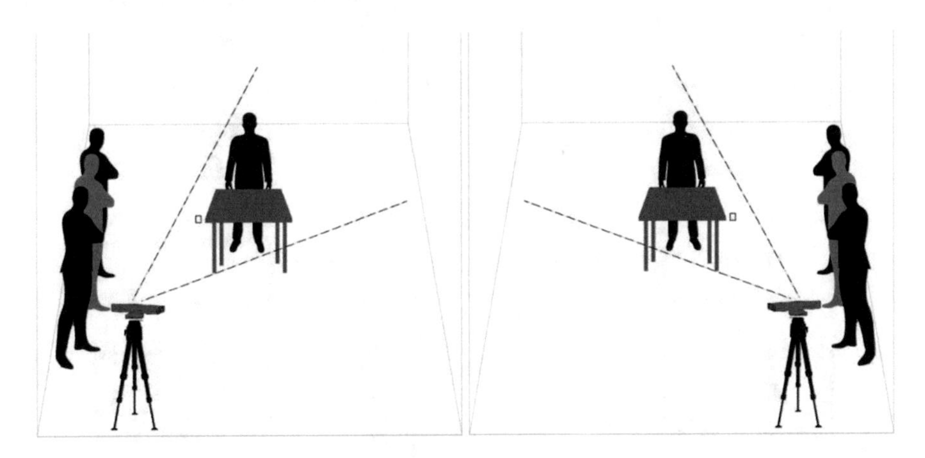

Fig. 1. Position of the sensor in the workspace.

2.5 Evaluation Procedure with SEEK V2

The SEEK V2 [20] is a system developed by the authors in previous investigations in which the programming, sensor capture method and the results found are shown. To capture the data through the Kinect, the depth camera is used as the main sensor. For this reason, the system obtains the angles in different positions using mathematical methods such as Euler angles and processing through the rotation matrix [21].

The Kinect is positioned at a height of 1.25 m and a distance of 1.5 m on the right (or left) diagonal of the cutting table. In order to facilitate evaluations in industry and other fields of study the SEEK V2 has the ability to evaluate in a range of 1 to 1.5 m in height and in a range of 1 to 2 m in lateral distance. In this study, the sensor was placed at the optimum system development distances for validation purposes.

There should be no objects or other people crossing between the Kinect and the person being tested. The processing of the sensors with respect to the evaluation system is an internal architecture where each participant generates information that the software processes through a programming for calculation of angles and points for each end evaluated (see Fig. 2).

Previous studies have found that Kinect's detection points are severely out of phase or cannot be located because clothing does not show the joints. The system proposed in this work can find the joints in people with clothing because it detects a limb and in the smaller segment where the skin is shown a point of union is located and takes it as a reference to locate a point of intersection. The correct functioning of this has been validated in an adjacent study.

The evaluations with the SEEK V2 were carried out offline, in a controlled environment where data was recorded with the sensor and after the execution was analyzed with the developed system. The objective is to obtain data a main study thrown by the direct

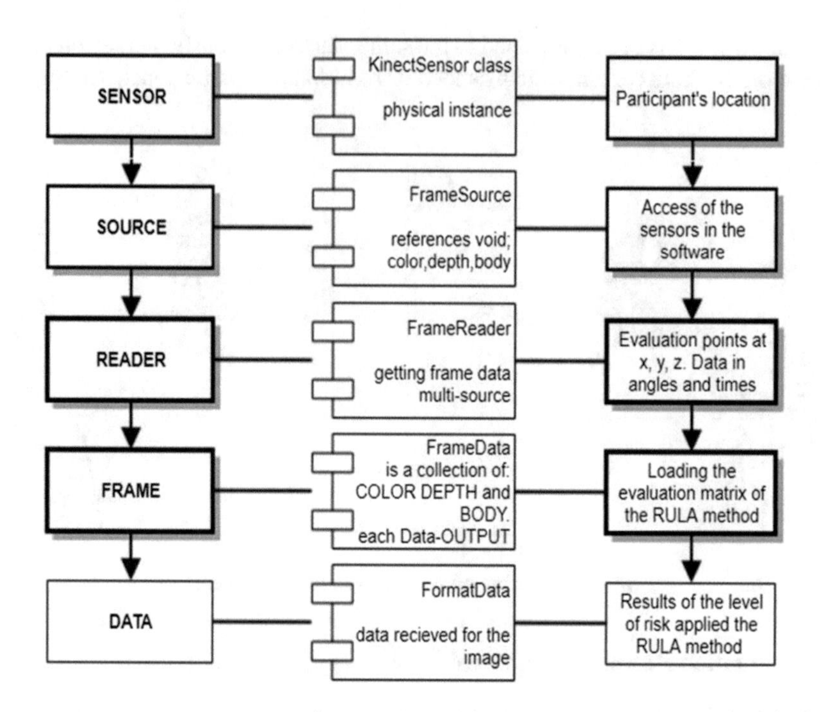

Fig. 2. Sensor processing with evaluation system.

method which allows us to compare with the obtained within the system once already evaluated.

2.6 Angular Measurement Procedure with the Direct Method

The angular values of flexion and abduction measured in the parts of the body (head, trunk, arms and forearm) taken with a goniometer are based on the RULA methodology [5, 24]. It should be indicated that the participants will carry out a task of marking the contour of a mold to simulate the manual cutting activity.

The following standardized verbal instructions are given to the participants on how to perform the manual cutting simulation activity, such as: seeking comfort to perform the task, emphasizing the edge of the cutting mold, and remaining immobile when listening to the evaluator's signal. When the most repetitive posture is determined, the participant must remain completely immobile, while measurements are taken using the anthropometric measurement points as a basis (see Fig. 3).

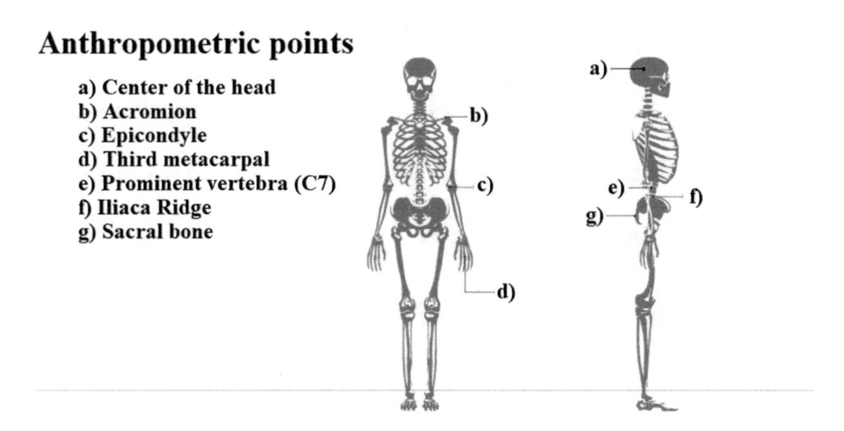

Fig. 3. Anthropometric measuring points.

The direct evaluation of the angles is done by experienced evaluators. The instruments used consist of the universal goniometer for angular joint measurements, a metric tape to define reference axes in the body and a plumb line to define the neutral position.

Arm: The angle formed by the end to the axis of the trunk shall be measured (see Fig. 4a).

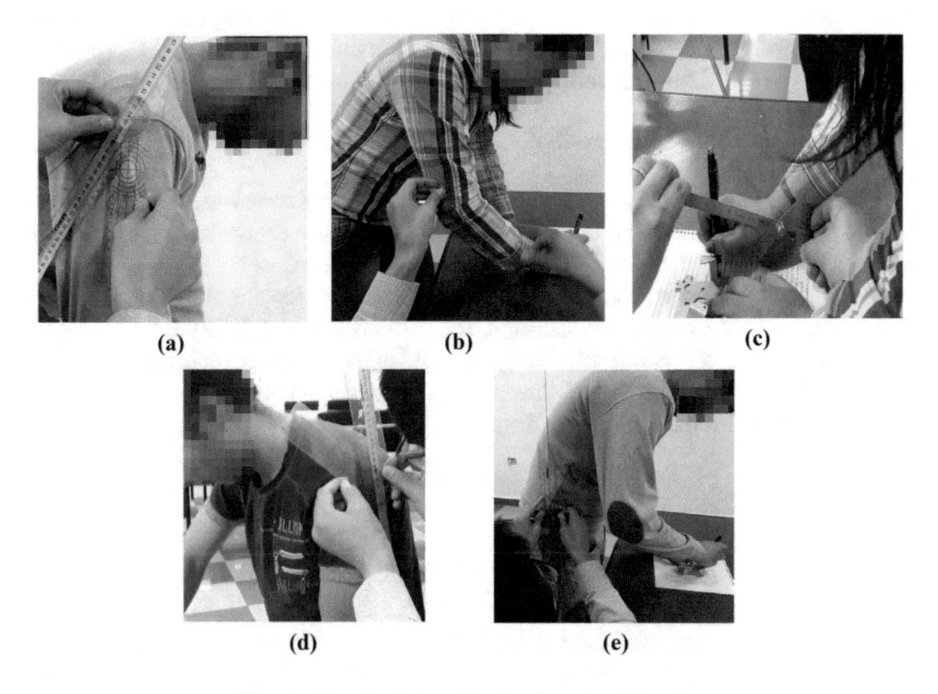

Fig. 4. Sample data collection from participants.

Forearm: The angle formed by the axis of the limb and the axis of the arm shall be measured (see Fig. 4b).
Wrist: The angle of flexion or extension is measured from the neutral position (see Fig. 4c).
Neck: The angle formed by the axis line of the head and trunk shall be measured (see Fig. 4d).
Trunk: The angle formed by the trunk and a vertical plane is measured (see Fig. 4e).

Participants were asked to wear clothes that fit their body, so methods will not be affected by the type of clothing used. Figure 3 shows the measurements taken using the direct method. Being the system able to evaluate a participant's exact points with or without appropriate work clothing.

2.7 Procedure for Statistical Analysis of Results

The absolute error is calculated from the measurements taken on the parts of the body evaluated such as the arm, forearm, wrist, neck and trunk of each participant. The direct method is considered as the standard measurement because it is the proven and validated method and the SEEK V2 is considered as an approximate value because it is the method whose uncertainty is to be estimated.

To calculate the expanded measurement uncertainty of the SEEK V2 with 95% confidence, the absolute error results (1) previously found for each joint evaluated are used. In addition, the typical uncertainty type A evaluation is used; the procedure outlined in the GUM guide is followed for the evaluation of measurement data [25].

$$|X_{Direct\,Method} - X_{SEEK\,V2}| = Absolute\,Error \tag{1}$$

It was applied the method of Bland-Atlman plots [26] of the measurement differences between the angles emitted by the SEEK V2 and those taken using the direct method of each part of the body evaluated are made using spreadsheets.

For the calculation of Cohen's Kappa concordance coefficient, the SPSS Software Version 23.0 (IBM Corporation, Chicago, IL, USA) is applied; the RULA final scores are considered as categories (see Table 1), exposed in the generated reports by the SEEK V2 and the scores reported by the direct method of postural evaluation, to interpret the degree of agreement the Landis and Koch [27] scale is used.

Table 1. Final scores RULA

Final score	Level of performance
1 o 2	Acceptable risk activity can be maintained
3 o 4	More study needed; changes in activity may occur
5 o 6	Redesign and more homework research coming soon
7	Urgent intervention in the activity

3 Results

The angular values taken from the neck, trunk, arm, forearm, wrist, and neck joints using the direct method and the reports issued by the SEEK V2 which is shown in Table 2.

Table 2. Angular values of the participants in each joint measured with direct method.

Participant number	Trunk (°)		Neck(°)		Arm (°)		Forearm (°)		Wrist(°)	
	S	D	S	D	S	D	S	D	S	D
1	8	8	21	20	11	11	70	69	42	43
2	38	38	29	29	52	52	40	39	50	52
3	31	31	54	54	25	25	58	57	14	15
4	27	26	40	40	48	50	65	64	23	28
5	19	18	37	37	14	14	39	40	33	36
6	13	13	30	30	24	24	27	26	39	46
7	23	23	41	40	34	34	38	38	15	18
8	11	10	39	40	14	14	48	46	52	57
9	44	44	29	28	67	66	66	66	16	17
10	30	30	40	41	44	44	48	48	25	26
11	43	44	29	30	51	52	62	61	23	26
12	28	30	50	50	52	52	63	64	26	25
13	46	46	29	27	46	44	66	66	14	16
14	13	15	34	34	24	24	62	63	12	12
15	8	6	82	84	13	14	42	42	26	27
16	30	30	68	70	64	63	61	60	13	13
17	28	30	55	55	21	22	80	80	24	24
18	49	48	18	18	49	50	70	70	6	8
19	41	39	58	58	54	54	69	70	22	23
20	33	33	55	53	58	58	65	66	1	1
21	52	51	36	36	40	40	96	94	20	21
22	31	31	62	61	38	38	74	74	22	23
23	28	28	58	57	30	30	61	61	25	26
24	58	58	44	44	62	60	92	92	7	8
25	62	62	26	28	92	96	88	88	5	5
26	42	42	16	15	72	74	55	55	27	28
27	46	45	26	26	74	73	59	59	8	9
28	30	30	44	44	24	26	64	64	35	35
29	35	34	30	30	38	40	54	54	1	0
30	22	22	38	40	26	26	65	65	17	20

*S: Angle taken by the SEEK V2.
*D: Angle taken with the direct method.

In the Bland-Altman plots with Average Mean Differences (AMD) and 95% Limits of Agreements (LOA), the joint measurements acquired by the SEEK V2 and by the direct postural evaluation method for all parts of the body evaluated (see Fig. 5).

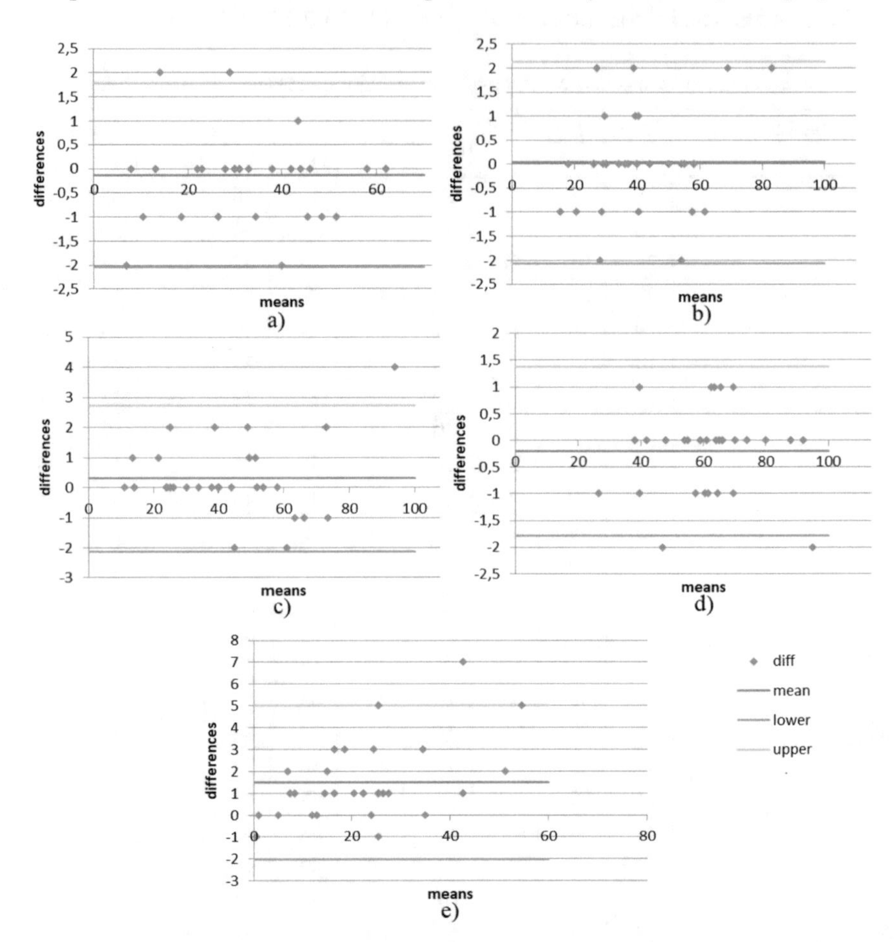

Fig. 5. Bland Altman plots (a) comparison of angles between SEEK V2 and direct method on the trunk, (b) comparison of angles between SEEK V2 and direct method on the neck, (c) comparison of angles between SEEK V2 and direct method on the arm, (d) comparison of angles between SEEK V2 and direct method on the forearm, (e) comparison of angles between SEEK V2 and direct method on the wrist.

Bland-Altman diagrams revealed agreement in the AMD between the direct methods and SEEK V2 on the trunk (−0.13°, LOA −1.77 to 2.04; see Fig. 5a), neck (0.33°, LOA −2.06 to 2.12; see Fig. 5b), arm (0.30°, LOA −2.12 to 2.72; see Fig. 5c), forearm (−0.20°, LOA −1.78 to 1.38; see Fig. 5d) and wrist (1.50°, LOA −2.02 to 5.02; see Fig. 5e). It is observed that the LOA is small in all parts of the body, which implies that there is not much variation between the measurements produced by both methods. AMDs are close to zero, which means that similar results are produced in the angular measurements between the two methods. The variability of the data is consistent with the fact that no measured value exceeds the limits of agreement in the neck, only 1 measured value exceeds the upper limit in the arm and wrist, 2 measured values exceed the upper limit in the forearm and trunk. The data found are found on the trunk (6° to 62°), neck (15° to 84°), arm (11° to 96°), forearm (26° to 96°) and wrist (0° to 57°) (see Table 2).

Table 3 shows the data of average, maximum variation and uncertainty ($\bar{X}$, S, ± U), which has the SEEK V2 with respect to the measurements taken with the direct method for each part of the body of the participants.

Table 3. Comparative statistical results of angular measurements.

	Average $\bar{X}$(°)	Maximum variation M(°)	Expanded uncertainty U(°)
Arm	0.76	4	0.36
Forearm	0.53	2	0.22
Wrist	1.63	7	0.59
Trunk	0.60	2	0.27
Neck	0.70	2	0.28

The results of the angular measurements of the different parts of the body are: arm ($\bar{X} = 0.76°$, $M = 4°$, U = ±0.36°), forearm ($\bar{X} = 0.53°$, $M = 2°$, U= ±0.22°), wrist ($\bar{X} = 1.63°$, $M = 7°$, U= ±0.59°), trunk ($\bar{X} = 0.60°$, $M = 2°$, U= ±0.27°) and neck ($\bar{X} = 0.70°$, $M = 2°$, U= ±0.28°) (see Table 3). It is appreciated that the wrist is the joint with the greatest variation in the data compared between the methods of postural evaluation, this is due to the fact that it is the most difficult joint to detect for the Kinect and complex to measure with the direct method because of the movements required by the task with which the data is corroborated, that is, it is the point of measurement of the lowest reliability that the system gives, followed by the arm, although with greater confidence. The other joints have a much smaller variation in results, which is believed to be due to their shapes and sizes, which makes them easy to detect for the sensor and to measure for the evaluators.

In Fig. 6 it shows the number of participants who obtained final RULA scores of 4 to 7 (see Table 1), comparing the SEEK V2 with the direct measurement method.

For the direct postural evaluation method and for the SEEK V2, five participants obtained a RULA final score of 4, both methods found 5 participants in a RULA final score of 5, the direct method found 13 participants in a RULA final score of 6, while the SEEK V2 method found 7 participants in a RULA final score of 7, while the SEEK V2 method found 8 participants in a RULA final score of 7.

The final RULA scores in the direct and SEEK V2 methods are similar in both evaluations, the existing variations are 1 point (see Fig. 6) and are due to the wide ranges that are considered in the ergonomic method applied.

Cohen's Kappa concordance coefficient with an alpha of 5% gives a result of 0.953, according to the Landis and Koch scale [26] there is a very good degree of agreement between the final RULA scores of the direct ergonomic evaluation method and those issued by the SEEK V2.

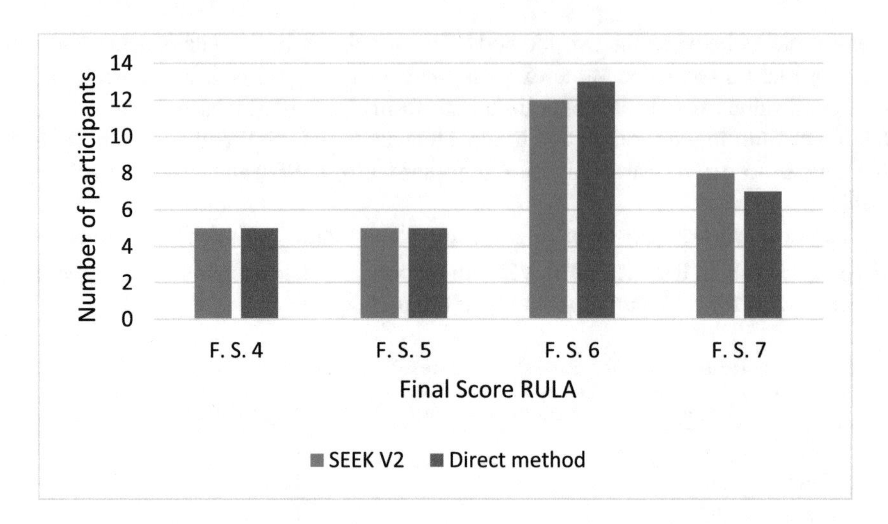

Fig. 6. Final RULA Direct Method vs SEEK V2 RULA Scores.

4 Discussion

In the literature consulted Manghisi et al. [7] presented a semiautomatic RULA evaluation system with Kinect v2 and validated it by obtaining an almost perfect Kappa (0.84) agreement between the system and an expert RULA evaluator. Additionally, it presented a lower Kappa statistical agreement (0.34) when it compared to commercial software based on Kinect v1, for 30 non-ergonomic positions. These postures were static and with the frontal sensor to the person, the results found are consistent with this article in which the Kappa coefficient (0.953) also has an almost perfect concordance between the proposed system and the evaluations made by the direct method. The difference with the present article, is the participants in this study perform diverse and complex movements because they had to perform a thorough task of tracking the mold in which the position of the wrist was difficult to detect for the Kinect and visually.

The angular measurements obtained from the evaluated joints of the body show an expanded uncertainty of less than 0.5° except for the wrist which is 0.59° the highest variation found. It is believed that this is due to the distance to which the Kinect should be placed. It is difficult to determine the small movements that the wrist makes, this distance is inevitable to evaluate all the necessary body parts. This agrees with the evaluation of the human body carried out by Alabbasi et al. [12] where it shows the Kinect V2 is robust in capturing complex movements, but it is not perfect, as the complexity of the movement increases the sensor capture errors increase too.

According to the RULA method and the different methods of ergonomic evaluation based on joint measurements, a measurement of precision is not needed, but a measurement that allows to determine between the ranges given by the method in whole angular values and to qualify the joint in one or another risk score.

A future solution to achieve more precise result in the wrist studies would be to implement a second Kinect, it should be exclusive for this articulation, this is a valid option because the sensor is only used to acquire angular data which is process by programming the SEEK V2.

During the taking of direct joint measures, the wrist was the most complicated joint to evaluate, for example: in Table 2 the data of participant number 6 is shown who obtained values of 39° with the system and 46° with the direct method on the wrist. This variation of 7°, influences by the distance exist from the Kinect to the person. Furthermore, it should be considered the perception of evaluators [28] also plays an important role to have greater uncertainty in this articulation and not only due to sensor accuracy problems.

As shown in Table 3, the remaining arm, forearm, trunk and neck joints are more reliable in terms of angular measurements, this could be since when the participant uses movements to perform the activity, the sensor is diagonally easy to grasp all these joints.

It is believed that the measurements have a strong reliability because the results of the Bland-Altman diagrams revealed an AMD agreement closer to 0 in all cases, besides the LOA values do not have an important amplitude, which provides confidence in the results obtained.

The proposed system has the advantage of evaluating with the sensor on the side of the person and with suitable garments as shown in Fig. 3, which is believed to be the main contribution of the project, as the RULA method mentions, the best evaluation plane is the sagittal [5] because it is where all the necessary articular measures are captured. SEEK V2 presents a viable contribution and alternative for the industry. As stated, it is intended to carry out future studies in footwear companies of the region under real working conditions.

5 Conclusions

The semiautomatic system developed of ergonomic evaluation with Kinect V2, demonstrates to be a non-invasive method which allows the worker to carry out their activities in a natural way making the data more real and reliable. Furthermore, it's allowing the discovery of the ergonomic risk associated with the postures in a precision task, thus decreasing the estimation errors of measurements taken by the evaluators with a direct measurement. As well, it considerably reduces the time in the development of an ergonomic evaluation.

The reliability of the SEEK V2 is evidenced by the angular values of the upper extremities emitted by the system, contrasted with those taken with the direct postural evaluation method, obtaining results of expanded uncertainty of 0.36° for the arm, 0.22° for the forearm, 0.59° for the wrist, 0.27° for the trunk and 0.28° for the neck. The wrist is the joint that shows the greatest variation but does not affect the results of the RULA method. On the other hand, the forearm shows more accurate results.

The correlation between the evaluations RULA of the SEEK V2 and performed with the direct method, provides a Cohen Kappa correlation coefficient of 0.953, which shows under controlled conditions that the RULA final scores of the SEEK V2 are in perfect agreement with a direct evaluation.

In the future research, it is suggested to apply the system in real working conditions to enhance it with the use of a new sensor centered on the wrist joint.

Acknowledgments. The authors would like to thanks to the Technical University of Ambato (UTA) for financing the project ''System of Evaluation of Postural Risk Using Kinect 2.0 in the activity of Cutting of the Production of Footwear for the CALTU Ambato", for the support to develop this work.

References

1. Geraldo AP (2015) Ergonomía y Antropometría Aplicada con Criterios Ergonómicos en Puestos de Trabajo en un Grupo de Trabajadoras del Subsector de Autopartes en Bogotá, D.C. Colombia. Rev Repub 135–150
2. Jesús M, Rojo F, Canga A, Ferrer AP, José P, Quintana MF (2000) Manual Básico de Prevención de Riesgos Laborales: Higiene industrial. Seguridad y Ergonomía. Imprenta Firma, S. A., Mieres
3. Cursoforum S.L.U: Riesgos Ergonómicos y Medidas Preventivas en las Empresas Lideradas por Jóvenes Empresarios., Madrid
4. Agencia Europea para la Seguridad y la Salud en el Trabajo: Introducción a los trastornos musculoesqueléticos de origen laboral - Salud y seguridad en el trabajo - EU-OSHA,https://osha.europa.eu/es/tools-and-publications/publications/factsheets/71/view
5. Asensio-Cuesta S, Bastante-Ceca J, Diego-Más JA (2012) Evaluación Ergonómica de los Puestos de Trabajo. Paraninfo, SA, Madrid
6. David GC (2005) Ergonomic methods for assessing exposure to risk factors for work-related musculoskeletal disorders. Occup Med (Chic. Ill) 55:190–199
7. Manghisi VM, Uva AE, Fiorentino M, Bevilacqua V, Trotta GF, Monno G (2017) Real time RULA assessment using Kinect v2 sensor. Appl Ergon 65:481–491
8. Lite AS, García García M, Ángel M, Del Campo M (2007) Métodos de evaluación y herramientas aplicadas al diseño y optimización ergonómica de puestos de trabajo
9. Dockrell S, O'Grady E, Bennett K, Mullarkey C, Mc Connell R, Ruddy R, Twomey S, Flannery C (2012) An investigation of the reliability of Rapid Upper Limb Assessment (RULA) as a method of assessment of children's computing posture. Appl Ergon 43:632–636
10. Rodríguez-Ruíz Y, Guevara-Velasco C (2010) Empleo de los Métodos ERIN y RULA en la Evaluación Ergonómica de Estaciones de Trabajo. Ing Ind 32:19–27
11. Garrido-Castro JL (2014) Diferencias en el análisis cinemático 2D/3D de los parámetros utilizados para la evaluación del pedaleo en el ciclismo. XXXVII Congr. LA Soc. Ibérica Biomecánica Y Biomater. pp 3–4
12. Alabbasi H, Gradinaru A, Moldoveanu F, Moldoveanu A (2015) Human motion tracking & evaluation using Kinect V2 sensor. In: 2015 E-Health and Bioengineering Conference (EHB). IEEE, pp 1–4
13. FLores-Bazán, R.: Análisis de la relación entre ergonomía, calidad de vida y eficiencia de la producción en la industria maquiladora de Tamaulipas. México, DF. In: ANFECA (2012)
14. Teschke K, Trask C, Johnson P, Chow Y, Village J, Koehoorn M (2009) Measuring posture for epidemiology: Comparing inclinometry, observations and self-reports. Ergonomics 52:1067–1078

15. María V. Posturas de Trabajo: Evaluación del Riesgo, http://www.insht.es/MusculoEsqueleticos/Contenidos/Formacion divulgacion/material didactico/Posturas trabajo.pdf
16. Plantard P, Auvinet E, Pierres A-S, Multon F (2015) Pose estimation with a kinect for ergonomic studies: evaluation of the accuracy using a Virtual Mannequin. Sensors 15:1785–1803
17. Greene RL, Azari DP, Hu YH, Radwin RG (2017) Visualizing stressful aspects of repetitive motion tasks and opportunities for ergonomic improvements using computer vision. Appl Ergon 65:461–472
18. Spielholz P, Silverstein B, Morgan M, Checkoway H, Kaufman J (2001) Comparison of self-report, video observation and direct measurement methods for upper extremity musculoskeletal disorder physical risk factors. Ergonomics 44:588–613
19. Clark RA, Pua Y-H, Fortin K, Ritchie C, Webster KE, Denehy L, Bryant AL (2012) Validity of the Microsoft Kinect for assessment of postural control. Gait Posture. 36:372–377
20. McAtamney L, Nigel Corlett E (1993) RULA: a survey method for the investigation of work-related upper limb disorders. Appl Ergon 24:91–99
21. Dutta T (2012) Evaluation of the Kinect™ sensor for 3-D kinematic measurement in the workplace. Appl Ergon 43:645–649
22. Alba-Martín R (2016) Fiabilidad y validez de las mediciones en hombro y codo: análisis de una aplicación de Android y un goniómetro. Rehabilitacion 50:71–74
23. Marino C, Vargas J, Aldas C, Morales L, Toasa R (2018) Non-invasive monitoring environment: Toward solutions for assessing postures at work. In: 2018 13th Iberian conference on information systems and technologies (CISTI). IEEE, pp 1–4
24. Diego-Más JA. Evaluación postural mediante el método RULA. https://www.ergonautas.upv.es/metodos/rula/rula-ayuda.php
25. Hipp J, Hill C, Hall M (2007) Evaluación de datos de medición Guía para la Expresión de la Incertidumbre de Medida
26. Martin Bland J, Altman D (1986) Statistical methods for assessing agreement between two methods of clinical measurement. Lancet 327:307–310
27. Landis JR, Koch GG (1977) The measurement of observer agreement for categorical data. Biometrics 33:159
28. Escamilla Esquivel A (2000) Metrología y sus aplicaciones. Larousse - Grupo Editorial Patria

Improving the Design of Virtual Learning Environments from a Usability Study

Germania Rodriguez Morales[1,2(✉)], Pablo Torres-Carrion[1], Jennifer Pérez[2], and Luis Peñafiel[1]

[1] Universidad Técnica Particular de Loja, San Cayetano Alto, Loja, Ecuador
{grrodriguez,pvtorres,lfpenafiel}@utpl.edu.ec
[2] Universidad Politécnica de Madrid, Madrid, Spain
jenifer.perez@etsisi.upm.es

Abstract. Usability as an area of knowledge of Computer Science closest to the user, has seen enhance their interest and development in recent times, due to the diversity of devices and forms of interaction. At the same time, educational institutions, aware of the potential existing in new technologies, and the disruption caused, invest their resources in the implementation of virtual learning platforms. In this investment, they seek diversify in the tools that allow the development of new teaching strategies, which are ubiquitous and continuous, adaptable to the needs of the user. In this context, a usability study is carried out, sustained in the method of inquiry with the questionnaire technique, following the proposal of Ferreira & Sanz in 2009, measuring the parameters of Satisfaction, Learning, Operability, Attractiveness, Content and Communication. The best results are obtained in the Learning and Content indicators, and with lower scores the Operability, highlighting in the latter the low value in the Accessibility and Availability indicators. According to the results obtained, a proposal for improvements is proposed, in order to achieve a greater degree of usability in the mentioned environment. Finally, these improvements are implemented in the case study, making a new evaluation of users in order to validate the improvements made.

Keywords: Usability · Higher education · LMS · Design

1 Introduction

Several of the human activities require the management of large amounts of information in relatively short times, as well as the constant decision making, to be quickly and efficiently allow the development and advancement of all types of entities around the world for the benefit of humanity. The highlight of this accelerated change is the migration of the various services to Inter-net; supported by technological platforms that eliminate spatial and temporal barriers allowing access to millions of users to the information they contain. One of the fields that since many years ago it has adopted as a strategic ally to new technologies, is the educational field, which through large platforms supports the teaching-learning process.

M. Botto-Tobar et al. (Eds.): TICEC 2018, AISC 884, pp. 100–115, 2019.
https://doi.org/10.1007/978-3-030-02828-2_8

Therefore, the need to develop computer applications centered on the user is evident. In this sense, usability has become a factor with a more important role even than the information architecture itself or content management and increasingly arouses greater interest in the community of developers, who emphasize rigorous processes of evaluation during and after development. Based on these aspects, the research aims to: a) Carry out a usability study of the UTPL Learning Management System (LMS) that was based on different parameters that allow identify the aspects that negatively affect the usability conditions of the platform; b) Formulate alternative solutions for the LMS platform to guarantee, favor and facilitate the teaching-learning process for teachers and students; c) Implement the improvement proposal in a test site to determine the feasibility of these being moved to the real operating environment.

The report begins with a general theoretical revision and state of the art on the Usability in higher education fields, with several LMS platforms, with greater emphasis on the MOODLE platform. It also analyzes emerging methodologies, considering the methods, techniques and parameters used by researchers in the area to carry out this type of evaluation. Next, the study conducted with the specific EVA of the UTPL is presented as a case study, from the model proposed by Ferreira and Sanz [1, 2] at the Universidad de la Plata. Then, the proposal of improvement of conformity to the representative problems revealed during the evaluation is shared, with the purpose of reaching a greater degree of usability within the environment. At the end, its implementation is detailed in the Moodle 1.9 platform hosted on a test server.

2 Usability

2.1 Theoretical Framework

Usability "studies the essential characteristics that a software or computer application must have in order to guarantee easy access to any person regardless of the technology available, and the knowledge they have about the use of ICT [3]. Bevan [4] shares a fairly specific definition, explaining how the usability and acceptability of a system or product for a particular class of users that perform specific tasks in a specific environment, where ease of use affects performance and user satisfaction and acceptability affects whether the product is used or not". Torres-Carrión [5], explains the great coverage of Usability, which evolves as a sub-area of UX, with great consideration in the scientific field for the design of quality; explains in [6] hat the study of Usability depends not only on the interaction system, but on how it has been validated and what the objectives were, and in the educational field it also has as premise that the learning attributes will vary between people, and this variable extends its value if we also consider the context from which it is evaluated.

The International Organization for Standardization (ISO) has also been concerned with providing technical and scientific coverage to this area of science. In the general field of quality in the Software industry, the ISO/IEC 9126-1:2001 standard is presented, later revised in ISO/IEC 25010:2011 (updated to 2017), where the key quality factors are established, including among them usability, and it is defined as a *subset of quality in use consisting of effectiveness, efficiency and satisfaction, for consistency with its established meaning* [7]. In the field of user ergonomics, as subfield of Human Computer Interaction (HCI), the ISO/DIS 9241-11:2017 standard is presented, where Usability is not defined like as an attribute of a product, although appropriate product attributes can contribute to the product being usable in a particular context of use, it is a more comprehensive concept than is commonly understood by "ease-of-use" or "user friendliness" [8]. Usability is not just another attribute of a system, but one of the key attributes that can make the difference between success and failure of the product; because *it is intimately related to the user's perception of the quality of the system; the internal algorithms or the definition of the architecture can be excellent, but the user does not have visibility of that, but of the interface with which he interacts* [9]. For the context of this study, the definition closest to the User Centered Design, framed in the HCI field, is considered as close.

2.2 Methods for the Evaluation of Usability

The main methods for the evaluation of usability derive from *User Testing* [10, 11], *Heuristic Evaluation* [12] y *model-based methods* [13]; those related to the first case, refer to empirical evaluations from the hypothesis, which implies the detail of dependent variables and indicators.

The three measurable usability attributes defined by ISO [8] are:

- **Effectiveness**: accuracy and integrity with which users achieve the specified objectives.
- **Efficiency**: resources spent in relation to the accuracy and integrity with which users reach the objectives.
- **Satisfaction**: absence of discomfort and positive attitudes towards the use of the product.

According to the model of measurable attributes of Nielsen [14], usability has the following attributes: ease of learning, efficiency, ease of recall, errors and subjective satisfaction. Rober Stake [15] proposes nine evaluation methods grouped by purpose: exploratory, predictive, formative and summative. The methodology proposed by Ferreira and Sanz [1, 2] belongs to this field, and that is the basis for the development of this proposal.

Heuristics usability are a widely accepted method. Mtebe and Kissaka [16] presents a heuristics usability that consolidates interface usability, didactic effectiveness and motivation to learn. Other researchers design an instrument on the basis of the general criteria for the heuristic evaluation proposed by Nielsen, as well as on international standards, guides, and recommendations for software quality (ISO 9241 and ISO 9126), with encouraging results for applying the usability evaluation instrument to Metacampus, an LMS developed by and used at the Virtual University System in University of Guadalajara [17]. Almarashdeh et al. [18] evaluate their prototype by six experts whom have good experience in different fields (LMS development, system quality, software engineering and distance learning) to show their satisfaction with features and requirements.

The User Affective Experience (UAX) is considered an emerging field in HCI [6, 19], with extended metrics from UX and Usability that take interaction data from various sources: observational, sensors and algorithms. Ulbricht et al. [20] evaluate the emotion component on usability human computer interface. To evaluate LMS's HCI describe methods and techniques used to evaluate HCI using Usability Tests with emotions and then validate them in the WebGD LMS. Pireva et al. [21] analyze user behavior in LMS and MOOC, using facial expression software to find seven emotional engagement attributes and three sentiment engagement attributes; as a methodology, they design an experiment with different tasks that each subject has to perform. Phongphaew and Jiamsanguanwong [22] measure 5 usability attributes; learnability, efficiency, effectiveness, memorability, and satisfaction, as a relationship of each usability attributes and emotional responses of user in order to understand the relationships of these two evaluation methods.

Artificial Intelligence (AI) is also making its contribution to this field of science. So, using data mining from 421 comments written by university students who frequently use an LMS, Jiménes et al. [23] propose an approach based on text mining techniques, which allows quick identification of usability and functionality issues. It has mentioned the most relevant methods currently applied in the study of Usability. There are several emerging methods that make use of look tracking technologies, counting clicks, gamification, degree of user attention in each option, among others, which continue to give dynamism to this area of science.

2.3 Related Research

HCI is an area of multidisciplinary science, and therefore it is necessary to carry out a broad study from different sciences; this has been inherited in Usability as one of its subfields. Freire [24] details a great diversity of points of view, including researchers from different scientific areas such as Ergonomics, Computer Science, Design and Education. Psychology is also an area of science to consider, given the imminent user behavior, as a center of the study of Usability.

Systematic Reviews of Literature About Usability in LMS

The efforts made in the search for Usability studies in LMS [24–26] have different approaches and give us a first view of the scope of this field of science. Freire [24] embraces three dimensions, namely the methods, models and frameworks that have been applied to evaluate LMS, including also the main usability criteria and heuristics used; their results show a notorious change in the paradigms of usability. The main criteria adopted in E-learning evaluations originated in criteria already researched by Informational Ergonomics, and the three main methods and techniques used (in the last 30 year <= 2012) are: (i) system performance evaluation, (ii) of user performance and (iii) the evaluations of the dialogs between users and systems.

Aydin [26] explore trends, gaps, and issues in the literature of the usability of Learning Management Systems (LMS). Their analysis revealed several gaps: *(1) engineering students have not been the main focus of research in any studies, (2) there is no research that compares usability of LMS between different academic disciplines, (3) there is no modeling effort for understanding if engineering students and instructors need different LMS design than other disciplines, (4) primary framework development for evaluating LMS has declined, (5) discount usability methods (heuristics) have been mostly preferred for the evaluation of LMS ignoring effectiveness and efficiency performance measures related to LMS usage, (6) there are very limited studies incorporating usability design with instructional and accessibility design, (7) there are very limited studies investigating LMS usability with regards to occupational training, (8) there are many researchers who mentioned the significance of research on usability of mobile e-learning platforms*. These results support the problems raised in this research.

In addition, Nakamura et al. [25] characterize the usability and UX evaluation techniques in the context of LMSs, checking a total of 62 publications (2004–2016), in which they identify the techniques used to evaluate the usability and UX of LMSs and their characteristics such as its origin, type, performing method, learning factors, restriction and availability. From the research question "*Which usability and UX evaluation techniques were applied on Learning Management Systems and how have they been used?*" propose 11 sub-questions, where they stand out for our purpose: Technique type (Inquiry 51,92%; Testing 33,65%; Inspection 27,98%), Performing method (manual 90,38%; automatic 7%), Evaluation focus (Usability 69,23%; Usability and UX 28,85%; UX 1,92%); Feedback (No 100%, Yes 0%), Investigation type (Case Study 25,86%; Survey 25,86%; Controlled Experiment 17,24%), Platform used (Desktop/Web 79,31%; Mobile 17,24%). Contrary to what we have seen in this study, changes in the platform are presented here and a complementary study is presented to support the validity of the changes.

Usability for Courses Design

Pástor et al. [27] proposes a methodology for creating e-learning design patterns, development a catalogue of ten design patterns for the creation of online courses in a VLE, articulating pedagogical methodologies used in virtual education. To validate, two evaluation processes were carried out: *the first one to measure ability to design online courses with the teachers and the second one to measure the usability of the online courses with the students*. Lai and Lin [28] analyzing the difficult tasks for expert (8 teaching assistants who have past Moodle teaching experience) and novice users (8 teaching assistants with non-Moodle experience); they carried out 18 usability tasks, post-task interviews, and finally a survey with 25 questions. Qualitative research provides a significant contribution, supported by techniques such as questionnaires, interviews, systematic observation and study of historical documents related to processes and the user.

Results gathered in usability research conducted among students confirm that development of eLearning systems needs to have learner in the center of development process [29]. In an empirical-based study, Harrati et al. [30] explore how university lecturers interact with an e-learning environment based on a predefined task model describing low-level interactions. From user feedback, via experimental System Usability Scale, results reveal that the evaluation must be fulfilled in tandem with analyzing the usage metrics derived from interaction traces in a non-intrusive fashion, and that these are not a sufficient measure to express the true acceptance and satisfaction level of lecturers for using the e-learning systems.

Usability in LMS for Higher Education

Higher Education institutions have been the pioneers in implementing interaction platforms, so it is essential to know the actual work related to the area. Thus [31] assess the usability of the Jusur Learning Management System (LMS) that is used in higher education in Saudi Arabia. Nine factors have been incorporated into a survey to evaluate the system: content, learning and support, visual design, navigation, accessibility, interactivity, self-assessment, learnability, and motivation. Kurata et al. [32] evaluate the effectiveness of the LMS, considering variables in the LMS design model, from the pedagogical approach, the usability and the satisfaction aspect of the user interface; the result shows that LMS is an effective tool to facilitate learning in an undergraduate engineering program in the Philippines, and that it could be made more efficient by adding collaborative learning tools for students.

As one of the emerging areas of Higher Education, there is Distance Education, now closely related to Online Education, which has been boosted since 2011 in an open manner with MOOC and REA. This field [33] describe the process of developing and evaluating the Moodle-based mobile Learning Management System (LMS) application called Student Centered e-Learning Environment (SCeLE). They implement the application with user-centered design basis, at the last phase we evaluated the application using usability testing and system usability score (SUS). Almarashdeh et al. [18] measure the usability of a DLMS prototype based on user's requirements, applying verification method using heuristic evaluation (experts review) to evaluate the prototype interface.

Usability in Several LMS

The availability of LMS in the market is increasing. This has required area researchers to evaluate various LMS platforms globally. Thus, Phongphaew and Jiamsanguanwong [34] identify the interface issues of myCourseVille by using usability evaluation method associated with student's interface and teacher's interface. The myCourseVille is a LMS that use in many universities in Thailand to support teachers and students to manage their class activities. Lalande and Grewal [35] from usability metrics compare two Learning Management Systems (LMS): Blackboard and Desire2Learn; they have as input the number of mouse clicks necessary, number of pages traversed, and data fields inputted. Okike and Morogosi [36] evaluate the use of Moodle and Blackboard from academic disciplinary domain perspectives, through quantitative research approach to conduct a survey study using 67 university lecturers (43 use of Blackboard, 11 make use of Moodle, and 13 do not use either platforms); both platforms are convenient for the creation and distribution of teaching materials, course assessment, communication and collaboration between users. In addition, from a logistic regression analysis, users have a greater probability of using Blackboard (Odd Ratio = 4.96) than Moodle (Odd Ratio = 1.00).

Usability for MOODLE

This is the base platform on which the proposed study focuses. For this LMS, Karagiannis and Satratzemi [37] present a usability study on a dynamic framework that adapts to the user, showing improvements in this adaptation in Moodle with respect to the standard version. Ifinedo et al. [38] explore the effects or roles of usability factors and external support (i.e., teacher and peer support) on undergraduates' use outcomes of Moodle in a blended learning environment. For this purpose they conducted a cross-sectional survey and collected data from 126 undergraduate students attending a university in the Maritime region of Canada; the result show that usability factors have positive effects on students' use outcomes; contrarily to predictions teacher and peer support did not. These results are relevant in terms of the contribution that Moodle as a tool can give to the didactic process in the classroom.

3 Evaluation Methodology of Usability for Study Case

The evaluation must be a planned and carefully applied process; it can go from the application of empirical methods as discussed in Sect. 2.2 with real users, to other more refined methods with domain experts, as established in Fig. 1.

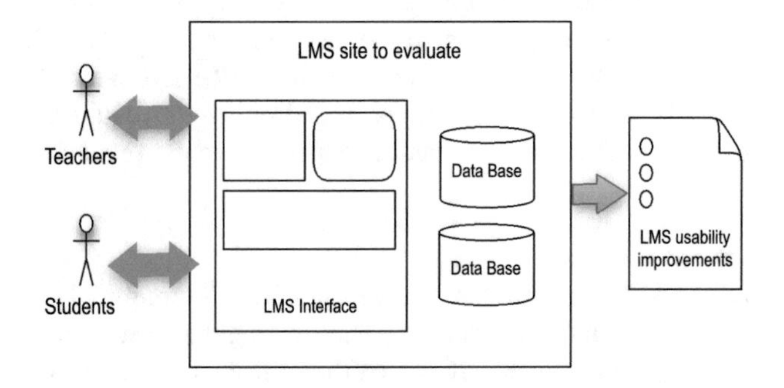

Fig. 1. Participants in the evaluation process

As it is an evaluation of usability from the user's perspective, two categories of users have been considered, who normally maintain a constant interaction with the UTPL LMS: **Teachers** and **Students**, with whom prior contact was established, obtaining their consent and commitment to participate in the study.

3.1 Definition of Methods-Techniques and Evaluation Parameters

Taking into account the restrictions and the variety of tasks that are carried out through the LMS, the evaluation is based on the study of the tasks considered most representative. The method of inquiry supported by the questionnaire technique was used.

In the evaluation of usability are important aspects such as *ease of use, effectiveness, efficiency, ease of learning, user satisfaction, accessibility, consistency, among others. All these concepts are related to the usability of the Web site* [2]. According to this description the parameters used to measure the LMS usability through a questionnaire are shown in the Table 1.

Table 1. LMS usability evaluation parameters

Parameter	Description
Satisfaction	Utility Speed
Learnability	Ease of learning Help
Operability	Ease of Navigation Ease of use Availability Error tolerance Accessibility
Attractiveness	Attractiveness of the interface Personalization
Content	Credibility Reach
Communication	Control of communication Forms of messages

3.2 Data Collection

The data collection was carried out through the application of the online questionnaires to students and teachers, through Google Drive forms. In the case of students in the face-to-face mode, the application was carried out directly. It is important to emphasize that the representative sample to which the questionnaires were applied consists of 160 students and 65 teachers. The information obtained from the evaluation process is the basis for carrying out the corresponding improvement proposal, in the aspects that reflect LMS usability problems and its implementation is feasible.

3.3 Results Measurement

The nature of the items of the questionnaires used allows obtaining information in terms of percentages, for this reason the interpretation and presentation of the results is based on an assessment scale as indicated in Table 2. These items allow determining the level of compliance of the attributes evaluated. The final score of each attribute is obtained by calculating the average weight of the questions involved with the evaluated attribute. In case the score has a fractional part, rounding is applied.

Table 2. Scale of assessment of results

Indicator	Weight	Range
Optimum	4	76% - 100%
Acceptable	3	51% - 75%
Improvable	2	26% - 50%
Critical	1	0% - 25%

According to this scale, an indicator of a good level of compliance is Acceptable or Optimal and reveals the need to make an improvement in cases in which it is located in Critical or Improvable.

4 Results

Once the surveys for the data collection were applied, the tabulation and presentation of the results were carried out (see Table 3) in which the different parameters selected for the evaluation can be differentiated and their location according to the rating scale proposal in Table 2.

Table 3. Results of the usability evaluation of the UTPL LMS

Parameter		Score	Assessment
Satisfaction	Utility	3,5	Optimun
	Speed	2	Improvable
Learning	Learnability	3,2	Acceptable
	Help and Documentation	3,4	Acceptable
Operability	Ease of Navigation	3,67	Optimun
	Easy to use	3	Acceptable
	Availability	2	Improvable
	Error tolerance	3,5	Optimun
	Accessibility	2	Improvable
Attractiveness	Attractiveness of the interface	4	Optimun
	Personalization	1	Critical
Contents	Credibility	2,67	Acceptable
	Scope	3	Acceptable
Communication	Communication control	2,43	Improvable
	Forms of messages	2,5	Acceptable

5 Proposal

From the results obtained, it is clear that the aspects related to usability that should be considered in the improvement proposal are the following:

- Communication control (improvable)
- Interface customization (critical)
- Accessibility (improvable)
- Speed (improvable)
- Availability (improvable)

According to certain aspects that have been determined in the investigation, these parameters are considered as problems due to certain factors. The proposal focuses on incorporating functionalities (use cases) into the platform to solve the usability problems revealed in the research. Table 4 establishes the relationship of the problems identified with the proposed solution.

Table 4. Problems and solution alternatives

Problems	Use Case	Solution
Communication	Use chat	Integrate Moodle chat activity and / or plugin for the Chat Console block.
	Format message text	Configure the use of the HTML editor.
	File upload	Upload files to the server and access through hyperlinks
Accessibility	Customize environment	Integration of the plugin for the Accessibility Block
Personalization	Customize environment	Allow customization using themes for the user interface.
Speed		Services in the cloud, ongoing project by the UTPL.
Availability		Services in the cloud, ongoing project by the UTPL.

5.1 Validation of Improvement Proposal

For the implementation of the mentioned usability improvements it was previously necessary to execute the following tasks:

- Configuration of a test environment, consisting of the Web Server, DBMS and PHP and installation of the Moodle 1.9 platform, because it is the version that currently supports the UTPL LMS.
- Analysis of the operation of the Moodle platform and comparison with the LMS of the UTPL, in the administration options of the site to determine the appropriate configurations.

- Creation of a course and test users that allow operating in the Moodle environment.
- Implementation of the necessary configurations in the platform to achieve the objectives of the proposal.

At the date of the study, the UTPL uses the Moodle 1.9 platform as the basis of its Learning Management System. Many adaptations have been made in the original site, although several Moodle configurations are preserved, such as: the navigation bar, the components in panels located in the sidebars.

The next step, before establishing a configuration to carry out the implementation was an analysis of the structure of the platform, fundamental knowledge for its operation, following the structure shown in Fig. 2.

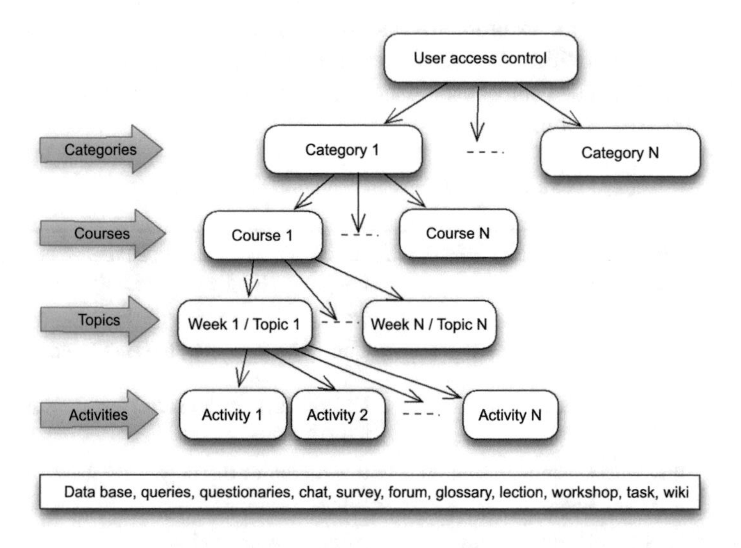

Fig. 2. Structure of the Moodle platform. Adapted from González [39]

In order to demonstrate the feasibility of operating with the functions that users consider necessary to facilitate their work within the LMS, it is necessary to configure an extended test version from the running version in the UTPL LMS, with full access to configuration, and access roles both administration and operation. As part of these activities, these tasks were carried out:

- Creation of the course
- User registration on the site and definition of roles
- Verification of the operation of the test course

Once the test operating environment for the installation, review, implementation and testing of the Moodle platform has been configured, the functionalities are incorporated and the appropriate configurations that satisfy the solutions proposed in Table 4 are established.

5.2 Test Results

The test plan fundamentally establishes the objectives that need to be achieved in accordance with the proposed solutions, the scope of the tests, the elements that will be tested on the platform, the requirements, procedures for registering the tests and the restrictions to which hold on. According to the plan, the testing process has been carried out in 3 stages: functional tests by the site administrator, functional tests by the users and acceptance tests. The population selected for this evaluation is made up of: Administrators of LMS (5), Teachers (10), and Students (20). The results of the functional tests and acceptance tests performed according to the plan established for the effect are presented.

In the functional tests, the obtained results can be evidenced that in regard to the access and use of the environment, as well as in the Communication, no functional problems have been detected. With regard to personalization and Accessibility, it was found that the functionality is somewhat affected by the response times of the site. However, the general results provided values that are located in each case, in operating percentages of 80% to 100%, and 20% or less error that have served to perform a general review of the platform (Fig. 3).

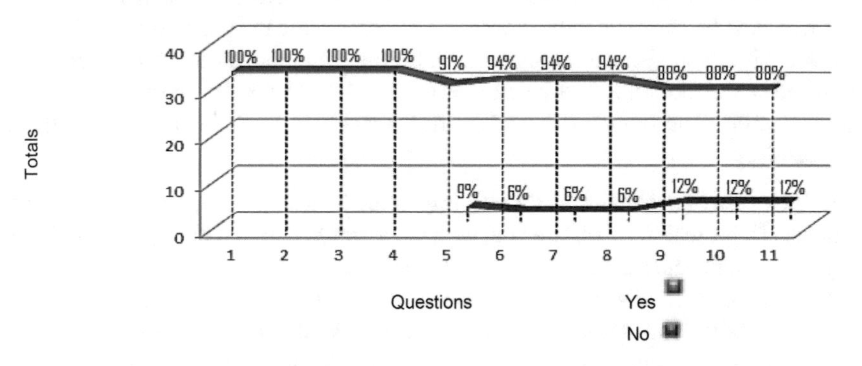

Fig. 3. Overall results of students functional tests

The results of the acceptance tests show a high degree of acceptance. Regarding Communication, the Messaging service with HTML has an acceptance of 94%, while chat sessions are 88% level and 85% for chat console, because some teachers consider it a distraction. The integration to the platform of the accessibility block; it has an acceptance level of 88%. In this case, the values are between 85% and 100% acceptance, demonstrating that it is justified that these functionalities are integrated into the LMS (Fig. 4).

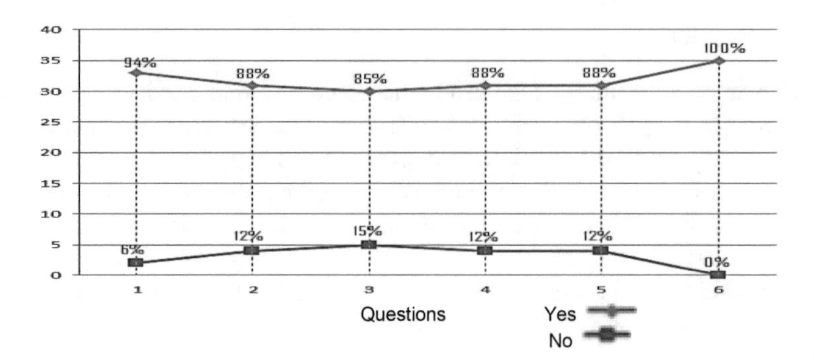

Fig. 4. Overall results of teachers functional tests

6 Conclusions

The development of this research work has allowed the acquisition of numerous knowledge and valuable experiences related to usability, a topic of interest at present in which humanity is virtually daily connected to the Internet to perform various activities. The objectives and level of knowledge of the users complicate the panorama of this technological environment. This is enough reason for the development of websites based on a user-centered methodology to be an alternative, a topic that has merited attention in various sectors. The study conducted on the subject has allowed to reach the following conclusions:

- A study case of evaluation of the usability of the Learning Management System LMS of the UTPL, which was carried out with the participation of real users of the platform (teachers and students) in order to obtain real results.
- A set of metrics and a scale was define that mark an initial diagnosis of usability for this type of platforms quantitatively, which was apply and validate in a study case.
- The improvement proposal was raised based on the analysis of the results obtained, the same ones that reveal those attributes that are located at an improvable or critical level, such as Communication, Personalization, Accessibility, Availability and Speed, which can be solved with a appropriate configuration of the platform and in the last two cases with the intervention of the UTPL through the cloud services project.
- The implementation allowed a post-study of the platform, determining that it offers a variety of alternatives to integrate sufficient and necessary functionalities to improve usability; this is corroborated by post-implementation tests that demonstrate its functionality with results above 80% and acceptance levels above 85%.

The favorable results of the usability study do not depend only on the platform available in the institutional web portal. Although it is true that it is the main factor, in platforms with a high degree of parameterization, it also depends to a large extent on the policies established by the Educational Institution that adopts it, the administrators, the teachers and the students. In this way, the range of possibilities offered by the LMS can be adequately and converted into an efficient and effective environment that guarantees satisfaction in its use.

References

1. Ferreira Szpiniak A, Sanz CV (2009) Un modelo de evaluación de entornos virtuales de enseñanza y aprendizaje basado en la usabilidad. In: IV Congreso de Tecnología en Educación y Educación en Tecnología, pp 382–392
2. Ferreira Szpiniak A (2013) Diseño de un modelo de evaluación de entornos virtuales de enseñanza y aprendizaje basado en la usabilidad. https://postgrado.info.unlp.edu.ar/wp-content/uploads/2017/11/2013_Ferreira_Szpiniak_Ariel.pdf
3. Alfonso Cuba IM (2012) Usabilidad en la Educación: Garantía de la calidad de la Educación Virtual. Editorial Universitaria, La Habana, Cuba
4. Bevan N (2001) International standards for HCI and usability. Int J Hum-Comput Stud 55:533–552
5. Torres-Carrión P (2018) Fundamentos de Interacción Humano-Computador. Texto-Guia. Universidad Técnica Particular de Loja, Loja-Ecuador
6. Torres-Carrión P (2017) Metodología HCI con análisis de emociones para personas con Síndrome de Down. Aplicación para procesos de aprendizaje con interacción gestual
7. International Organization for Standardization (2011) ISO/IEC 25010:2011(en) Systems and software engineering — Systems and software Quality Requirements and Evaluation (SQuaRE) — System and software quality models. https://www.iso.org/standard/35733.html
8. International Organization for Standardization (2016) ISO/DIS 9241-11.2 Ergonomics of human-system interaction – Part 11: usability: definitions and concepts. https://www.iso.org/obp/ui/#iso:std:iso:9241:-11:dis:ed-2:v2:en
9. Fagalde P, Fontela C (2011) Artefactos de especificación de requerimientos de usabilidad. Universidad de Buenos Aires, Buenos Aires
10. Dumas J, Fox J (2007) Usability testing. In: The human-computer interaction handbook. CRC Press, pp 1129–1149
11. Dumas JS, Redish JC (1999) A practical guide to usability testing. Intellect Books, Exeter
12. Nielsen J (1994) Enhancing the explanatory power of usability heuristics. In: Proceedings of the SIGCHI conference on human factors in computing systems. ACM, New York, pp 152–158
13. John BE, Kieras DE (1996) Using GOMS for user interface design and evaluation: which technique? ACM Trans Comput Interact 3:287–319
14. Nielsen J (1993) Usability engineering. Morgan Kaufmann Publishers. Inc., San Francisco
15. Stake RE (1976) Evaluating educational programmes: the need and the response
16. Mtebe JS, Kissaka MM (2015) Heuristics for evaluating usability of learning management systems in Africa. In: Cunningham P, Cunningham M (eds) IST-Africa Conference, IST-Africa 2015
17. Medina-Flores R, Morales-Gamboa R (2015) Usability evaluation by experts of a learning management system. Rev Iberoam Tecnol Del Aprendiz 10:197–203
18. Almarashdeh IA, Sahari N, Zin NAM (2011) Heuristic evaluation of distance learning management system interface. In: 2011 international conference on electrical engineering and informatics, ICEEI 2011, Bandung
19. Torres-Carrion P, Gonzalez-Gonzalez CS, Barba-Guamán R, Torres-Torres AC (2017) Experiencia Afectiva de Usuario (UAX): Modelo desde sensores biométricos en aula de clase con plataforma gamificada de Interacción Gestual. In: Actas del V Congreso Internacional de Videojuegos Educativos (CIVE 2017). Grupo ALFAS, Puerto de la Cruz - Tenerife - España

20. Ulbricht VR, Berg CH, Fadel L, Quevedo SRP (2014) The emotion component on usability testing human computer interface of an inclusive learning management system. In: 1st international conference on learning and collaboration technologies, LCT 2014, Heraklion, Crete, pp 334–345
21. Pireva K, Imran AS, Dalipi F (2016) User behaviour analysis on LMS and MOOC. In: IEEE conference on e-learning, e-management and e-services, IC3e 2015, pp 21–26
22. Phongphaew N, Jiamsanguanwong A (2016) The usability evaluation concerning emotional responses of users on learning management system. In: 2016 6th international workshop on computer science and engineering, WCSE 2016, pp 43–48
23. Jiménes K, Pincay J, Villavicencio M, Jiménez A (2018) Looking for usability and functionality issues: a case study. In: Rocha Á, Guarda T (eds) International conference on information technology and systems, ICITS18, pp 948–958
24. Freire LL, Arezes PM, Campos JC (2012) A literature review about usability evaluation methods for e-learning platforms. Work 41:1038–1044
25. Nakamura, WT, De Oliveira, EHT, Conte T (2017) Usability and user experience evaluation of learning management systems a systematic mapping study. In: Proceedings of the 19th International Conference on Enterprise Information System, ICEIS 2017, pp 97–108. For CrossRef propouse it's available in http://www.scitepress.org/PublicationsDetail.aspx?ID=fMKcrDdaCvI=&t=1
26. Aydin B, Darwish MM, Selvi E (2016) State-of-the-art-matrix analysis for usability of learning management systems. In: 123rd ASEE annual conference and exposition
27. Pástor D, Jiménez J, Arcos G, Romero M, Urquizo L (2018) Design patterns for building online courses in a virtual learning environment. Ingeniare 26:157–171
28. Lai L-L, Lin S-Y (2017) An analysis for difficult tasks in e-learning course design. In: Nah FH, Tan CH (eds) 4th international conference on HCI in business, government and organizations, HCIBGO 2017, held as part of the 19th international conference on human-computer interaction, HCI 2017, pp 171–180
29. Minović M, Štavljanin V, Milovanović M, Starčević D (2008) Usability issues of e-learning systems: case-study for moodle learning management system. In: Meersman R, Tari Z, Herrero P (eds) International conference on On the Move to Meaningful Internet Systems, OTM 2008 and held ADI 2008, AWeSoMe 2008, COMBEK 2008, EI2N 2008, IWSSA, MONET 2008, OnToContent 2008, QSI 2008, ORM 2008, PerSys 2008, RDDS 2008, SEMELS 2008 and SWWS 2008, pp 561–570
30. Harrati N, Bouchrika I, Tari A, Ladjailia A (2016) Exploring user satisfaction for e-learning systems via usage-based metrics and system usability scale analysis. Comput Hum Behav 61:463–471
31. Althobaiti MM, Mayhew P (2016) Assessing the usability of learning management system: user experience study. In: Vincenti G, Bucciero A, Vaz de Carvalho C (eds) 2nd international conference on e-learning, e-education, and online training, eLEOT 2015, pp 9–18
32. Kurata YB, Bano RMLP, Marcelo MCT (2018) Effectiveness of learning management system application in the learnability of tertiary students in an undergraduate engineering program in the Philippines. In: Andre T (ed) AHFE 2017 international conference on human factors in training, education, and learning sciences, 2017, pp 142–151
33. Banimahendra RD, Santoso HB (2018) Implementation and evaluation of LMS mobile application: scele mobile based on user-centered design. In: 2nd international conference on computing and applied informatics 2017, ICCAI 2017
34. Phongphaew N, Jiamsanguanwong A (2018) Usability evaluation on learning management system. In: Ahram T, Falcão C (eds) AHFE 2017 international conference on usability and user experience, 2017, pp 39–48

35. Lalande N, Grewal R (2012) Blackboard vs. Desire2Learn: a system administrator's perspective on usability. In: 2012 international conference on education and e-Learning innovations, ICEELI 2012, Sousse
36. Okike EU, Morogosi M (2018) Measuring the usability probability of learning management software using logistic regression model. In: 2017 SAI computing conference 2017, pp 1217–1223
37. Karagiannis I, Satratzemi M (2017) Enhancing adaptivity in moodle: framework and evaluation study. In: Auer M, Guralnick D, Uhomoibhi J (eds) 19th international conference on interactive collaborative learning, ICL 2016, pp 575–589
38. Ifinedo P, Pyke J, Anwar A (2018) Business undergraduates' perceived use outcomes of Moodle in a blended learning environment: the roles of usability factors and external support. Telemat Inform 35:93–102
39. González de Felipe, AT (2009) Guía de apoyo para el uso de Moodle 1.9. 4. Usuario Profesor. Universidad de Oviedo, Oviedo

The Digital Preservation in Chimborazo: A Pending Responsibility

Fernando Molina-Granja(✉)

National University of Chimborazo, Riobamba, Ecuador
fmolina@unach.edu.ec

Abstract. Currently, every public or private institution generates digital information that by legal mandate, social responsibility, cultural and historical value must be preserved in the long term by means of techniques, methods or appropriate models that allow a technical way to have digital information accessible and informationally useful in a near and far future. In the world and in Ecuador there is a legal basis that motivates and demands that this responsibility is fulfilled, as well as several models of digital preservation that could be applicable. This research intends to know the current state of preservation of digital information of public and private institutions of Ecuador, specifically Chimborazo, valued by means of a validated survey of 68 items, applied to 63 public and private institutions, which measures 4 aspects globally accepted and of fundamental compliance, these are: (A) Organizational infrastructure, (B) Administration of digital objects, (C) Infrastructure management and security risks, and (D) Management of aspects of integrity in institutions, aspects that allow a guarantee of an acceptable level of integrity and security of digital information and its correct preservation. After a statistical analysis it is determined that there is a very low percentage of companies that apply some formal method and in the same way a very low compliance with the minimum aspects necessary for an adequate digital preservation.

Keywords: Digital preservation · PREDECI · Digital information

1 Introduction

Currently, any digital device that is part of the life of a person or institution is capable of generating information that may become necessary in the future as a source of consultation, as a record of the actions taken and even as evidence in case of an incident of security; whether among many others in the form of a photograph, document, geolocation record, text message, email or even a phone number registered as part of a call, the problem is that many times the collection, handling and preservation of this information is not done properly [1].

M. Botto-Tobar et al. (Eds.): TICEC 2018, AISC 884, pp. 116–126, 2019.
https://doi.org/10.1007/978-3-030-02828-2_9

Table 1. Type and number of institutions analyzed

Institution	Quantity	Percentage
Public	42	67%
Private	21	33%

Source: Prepared by the author

Table 2. Number of institutions that apply some formal PD method

Institution	Quantity	Percentage
Public	7	11%
Private	56	89%

Source: Prepared by the author

Table 3. Institutional valuation by evaluated aspects

Aspect	ASSESSMENT About 5 points
A.- Organizational infrastructure	2,78
B.- Administration of digital objects	2,15
C.- Infrastructure management and security risks	2,35
D.- Management of integrity	1,87

Source: Prepared by the author

Within the processes of preservation of entities or institutions that have the responsibility of guarding long-term data, digital preservation is found, intervening techniques and skills in the field of information science and information technology. One of the benefits of digital preservation is to make the digital material of these entities informationally accessible over time, regardless of whether this digitally available material was created digitally from its origin or generated from the analog material.

Entities and persons that require preserving digital material may be hospitals, libraries, museums, prosecutor's offices, criminal investigation institutions, higher education institutions, or any other entity that has the responsibility or legal obligation to safeguard digital data, usually using digital repositories.

According to Ecuadorian regulations, in the Law on the Conservation of archives, it states that "It is the State's obligation to ensure the conservation of the country's historical and sociological sources, as well as to modernize and modernize the organization and administration of archives;" and in its articles 1 and 2, it states that "The basic documentation that currently exists or that is produced in the archives of all the institutions of the public and private sectors, as well as that of private individuals, constitutes State Patrimony. Said basic documentation shall be constituted by the following instruments: (a) Handwritten, typographic or printed writings, whether original or copies; (b)

Maps, plans, sketches, and drawings; (c) Photographic and cinematographic reproductions, be they negative, plates, films and cliches; (d) Sound material, contained in any form; (e) Cybernetic material; and, (f) Other unspecified materials.

The digital preservation within the historical implications, has direct affection to the cultural inheritance, because without information (digital or not) that is guarded, there is no inheritance nor culture, but without the technology that allows the preservation or recovery of said information, neither there would be.

In an institution that carries out digital preservation processes, if the adequate technology is not available to support the preservation of data, such preservation processes cannot be executed. In any case, there is not a generally accepted established technological model, although standards and practical criteria have been created [2].

The relationship between social, historical, economic and technological implications makes digital preservation a recognition as an important area of research, framed in the information and information sciences, due to the informational and technological implications that exist.

There are models of digital preservation applied to specific environments, some tend to focus on features and record events in a static way, others focus only on technical aspects, others are limited to certain types of digital objects, others focus on approaches to specific solutions, others describe functional interactions at a high level, and others tend to describe absolute solutions.

These models may not be applicable in Ecuadorian legislation, or in specific institutions, there are other models such as PREDECI [3], which can be applied to any reality.

Now, knowing how digital preservation is being applied in public and private institutions, and how digital information is administered, in Ecuador and specifically in Chimborazo, is a pending issue.

The purpose of this research is to determine the application of digital preservation techniques and models in public and private institutions in Ecuador and their impact in terms of organizational infrastructure, digital object management, Technical Infrastructure Management and Security risk, and integrity; and answer the following research question: "Is there a formal digital preservation mechanism applied in the public and private institutions of Chimborazo?"

This article is organized in the first part of an introduction, followed by the methodology, then a conceptualization is presented as a theoretical basis, an analysis of the threat of validity is included to guarantee the study, to finally expose the obtained results and with it the conclusions and future work.

2 Methodology

The research that was carried out was of exploratory and descriptive type [4, 5]. It is descriptive because there are studies carried out worldwide on digital preservation in society, and it is exploratory since it is a subject that has not been investigated in Ecuador and in Chimborazo in particular.

A survey is used as an instrument to collect the required information, based on the guide proposed by OAIS [6] and the questionnaire proposed by NESTOR [7] with the variations of integrity parameters proposed by Molina&Rodriguez in the PREDECI model [3]. For the validation of the instrument, the reliability analysis of the survey is carried out, applying the Cronbach's Alpha whose coefficient is 0.842, which provides an acceptable reliability of the consulting instrument.

The questionnaire evaluates: (a) the validity of the content - observing the ability of the instrument to measure what it has been built for; (b) its application.- with an analysis of the advantages and disadvantages and with the completion of a revision of the instructions for completing the instrument; (c) its structure.- with a review of the formulation of the questions, the proposed sequence and the response scale; and (d) its presentation.- in which the best characteristics in appearance and format for the instrument are identified [8].

These general criteria constitute the basis for cooperation and collaboration, the key premise underlying the fundamental requirements is that for repositories of all kinds and preservation activities they should be reduced to the needs and means of the defined community or designated communities, as well, in addition, it is evaluated in case of applying digital preservation techniques if: (a) The repository commits to continue maintaining digital objects for the community or communities identified. (b) Demonstrate organizational fitness (including personal financial structure and processes) to fulfill its commitment. (c) Acquires and maintains the necessary contractual and legal rights and fulfills the responsibilities. (d) Has an effective and efficient regulatory framework. (e) Acquires and ingests digital objects based on established criteria that correspond to their capabilities and commitments. (f) Maintains/guarantees the integrity, authenticity, and use of digital objects sustained over time. (g) Creates and maintains necessary metadata on the measures adopted during the preservation of digital objects, as well as on the relevant production, access support, and contexts of the process of use before preservation. (h) Meets the necessary diffusion requirements. (i) It has a strategic program for the preservation of planning and action. (j) It has the adequate technical infrastructure to continue the maintenance and security of digital objects.

The instrument is applied in different public and private institutions regardless of the type of business, located in Chimborazo - Ecuador, and applies to technical specialists responsible for the process of preservation and management of digital information, in a number of 63 respondents, number that obeys to the availability of the companies that generate digital information and that authorize the application of the survey. The questionnaires were answered anonymously, although there is an additional document indicating that they carried out the survey.

For the statistical treatment of the data, an electronic spreadsheet was used, and Fisher's test was applied in Tukey's analysis of variance and separation of means, the standard error was analyzed and the percentage of good classification and the coefficients were obtained. With its corresponding exponentials and confidence intervals for them to 95% [9].

Finally, a statistical analysis and interpretation of the results obtained is carried out.

3 Conceptualizations

UNESCO states that "Digital heritage is not subject to temporal, geographical, cultural or format limits. Although specific to a culture, anyone in the world is a potential user. Minorities can address the majorities and individuals to a global audience. We must preserve and make available to anyone the digital heritage of all regions, nations, and communities in order to promote, over time, a representation of all peoples, nations, cultures, and languages " [10].

The National Digital Management Agenda NASD - Consortium of institutions that are committed to the long-term preservation of digital information - in 2014, mentions that effective digital preservation is vital to maintaining the necessary records to understand and evaluate government actions, the basis of scientific evidence to replicate the experiments, based on prior knowledge and preservation of the nation's cultural heritage [11]. Substantial work is necessary to ensure that today's digital content remains accessible, useful and understandable in the future, regardless of the way it is created.

Ferreira defines it as "the ability to ensure that digital information is maintained with accessible and sufficient qualities of authenticity, which can be interpreted in the future with the use of a different technological platform used at the time of its creation", this means that the Digital objects can be adapted to technological advances over time "in hardware, networks, software architectures, management systems, plans and needs, and even changes of custodian responsible for keeping records [12].

The increasing use of digital evidence in processes offers new opportunities and challenges. An e-mail, an interception of satellite communications, or a digital recording of live events can help establish a link of evidence between the accused and the commission of an international crime. Depending on the authenticity of the data, the digital evidence can also provide information about the time, place and form of an event to complement the living voice evidence or living testimony.

However, digital information can also be altered or degraded. In addition, the digital information at the moment of generating it was separated from its source; For example, a photograph captures a single point of view of a place at a particular time, and, similarly, a mailing address does not capture the author's poise or tone of voice.

In recent years, digital documents have been multiplied, their use among different sectors of society and the type of documents, and different proposals emerged, although international libraries did not agree on the most appropriate method, if they agreed to Regard the need to share some basic rules and a list of appropriate tasks for the implementation of a file digitization program, the Open Archival Information System (OAIS) model emerges [6], as a model of conceptual reference that establishes the six functions of an integral system of preservation of digital documents in the long term [13].

The importance of the publication of this model lies in its standardization in ISO 14721: 2003 [14], that would begin to materialize in specific projects, although the main objective is long-term preservation, the scientific community has also considered other associated problems: guarantee the principles of originality (the uniqueness of the document) and integrity (that the document has not been modified or altered) of the object, and of the information itself when it is migrated, modified or changed support.

Nowadays, the most applied options within digital preservation are the preservation of technology, the migration of old formats to newer ones, emulation through current software of previous software, metadata and, the simplest and least expensive, the replication or copying of the information. None of them imposes on others since preservation is a field still in development and not all entities have the same needs [15].

With each of these methods, different projects also emerged, which the different national or university libraries set in motion when it became evident that the volume of digital information would continue to grow and it would be necessary to preserve it. Among them stood out as pioneers InterPARES, PREMIS, PADI, PLANETS, LOCKSS, and PREDECI. It has not been defined yet what resources should be preserved, it will depend on the institution and the regulation of the country, as well as that of the author's rights in the digitally born document since there is still no definitive solution [16].

According to Torres, the tools necessary to build a long-term digital preservation system have not been developed, although digital preservation strategies have been carried out that is not the definitive solution. He adds that information specialists do not have the necessary skills to develop these tools and need to "ally in a much more enthusiastic way with the computer" so that digital preservation can be a fact. Digital information has an important economic value as a cultural product and as a source of knowledge [17].

It also plays a very significant role in sustainable development at the national level, taking into account that usually personal, governmental and commercial data are created in digital form. The disappearance of this heritage will generate economic and cultural impoverishment and hinder the advancement of knowledge [18].

It also indicates that no progress has been made in the implementation of methods for digital preservation and scientists are still working on this theory. "There is no clear solution or set of solutions to confront the challenges of digital preservation. The unpredictable technological development (…) and the political environment (…) contribute to the challenge of proposing the course that will follow[1]".

Another critical issue to consider is the lack of adequate storage for the protection of this type of evidence, based on the special care and physical space required by the technological elements, and it is usual for typical kidnapping offices do not meet the minimum requirements for guarantee the correct preservation of digital evidence sources [19]. It is further indicated that a direction for the investigation provides insufficient depth for a detailed evaluation.

Those institutions that currently make proposals for digital preservation use large and expensive systems. The vast majority of them apply the Open Archival Information System (OAIS) model for the preservation of their digital data [20]. A relevant aspect of this research is to study if the current OAIS and PREDECI reference model is adaptable to any institution. From the scientific community itself, this need for complementary documentation for small institutions under the name of OAIS-LITE has been raised. [21].

[1] "Preserving Our Digital Heritage: Plan for the National Digital Information Infrastructure and Preservation Program" 2002, p.19.

4 Results

The actors carried out a previously designed questionnaire, consisting of 68 items or variables, which measures the level of importance of the aspects that institutions should consider when applying digital preservation techniques or models in their information. The level of importance is measured on a Likert scale, with mathematical weights; 5 indicates “High importance”; 4 indicates “moderate importance”; 3 indicates “Average importance”; 2 indicates “Low importance”; 1 indicates “insignificant”.

The questionnaire was divided into the following four parts to verify the importance of the aspects. (A) organizational infrastructure, to determine if the preservation policies act within an organizational framework that is determined by the defined objectives, the legal conditions and the available financial and personnel resources, (B) administration of digital objects, where it is intended to verify if The application policies of digital preservation techniques or models analyze the objectives and strategies, and specify all the requirements related to the object for the management of digital objects during the life cycle of the objects in the preservation. (C) management of infrastructure and security risks, to analyze the technical and security aspects of the general system, and (D) aspects of management of the integrity of criminal investigation institutions; each of these aspects with the variables that evaluate for this work.

After the consolidation and statistical analysis of the data, we obtain:

Analysis and Interpretation
It is determined that 63 institutions were analyzed and evaluated in terms of digital preservation, thus, 67% belong to the public sector and 21% to the private sector (see Graph 1).

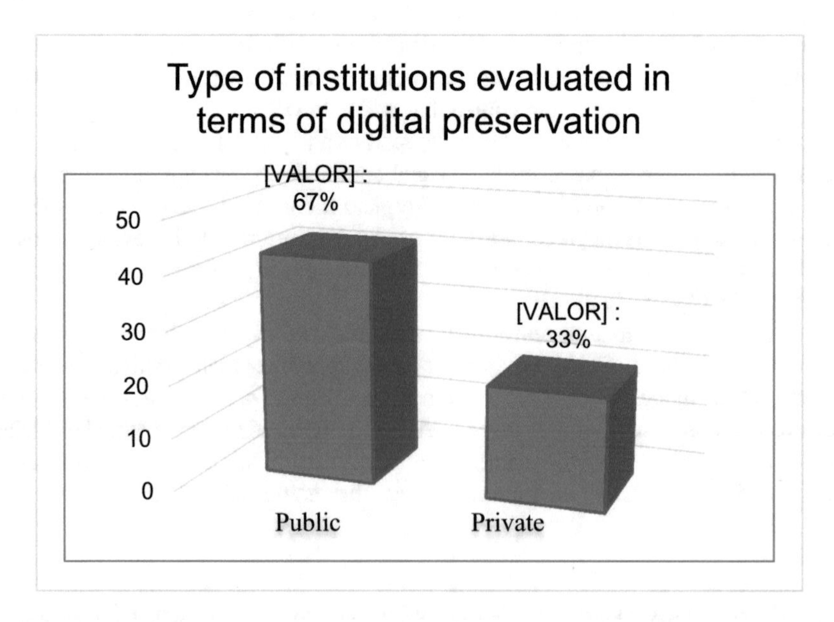

Graph. 1. Type of institutions analyzed Source: Table 1, Prepared by the author

Analysis and Interpretation

It is determined that only 11% of institutions evaluated have any applied any formal method of digital preservation, such as digital repositories or maintain a digital file and 89% of institutions only store digital information. (See Graph 2).

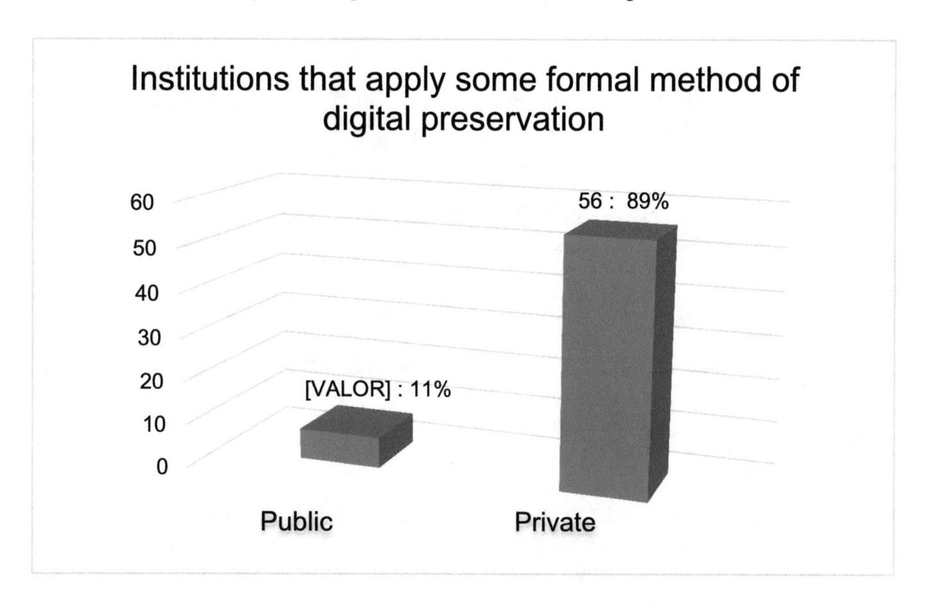

Graph. 2. Number of institutions that apply some formal method of PD Source: Table 2, Prepared by the author

Analysis and Interpretation

It can be determined that a value of 2.78 is obtained for the aspect of Organizational Infrastructure, for the aspect of Administration of digital objects a value of 2.15, for the aspect of Management of Infrastructure and security risk a value of 2.35 and for the aspect of Integrity Management, a value of 1.87 out of 5 points as the maximum compliance score. This assessment determines a low level of compliance with aspects of digital preservation, obtaining an average of 2.28 in the aspects analyzed (see Graph 3).

This result shows a low importance to processes of digital preservation, it may be due to the availability of resources, formal training or ignorance of the law; in any case, they are data that demonstrate the inadequate treatment of digital information, therefore a high possibility that in the long term that information is inaccessible, provoking from the cultural, historical and legal aspect a loss of information.

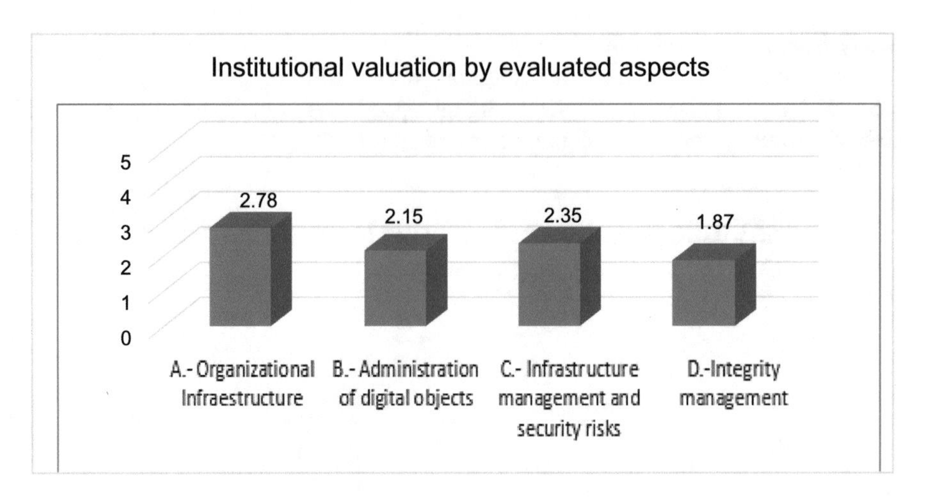

Graph. 3. Institutional valuation by evaluated aspects Source: Table 3, Prepared by the author

5 Threats of Validity

According [22], to diminish the threat to the validity of the statistical conclusion, aspects that are explained below are taken into account.

The questionnaires were answered with the same scale, in the same period of time and by all the actors involved, without there being a change of environment in that period of time. It was performed anonymously, however, there is an additional document that the actor has signed as proof of its application.

In the internal threat to the validity, the effect of history, maturation, and testing has not been considered, there are no threats of selection bias, because, in the institution where the survey is applied, no groups have been selected, and all the actors they have an interest in managing the preservation of evidence.

In the external threat of validity, there is no interaction risk, since the sample is the total population of the institution applied.

6 Conclusions

Digital preservation is a legal requirement in Ecuador and in most countries, which means a responsibility to fulfill at a personal and institutional public and private level, regardless of the type of business or activity performed. This digital preservation must be defined in the long term, and this time is the one that determines the legal basis of each country, in the case of Ecuador, as the case may be, it must be preserved for 5 years for some legal or patrimonial actions and without limit for others

There are several models of digital preservation, which are applicable to institutions such as hospitals, libraries, museums, prosecutors, criminal investigation institutions or any other entity that has the responsibility or legal obligation to safeguard digital data, however, they are not yet fully applicable in our country.

A critical level is determined in the application of techniques, methods or models of digital preservation in institutions that generate digital information, and even in an adequate management of some applied method; only 11% of institutions use a formal method of digital preservation. This percentage demonstrates a reality in which the guarantee and integrity of the digital objects generated by these institutions are accessible in the long term is diminished.

There is an average value of 2.28 out of 5 points in the digital preservation aspects, these aspects evaluate the adequate minimum process necessary for an adequate digital preservation, which means that, on average, of 63 public and private institutions evaluated, there is the need to apply adequate digital preservation techniques, methods and/or models to guarantee the legal mandate and the cultural, social and historical responsibility to preserve the digital information generated.

7 Future Work

Investigate the application of the PREDECI model as an alternative to digital preservation in the long term in public and private institutions

Investigate the admissibility of digital evidence in court, digital information from public and private institutions that apply models of digital preservation.

References

1. Allinson J (2006) OAIS as a reference model for repositories. An evaluation. http://eprints.whiterose.ac.uk/3464/. Accessed 29 Apr 2018
2. Ashoire AB (2013) An overview of the use of digital evidence in international criminal courts. In: Salzburg workshop on cyber investigations, Berkeley
3. CCSD (2012) Reference model for an open archival information system (OAIS) 650.0-M-2. Washington, DC, USA
4. CCSDS (2012) Reference model for an open archival information system (OAIS) 650.0-M-2. Washington, DC, USA: CCSDS. Obtenido de http://public.ccsds.org/publications/archive/650x0m2.pdf
5. Plaza LAC, Guadalupe JMP (2015) Diseño experimental en el desarrollo del conocimiento científico, vol 1. Riobamba, La Caracola Editores, Ecuador
6. Corral Y (2009) Validez y confiabilidad de los instrumentos de investigación para la recolección de datos. Revista Ciencias De La Educación 19(33):228–247
7. Ferreira M (2006) Introdução à Preservação Digital: Conceitos, estratégias e actuais consensos. https://repositorium.sdum.uminho.pt/bitstream/1822/5820/1/livro.pdf. Accessed 23 May 2018
8. Gladney HM (2009) Critique of Architectures for Long-Term Digital Preservation. HMG Consulting, United Kingdom
9. Gomez LS (2013) Delitos, prueba y evidencia digital. http://listas.hackcoop.com.ar/archivos/bla/attachments/20131026/b708d60b/attachment.pdf. Accessed 28 May 2018
10. ISO (2012) Space data and information transfer systems – Audit and certification of trustworthy digital repositories. ISO 16363:2012
11. Kotler P (2010) Fundamentos de Marketing. Prentice Hall, Mexico

12. LeFurgy W (2009) NDIIPP partner perspectives on economic sustainability. Libr Trends, 413–426
13. Molina F, Rodriguez G (2015) Digital preservation and criminal investigation: a pending subject. In: Molina Granja F, Rodriguez G New contributions in information systems and technologies. Springer, US, pp 299–309. https://doi.org/10.1007/978-3-319-16486-1_30
14. Molina F, Rodriguez G (2017) Model for digital evidence preservation in criminal research institutions – predeci. Int J Electron Secur Digit Forensics 9(2):150–166
15. Molina F, Rodriguez G (2017) The preservation of digital evidence and its admissibility in the court. Int J Electron Secur Digit Forensics 9(1):1–18
16. Morales P (2010) Investigación e innovación educativa. Revista Iberoamericana sobre Calidad, Eficacia y Cambio en Educación 8(2):47–73. http://www.redalyc.org/pdf/551/55114080004.pdf. Accessed 18 May 2018
17. National digital stewardship alliance (2015) National agenda for digital stewardship. The Library of Congress, Washington
18. Nestor working Group (2009) Catalogue of criteria for trusted digital repositories Version 2. Nestor working Group Trusted Repositories – Certification, Frankfurt
19. Martinez R (2012) La importancia de la evidencia y el análisis forense digital. Obtenido de http://www.bsecure.com.mx/opinion
20. Serra Serra J (2001) Gestión de los documentos digitales: estrategias. El profesional de la información, 4(18)
21. Shadish WR, Cook TD, Campbell DT (2002) Experimental and quasi-experimental designs for generalized causal inference. Hougthon Mifflin Company, Boston, New York
22. Térmens M (2009) Investigación y desarrollo en preservación digital: un balance internacional. El profesional de la información 4(8):4–18
23. Torres Freixinet L (2008) Preservacion Digital, el Reto del Futuro. Actas de las VIII Jornadas de Archivos Aragoneses. Gobierno de Aragón
24. Unesco (2004) Charter on the preservation of the heritage digital, Paris, France

Offensive Security: Ethical Hacking Methodology on the Web

Fabián Cuzme-Rodríguez[1(✉)], Marcelo León-Gudiño[2], Luis Suárez-Zambrano[1], and Mauricio Domínguez-Limaico[1]

[1] Carrera de Ingeniería en Telecomunicaciones, Universidad Técnica del Norte, Av. 17 de Julio 5-21 y Gral. José María Córdova, 100105 Ibarra, Ecuador
fgcuzme@utn.edu.ec

[2] Carrera de Ingeniería en Electrónica y Redes de Comunicación, Universidad Técnica del Norte, Av. 17 de Julio 5-21 y Gral. José María Córdova, 100105 Ibarra, Ecuador

Abstract. The implementation of security measures in IT directorates within Higher Education Institutions (IES) have increased in recent years due to a high rate of cyber attacks aimed at finding vulnerabilities in their Web services and communication networks, with an emphasis on government segments and strategic institutions such as HEIs. The objective of this research is to generate policies, protocols and an information assurance plan based on methodologies controlled in terms of security; As well as standards aimed at compliance with information security such as ISO 27001. For this purpose, a controlled scheme of attacks was established for the web server of the Universidad Técnica del Norte (UTN) in which the Offensive Security Methodology) For the execution of a Pentesting establishing improvements in the performance of the web service, as well as the assurance of the same web portal UTN managing to generate processes, policies insurance plans based on the norm ISO 27001 and the migration.

Keywords: Crime · Ethical hacking · Cyber attacks · Assurance of information

1 Introduction

According to the latest report of the General Prosecutor's Office of the State of Ecuador, about computer crimes, he mentions that since August 10, 2014, the date on which the Comprehensive Organic Criminal Code (COIP) came into force, crimes such as identity theft, espionage, fraud, among others [1]. In the Department of Technology and Computer Development (DDTI) of the Technical of the North University (UTN) have reported computer security problems such as: Access of unauthorized persons to the data network and denial of services, affecting the availability of systems such as the web server, which is exposed on the internet and visited by a large number of users daily.

It is necessary for organizations to take preventive actions against computer attacks, such as security audits. To do so, a methodology must be considered that adapts to the needs of the company; among the best known computer audit methodologies are: Open Source Security Test Methods Manual (OSSTMM), Security Information Systems

M. Botto-Tobar et al. (Eds.): TICEC 2018, AISC 884, pp. 127–140, 2019.
https://doi.org/10.1007/978-3-030-02828-2_10

Assessment Framework (ISSAF), Open Web Security Project Application (OWASP), Ethical Hacking Certificate (CEH) and Offensive Security (OS) [2]. Thus, companies or institutions, be they public or private, must have information security plans that allow them to secure their infrastructure and services from the internal and external point of view; This is where the application of standards, standards, reference frameworks and best practices such as ISO, COBIT, ITIL, among others, becomes indispensable, with reference to their application in some institutions today [3, 4].

Regarding the legality or illegality of ethical hacking in Ecuador is analyzed by Rojas [8], that is how, based on Ecuadorian legislation, he mentions: "*an offense is an unlawful, guilty or fraudulent action that is sanctioned with a penalty, and using this definition, ethical hacking is not punishable by Ecuadorian law, which explains in terms of intrusions and theft of information using tools such as spamming, pishing, tampering, among many others, sanctioning the fact of stealing information, swindling the client and break security causing harm to the victim, and in fact clearly defines the hacker term having the negative connotation directly, but does not contemplate an ethical hack to find errors and vulnerabilities was the definition of the word at first*".

2 Materials and Methods

This section establishes the materials and methods used in the development of the research, according to the scope established and that are adapted to the needs of the UTN Web Portal.

2.1 Methodology Offensive Security

Although there are some methodologies such as OSSTMM version 3 to perform a computer audit [2, 3, 5] applied to a public entity, the Offensive Security (OS) methodology is used for this case study since it allows verification of solutions, that is to say to verify if the controls that are applied will improve the security.

The purpose of the OS methodology is to perform penetration tests and study safety; because it is clearly intrusive, it is based on studying offensive security to exploit vulnerabilities [6]. In this case, the UTN web service is adapted to the needs of OS, since it fulfills the condition of Exploitation of real platforms; The results are not based on statistics generated by applied tools, but on results of the penetration test.

2.2 ISO/IEC 27001 Standard

The ISO/IEC 27001 standard has been in force since 2013, specifically focusing on Information Security Management Systems (ISMS); the standard determines the fulfillment of 130 requirements so that it is guaranteed.

On the other hand, as shown in Fig. 1, the standard is based on complying with the ISMS, which has the principle of offering Confidentiality, Integrity and Availability of information, thereby seeking trust between organizations and organizations and customers [7].

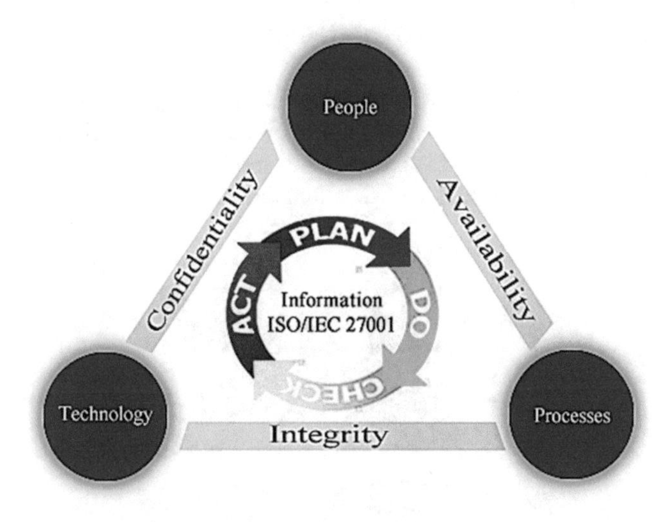

Fig. 1. Information security assessment areas.

Although ISO 27001 establishes some evaluation areas for the information assurance plan, which should be composed of security policies and processes, in this study the following areas are considered in order to present the security policies:

- General security policies.
- General policies to administrators.
- Security policies for infrastructure.
- Security policies for applications.

In terms of processes, the following evaluation areas are determined for the preparation of safety procedures manuals.

- Manual of procedures for the control of documentation.
- Procedures manual for internal audit.
- Procedure manual for corrective measures.
- Procedures manual for preventive measures.
- Technical procedures manual.

2.3 MSAT 4.0 Risk Analysis

The risk analysis consists of studying the possible threats and probable unwanted events and the damages that could be caused, in this case, the audit was made to the computer technology department of the UTN through the MSAT 4.0 (Microsoft Security Assessment Tool) tool. It is based on the ISO/IEC 27005 standard.

3 Implementation of the Methodology

The methodology Offensive Security (OS) explained in [9], mainly contemplates the methods for the development of security studies focused on offensive security, allows the independent exploitation of the indicators of risks and vulnerabilities, that is, it is carried out in real platforms with statistics tangible, and has five phases as shown in Fig. 2.

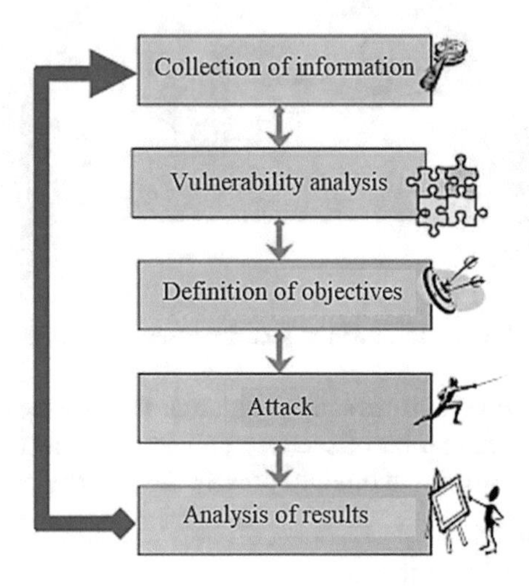

Fig. 2. Phases of the offensive security methodology.

3.1 Information Collection

In this stage an evaluation of the current situation of the company is carried out, for this case of study it is carried out with a risk analysis using the MSAT 4.0 tool.

3.2 Vulnerability Analysis

Once the risk analysis is carried out, the vulnerabilities of the system are obtained, in this phase all the possible faults in the institutional network must be looked for. To do this, it uses tools such as **Nmap** for port scanning; **Footprinting** to find critical information that is public to the user; **Banner grabbing** to find hosts on the network; and to scan vulnerabilities using **Nessus**.

3.3 Definition of Secondary Objectives

In most scenarios, the target of attack is not within reach or visible; that is, there are barriers or stages that must be passed to reach the main objective. For this, access measures must be determined and these obstacles must be overcome. In this research work, the following are established as secondary objectives: Wireless networks, wired networks, computers located in the internal network, switching and routing devices.

3.4 Attacks

In this stage, the exploitation of vulnerabilities to the network infrastructure begins; in this case the Kali Linux tool is used, which is a robust tool in terms of conducting computer audits; as well as additional tools to be the case, for example: Wireshark, FOCA, WinSCP, among others. This is explained in detail in the section Attacks on the Internal Network of the UTN.

Scanning. The scanning methodology consists of determining what equipment is available in a network, what services are offered, what operating systems are running, what filter or firewall is being used [10], the state of the ports, doing a Banner Grabbing, scanning vulnerabilities and perform the connection survey of the network diagram. For this the Kali Linux tool is used; Nmap to perform port scans with the command nmap sS <ip victim>, for vulnerability scanning Nessus is used and to create the topology of the victim Zenmap.

Phishing. The term Phishing refers to the impersonation, in this case the cloning of the UTN web portal will be carried out, which is shown in Fig. 3.

Fig. 3. Phishing in the portal of the UTN.

Beef is the abbreviation of The Browser Explotation Framework; This penetration testing tool comes in Kali Linux; Beef allows you to capture from the web what the victim types in the cloned web portal. The attack consists in creating a clone of the victim's website through the Setoolkit tool so that users can access the fake website, capture the username and password through the Beef tool, and then access the original website.

Extract Metadata, Snooping. This attack consists in extracting information that has inside the web site, FOCA is a tool that allows us to extract the metadata and analyze it; It has several parameters that it analyzes, such as the domain.

Brute Force Attack. It consists of creating a dictionary using the Crunch tool of Kali Linux, to then make the attack with the Hydra tool, which allows the dictionary to be executed. This attack consists of searching for possible users and authentication passwords, in order to access the institution's systems.

DoS. The Denial of Service Attack is that the network or service is not available to users. There is also the DDoS which is the distributed denial of service attack, which consists of two or more computers being controlled by a remote attacker, that is, said computers act in zombie mode.

This is done by running the command hping3 [11], which is a terminal application for Linux that will allow to easily analyze and assemble TCP/IP packets.

Injection SQL. It is a method of infiltration of intrusive code that uses a computer vulnerability present in an application at the level of validation of inputs to perform operations on a database. The SQL injection is intended to embed intruder SQL code and the embedded code portion [12].

The attack was done with the Sqlmap tool that comes integrated with the Kali Linux tool.

3.5 Analysis of Results

The risk analysis carried out with the MSAT 4.0 tool based on the ISO 27005 standard is taken as reference, the results are shown in Fig. 4:

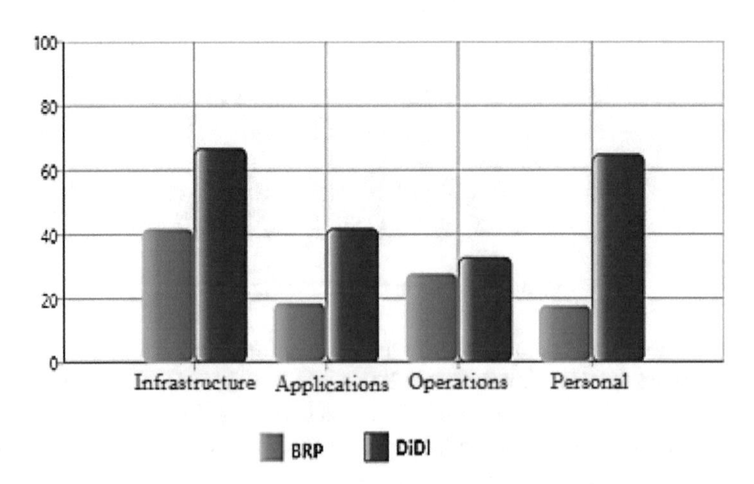

Fig. 4. Results through MSAT 4.0.

Risk profile for the company (BRP): Measure of the risk to which an organization is exposed, according to the environment and the sector in which it operates. The BRP score ranges from 0 to 100. A higher score represents greater risk to which the institution is exposed in this area of analysis. It is important to know that a score of 0 is not possible; any organization carries a level of risk, in addition there are commercial risks that can not be mitigated directly.

Defense in Depth Index (DiDI): Measurement of security defenses used in personnel, processes and technology to help reduce the risks identified in an organization. DiDI also has a score of 0 to 100. A higher score means an environment where more measures have been taken to implement DiD strategies in the area of specific analysis. The DiDI score does not indicate the overall effectiveness of security or even the amount of resources for it, but rather quantifies the overall strategy used to defend the environment.

4 Results

According to the critical vulnerabilities previously determined in the systems, solutions must be proposed that mitigate these problems. Next, the technical solutions proposed to improve the security system of the equipment of the Data Center on the UTN are described.

Table 1 shows a summary of the risks extracted from the MSAT 4.0 tool, for which a risk analysis was performed before the intrusion process and after implementing the improvements proposed to solve the security problems found in the institutional network.

Table 1. Risk analysis

Evaluated areas	Initial analysis	Final analysis
Infrastructure		
Applications		
Operations		
Personal		

Where the color code has the following interpretation:

It complies with the best recommended practices.
Needs improvement.
Severe deficiencies.

4.1 Bandwidth Exinda

Exinda is responsible for the monitoring and control of the network, it is a device that allows controlling the bandwidth of the computer equipment that is connected to the network.

In Table 2, the bandwidth of the internal requests is shown, the packets, the data, the maximum and minimum bandwidth, and the information flow are analyzed.

Table 2. Top 5 inbound URLs

Name	Packets	Data (MB)	Throughput (Average)	Throughput (Max)	Flow
download.microsoft.com	85852367	117607.210	7032.80	55586.32	74
edutrial.autodesk.com	74514366	101897.456	8651.60	58550.61	56
osxapps.itunes.apple.com	63442208	87002.200	6996.72	57666.98	31
mirror.espoch.edu.ec	51436395	70592.292	14852.55	66331.95	21
www.utn.edu.ec	263842460	38019.006	3.10	1146.38	892

Table 3 shows the bandwidth of the external requests, analyzes the packets, the data, the maximum and minimum bandwidth, and the flow.

Table 3. Top 5 outbound URLs

Name	Packets	Data (MB)	Throughput (Average)	Throughput (Max)	Flow
www.utn.edu.ec	321276096	321588.250	26.28	24019.21	892
repositorio.utn.edu.ec	3421310	4426.036	497.36	13939.87	21
rs.qustodio.com	2164357	2886.771	2896.65	4938.47	6
dfw.coolrom.com	50523481	2830.531	821.88	1125.62	2
10-img-fr.iloveimg.com	2064912	2709.750	1894.25	4352.53	5

4.2 Web Portal Security Levels

Next, Fig. 5 shows the security levels according to the attacks made to the system, this is done on a scale of 1 to 10, where number 1 represents the lowest level of security and 10 maximum.

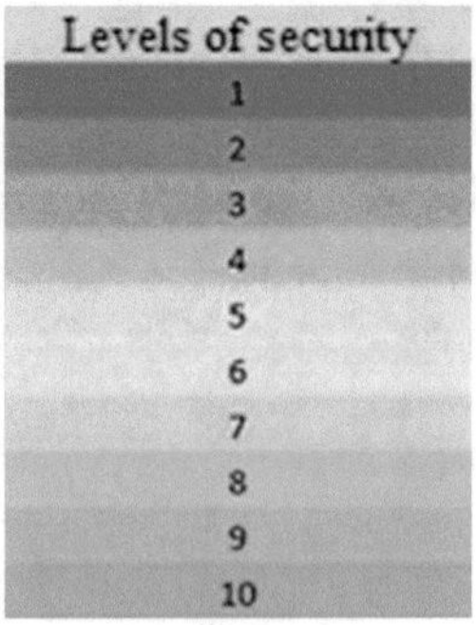

Fig. 5. Levels of security.

4.3 Summary of Attacks Against the Web Portal

Therefore, Fig. 6 summarizes the computer attacks carried out, the details, the tool, applicability, observations and the level of security.

Level	Name	Detail	Tool	Proposal submitted by the Author	Levels of security
Operating System	Ping	Site identification, discovery IP and domain.	Promt and SHODA	Ips firewall protection	9
Operating System	Scan Ports	Method to look for open ports	NMAP	Elminate unnecessary	7
Application	Phishing	Phishing	Beef and Kali Linux	Implementation SSL/TTL	1
Application	Extract Metadata	Extract Metadata of Portal Web	FOCA	Implementation SSL/TTL	3
Application	Snooping	Resolve domains of server DNS	FOCA	Implementation SSL/TTL	3
Operating System	Authentication attack	Remote authentication	WinSCP	System of difficult passwords	10
Operating System	Brute force attack	Obtaining server passwords	Hydra and Crunch	Difficult passwords	8
Operating System	Dos attack	Deny service of Portal Web with a computer	Kali Linux	Implement passwords with a high degree of dificulty and change them every so often	7
Operating System	DDoS attack	Deny service of Portal Web with multiple computer	Kali Linux	Protection of the DNS server port through firewall rules, improve the CPU capacity of the server	5
Application	Injection sql	Inject malicious code in the web server	Kali Linux	Exploit the benefits of Modsecurity	9
Application	Cross Site Scripting	Inject malicious scripts	Kali Linux	Exploit the benefits of Modsecurity	9

Fig. 6. Result of attacks on the Web Portal.

4.4 Technical Solutions

SSL/TLS Digital Certification, Web Portal with HTTPS Security. The Digital Certification TLS (Transport Layer Security) is responsible for providing security in the transport layer, its predecessor SSL (Secure Sockets Layer) secure ports layer is a protocol that allows applications to transmit information in a secure manner. Through this will provide greater security to all users of the UTN web portal against phishing attacks or identity theft.

The process of installing an SSL/TLS digital certificate in the CentOS operating system, which is where the Web server is hosted, is shown below. In this process a self-signed certificate is used. Within the proposed security polices it is proposed that it is necessary to use a digital certificate that is valid, which is reflected in the proposal provided to the DDTI department.

The purpose of HTTPS is to achieve safer connections on the web, in this way the sensitive information is encrypted, in case certain data are intercepted, however, it presents vulnerabilities when applied to publicly available static content.

To get the secure web page, the URL must start with https://, and port 443 must also be enabled, once this is done HTTPS will use SSL and TLS encryption. The HTTP protocol operates in the application layer of the TCP/IP model, but the HTTPS protocol works in the lower sublayer, encoding the HTTP message in the transmission and decoding the information before it arrives for the Reception.

Implementation of an IDS with SNORT. It is proposed to install an IDS server in the network of the Technical University of the North, with the purpose of having a computer that allows detecting attacks to the Data Center, which is the main objective for computer crimes. This server will allow detecting intrusions via addressing, service ports and interfaces where the attacks are made.

Installation of Antivirus on WEB and DNS Servers. The UTN has the Kaspersky antivirus license for computers, given the results obtained in the investigation, it is proposed to implement an antivirus on the servers since this measure will allow greater protection against malware that can be injected by means of computer attacks.

It should be noted that the equipment of these services are in production and serve a large number of users, so it should optimize the processing of this. What is proposed is for administrators to scan the entire system, so that possible malware is identified. For the verification and demonstration of the importance of an antivirus, tests were carried out using a free ClamAV antivirus.

Iptables in the WEB Server. A possible solution to the brute force attack is to restrict the number of parallel connections to a server by the client's IP address. You can use connlimit to create some restrictions, to allow 3 ssh connections per client the following rule is placed:

#iptables A INPUT p tcp -syn dport 22 m connlimit connlimit-above 3 j REJECT

On the other hand, to block and prevent DDoS attacks, the following rule is used:

#iptables A INPUT p tcp dport 80 m limit limit 25/minute limit-burst 100 j ACCEPT

To prevent a DDoS attack, a rule can be applied in the Firewall to limit the number of requests on port 80 that belongs to the Web server.

#iptables A INPUT p tcp dport 80 m limit limit 25/minute limit-burst 100 j ACCEPT

Firewall rules to block the ping command.

iptables -AINPUT -p icmp -icmp-type echo-request -j DROP
iptables -A INPUT -i eth1 -p icmp -icmp-type echorequest -j DROP

To hide open ports using firewall rules.

Syn, Ack, End, or Rst scan:
iptables -A port-scan -p tcp tcp-flags SYN, ACK, END, RST RST -m limit limit 1/s -j RETURN iptables -A port-scan -j DROP

Xmas, banner scan and others:

iptables-INPUT -p tcptcp-flags ALL NONE -j DROP
iptables-INPUT -p tcp tcpflags ALL END, URG, PSH -j DROP
iptables -AINPUT -p tcp tcp-flags SYN, RST SYN, RST -j DROP
iptables -INPUT -p tcp tcp-flags SYN, FIN SYN, FIN -j DROP
iptables-INPUT -p tcp tcp-flags ALL FIN -j DROP

Antifragments:

iptables -A INPUT -f -j DROP
Discard null packages:
iptables-INPUT -p tcp tcp-flags ALL NONE -j DROP

SSH Port Change. The change of port SSH increases the security in this type of connections since the default port is 22. Therefore, it is convenient to change this port for a higher than 1024 that complicates an attacker connection in this way.

Web Application Firewall (WAF). It is a web application firewall, responsible for filtering, monitoring and blocking HTTP traffic to and from a web application. WAF differs from a regular firewall because it is capable of filtering the content of specific web applications, while the regular firewall serves as a security gateway between the servers. The objective of WAF is to avoid attacks such as Cross Site Scripting (XSS), SQL Injection (SQLi), Remote File Inclusion (RFI), Local File Inclusion (LFI), poisoning, manipulation of headers, among others. Trying to protect what the IDS/IPS do not.

The areas of application of WAF are two, the first is responsible for denying all transactions, accepts those that believe are safe and the second accepts all requests, only denies those that are possible threats or real attacks. It depends on signatures and updates, so it does not make it so accurate.

There are two security models WAF: Aqtronix and modSecurity, it was determined to use Mod Security because it is adapted to the requirements to solve the problems found in the network of the UTN, since it has more security features for computers under GNU operating systems and Apache, since this avoids attacks such as Cross Site Scripting (XSS), SQL Injection (SQLi), Remote File Inclusion (RFI), Local File Inclusion (LFI), poisoning, manipulation of headers [13], which Unlike Aqtronix, it is the one for ISS (Secure Internet Service) and it is related to Windows servers.

4.5 Zabbix Administration Software Statistics

Zabbix allows to give reports regarding the loading efficiency of the UTN Web portal, with this tool the bandwidth consumed is determined as shown in Fig. 7:

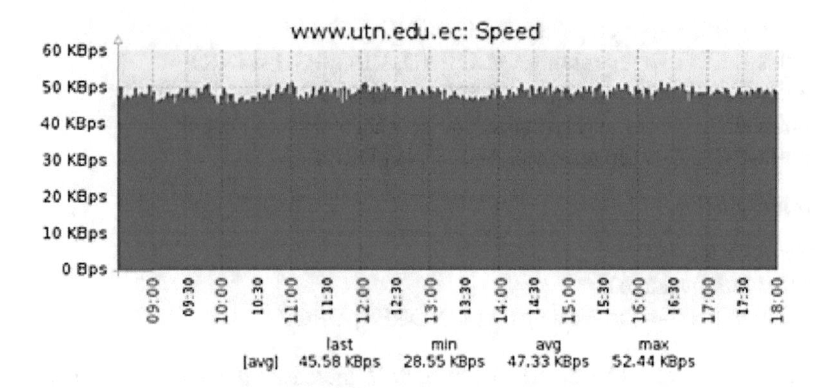

Fig. 7. Bandwidth consumption of the Web Portal.

The time it takes a user to access the UTN page before implementing securities, the following results were shown in Fig. 8:

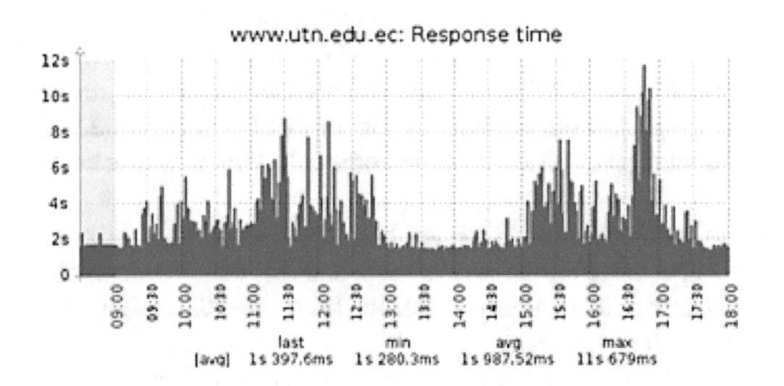

Fig. 8. Charge time before giving solutions.

With the improvement actions proposed, the Web portal has superior benefits for users in terms of response speed, this is verified in Fig. 9:

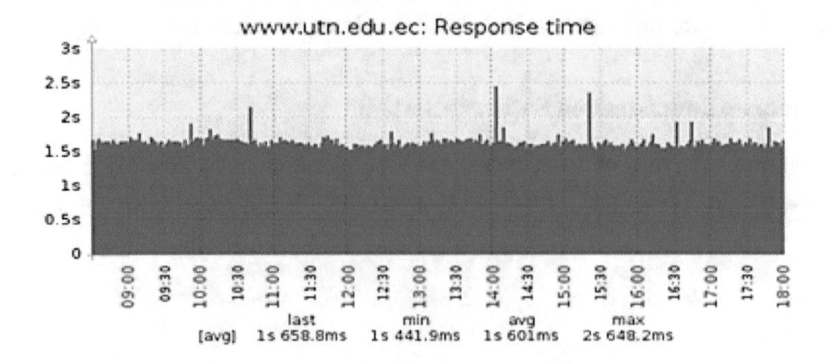

Fig. 9. Load time implemented solutions.

4.6 Organizational Solutions

The security of the systems can be improved, with an organizational system and constant training for users and administrators. There are security measures through the implementation of security policy and processes, for this we have the ISO 27001 guide for the preparation.

4.7 Presentation of the Documentation

It is important to submit safety reports to the institutions that are evaluated so that the corresponding measures are taken, for this a clear report must be made with emphasis on the critical problems, since they must be taken into account so that the administrators mitigate the flaws.

5 Discussion

With respect to other articles related to Ethical Hacking and Pentesting Methodologies in Web Applications, the importance of implementing preventive computer security measures was determined, unlike OSSTMM, which makes a report of the shortcomings and possible solutions; OFFENSIVE SECURITY performs a preliminary analysis to determine attacks according to the vulnerabilities found and allows to verify the results by means of a feedback of the audit process executing the stages from the beginning.

In the same way, the solutions presented in this article are consistent with the most common IT problems in the network security of the institutions, in addition to the results of the risk analysis prior to the exploitation of the vulnerabilities, it should be emphasized that for now are solutions, then they will be vulnerabilities that need better strategies to mitigate cyber attacks.

6 Conclusions

The Offensive Security methodology was chosen because, unlike other methodologies, it makes a feedback to verify the proposed solutions, the ISO/IEC 27001 standard is the guide for the realization of security manuals and policies, working together with the national legislation and international cybercrime, obtaining favorable results in terms of security in the Web Portal due to the implementation of the proposed solutions.

The risk analysis of the MSAT 4.0 tool and the vulnerability analysis of the Offensive Security methodology determined flaws in the UTN web portal, which were resolved through security policies and procedures.

Acknowledgment. To the Department of Technological and Computer Development of the Universidad Técnica del Norte for the trust of its leaders in allowing the development and implementation of research through Ethical Hacking.

References

1. del Estado FG (2015) Los delitos informáticos van desde el fraude hasta el espionaje - Fiscalía General del Estado, 13 May 2015. https://www.fiscalia.gob.ec/los-delitos-informaticos-van-desde-el-fraude-hasta-el-espionaje/. Accessed 09 Mar 2018
2. Bracho-Ortega C, Cuzme-Rodríguez F, Pupiales-Yepez C, Suárez-Zambrano L, Peluffo-Ordoñez D, Moreira-Zambrano C (2017) Auditoría de seguridad informática siguiendo la metodología OSSTMMv3 : caso de estudio, Maskana, vol. 8, pp 307–319
3. Cuzme-Rodríguez F, Suárez-Zambrano L, Bracho-Ortega C, Pupiales-Yepez C (2017) DISEÑO DE POLÍTICAS DE SEGURIDAD DE LA INFORMACIÓN BASADO EN EL MARCO DE REFERENCIA COBIT 5. In: Innovando Tecnología, UTN, Ibaquingo D, Guevara C, Arciniega S, Pusdá M, Granda P (eds) Ibarra, pp 129–137
4. Rocha Haro CA (2011) La Seguridad Informática. Rev Cienc Unemi 4(5):26–33
5. Valdez Alvarado A (2013) OSSTMM 3. Rev Boliv **8**:29–30
6. Valencia Blanco LS (2013) Metodologías Ethical Hacking. Rev Boliv 8:27–28
7. López Neira A, Ruiz Spohr J (2013) ISO27000.es - El portal de ISO 27001 en español. Gestión de Seguridad de la Información. http://www.iso27000.es/iso27000.html. Accessed 9 Mar 2018
8. Rojas D (2014) HACKEO ETICO EN EL ECUADOR "El Hacking Ético en el Ecuador, es legalmente posible"
9. Isaza Villar MA (2013) La Seguridad Informática Hoy, 19 February 2013. https://seguridadinformaticahoy.blogspot.com/2013/02/metodologias-y-herramientas-de-ethical.html. Accessed 12 Mar 2018
10. Domínguez HM, Maya EA, Peluffo DH, Crisanto CM (2016) Vulnerar servicios con métodos de autenticación simple 'Contraseñas', pruebas de concepto con software libre y su remediación, Maskana, vol. 6, pp 87–95
11. Kali Tools (2014) Fierce|Penetration testing tools. https://tools.kali.org/information-gathering/hping3. Accessed 25 Aug 2018
12. López de Jimenez RE (2016) Pentesting on web applications using ethical – hacking. In: 2016 IEEE 36th central American and Panama convention (CONCAPAN XXXVI), pp 1–6
13. Elhacker.net, Introducción a los Web Application Firewalls (WAF) - wiki de elhacker.net. http://wiki.elhacker.net/seguridad/web/introduccion-a-los-web-application-firewalls-waf. Accessed 9 Mar 2018

Identification of Skills for the Formation of Agile High Performance Teams: A Systematic Mapping

Héctor Cornide-Reyes[1(✉)], Servando Campillay[1], Andrés Alfaro[1], and Rodolfo Villarroel[2]

[1] Departamento Ingeniería Informática y Ciencias de la Computación, Universidad de Atacama, Avenida Copayapu 485, Copiapó, Chile
hector.cornide@uda.cl

[2] Escuela de Ingeniería Informática, Pontificia Universidad Católica de Valparaíso, Valparaíso, Avenida Brasil 2241, Chile

Abstract. The irruption and wide adoption of the agile methods have generated tremendous challenges to innovate in matters of education in Software Engineering. These pedagogical innovations seek to strengthen the skills of students to achieve optimum performance in the industry, however, recent research still points out differences with respect to what the industry requires.

This article's main objective, to collect and analyze scientific evidence on the skills required for the formation of agile high performance teams.

A systematic mapping of the literature was carried out to obtain a visualization of the scientific contributions existing in this topic.

Twenty-two primary studies were selected, which were classified according to the defined protocol. It was possible to identify a set of necessary skills and some methodological proposals aimed at stimulating and strengthening them.

The results obtained allowed to identify and classify the skills for the formation of agile teams, which will allow conducting future investigations on this subject.

Keywords: Skills · Agile · Education · Software Engineering
Systematic literature mapping

1 Introduction

At present, agile methodologies enjoy a greater and better presence in the software development industry. The scientific community has been very active in this issue, contributing publications based on improvement proposals for: their development processes, teaching methodologies, adoption experiences in the industry, key aspects for agile team training, among others.

The teaching of agile methodologies constitutes a tremendous challenge because the construction of true learning will be achieved through its use. The principles of agilism are the agile manifesto [4] that, in order to achieve this, requires that people think in a different way about the software development process.

M. Botto-Tobar et al. (Eds.): TICEC 2018, AISC 884, pp. 141–152, 2019.
https://doi.org/10.1007/978-3-030-02828-2_11

In matters related to education in Software Engineering, academics have turned to the search for new ways of teaching to strengthen student learning. This, because the software development industry increasingly requires a greater number of professionals with the skills demanded by the market [1].

There is a gap between what is taught in the classroom and what is required by software development companies. This situation has generated a reflection on academia, about the skills that must be developed in future professionals and discipline of software engineering to incorporate teaching strategies that can be used so that their learning experiences are highly influenced for the practices, techniques and work methods required by the development of industrial quality software [5].

In Chile, the technology industry is warning about the shortage of existing professionals in the sector, with a demand higher than the supply in the local environment, as indicated by the latest Cisco IDC Skills Gap 2016 study [1], carried out in ten countries of Latin-American. According to this research, in Chile there will be a 31% deficit of IT professionals for the year 2019 (19,513 vacancies).

According to this, it is important to maximize the efforts and resources in the training processes related to Software Engineering to reduce the gap between what has been learned in the University and what the industry requires; even more, when the demand for professionals will be above the offer.

This article is part of a broader research project that seeks to generate a deeper understanding of this topic, with the aim of proposing a framework that facilitates the traceability of skills for the formation of high performance agile teams.

In this first part, we conducted a Systematic Mapping of Literature (SML) to identify and analyze the existing contributions related to the identification of skills for the formation of agile teams.

This article is organized as follows: in Sect. 2, the methodology of the mapping performed is detailed. In Sect. 3, the applied research method is detailed, while Sect. 3.1 shows the results and answers to the research questions. Finally, Sect. 4 details the conclusions of the work carried out.

2 Mapping Methodology

SLM is a defined method for constructing classifications and conducting thematic analyzes in order to obtain a visual map of existing knowledge within a broad topic [6]. The analysis of the results was done by categorizing the findings and counting the frequency of publications within each category to determine the coverage of the different areas of a specific research topic.

The information generated can be combined to answer more specific research questions and save research time and effort. Therefore, systematic mappings should be of quality in terms of completeness and rigor. Systematic mappings allow having a general view on a subject when there are a large number of primary studies [7]. Another type of secondary study is the systematic reviews of the literature (SRL). These, unlike systematic mapping, are used to find, evaluate and aggregate all the evidence present in the relevant research articles regarding a specific research question. The objective is to ensure that the review of the literature is objective, rigorous and auditable [7]. However, its main drawback is the considerable effort required to carry it out [6]. It was decided to carry out a systematic mapping of the literature due to the need to obtain a general vision regarding the processes of traceability of requirements in agile methods. In [6] they suggest a procedure consisting of 5 stages (see Fig. 1):

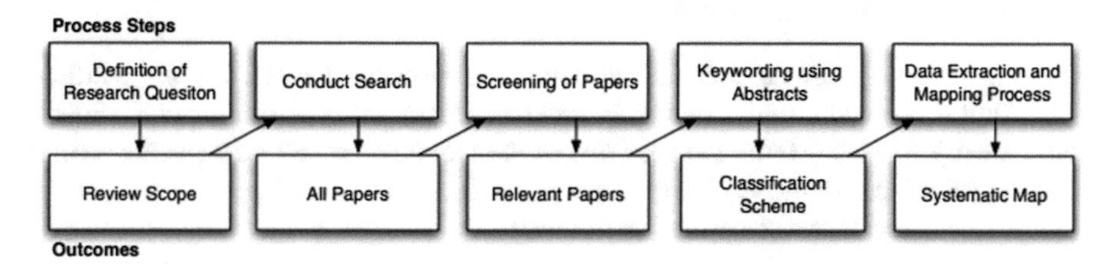

Fig. 1. Process for systematic mapping [6]

2.1 Research Questions

The objective of this research is to analyze scientific evidences that inform about the necessary skills to form agile software development teams.

This objective is divided into three specific research questions to obtain a more detailed knowledge and a comprehensive view of the subject. The research questions are the following:

- RQ1: What skills are necessary to be part of an agile development team?
- RQ2: What methodological proposals exist to develop the teaching/learning process on agile methods?
- RQ3: What evaluation methods will be adequate to measure achievement levels in the skills detected in RQ1?

2.2 Data Source and Search String

The search string is systematically determined by conducting various searches related to the research topic. In this way it was possible to refine the vocabulary and know the use of synonyms in order to generate more powerful chains. The citation databases used were: Web of Science (WOS) and SCOPUS, while the databases of scientific publications were: ScienceDirect and SpringerLink. The previous selection is due to the great reputation of these databases and because there was total access to the published material.

The search string was adapted to each engine considering the following: ((agile OR agil OR method *) AND skills) OR ((method * agil *) AND (skills OR competences)).

2.3 Criteria for Inclusion and Exclusion

All scientific publications related to Software Engineering will be eligible. If the engine allowed it, it was filtered by the Computer Science discipline and the Software

Engineering sub discipline. Only articles published after 2012 are considered to analyze the most recent publications in this area and, to limit the number of articles in the searches to be carried out. Articles that are not related to the formation of work teams in agile methods were excluded.

2.4 Data Extraction and Synthesis

The data extraction form was developed with the following fields: Title of the article, Year, Author; Type (Journal, Conference); No. of citations; Scope of the Study (Education, Industry, Other); Method used; Type of Results (Generic Skills, Specific Skills), Main contributions; Evidence RQ1; Evidence RQ3; Evidence RQ3.

The search conduct was defined as follows:

As a first filter (1F), we proceeded to review all the titles and keywords of the articles thrown by each database. Then we proceeded to eliminate the repeated items.
As a second filter (2F), we proceeded to read the abstract for all those articles that passed the first filter.
Finally, the selected articles were downloaded from the web, completely read and added to the information registration form created in Microsoft Excel according to the defined protocol.

3 Execution of Systematic Mapping

Once the research protocol was defined, the mapping was executed. As a result of the application of the search string, a total of 421 articles were obtained. In Fig. 2, it is possible to observe the process of execution of the mapping, finally selecting 22 primary studies.

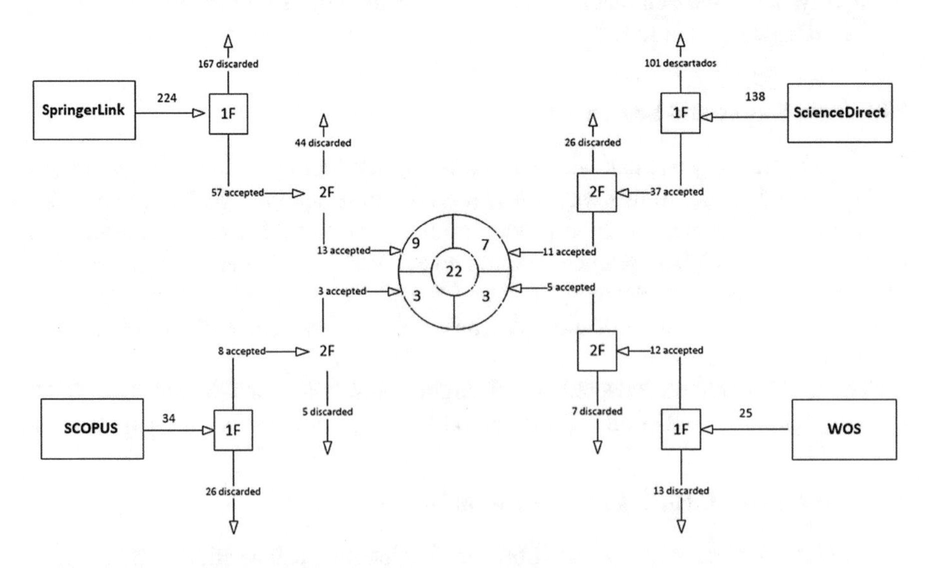

Fig. 2. Results of the systematic mapping process

In Table 1, shows the selected primary studies and identifies the scope of study, the research method and the type of result that each primary study delivers. This information will be used to answer the research questions that gave rise to the present investigation.

Table 1. List of selected primary studies

Authors	Year	Reference	N° citations	Scope of study	Methodological proposal	Type result
Steghöfer et ál.	2017	[11]	2	Education	Learning based on games	Generic skills
García et ál.	2017	[12]	8	Education	Learning based on games	Generic skills
Melo et ál.	2013	[3]	68	Industry	Study cases	Specific skills
Scott et ál.	2016	[21]	8	Education	Study cases	Generic skills
Holtkamp et ál.	2015	[13]	19	Education	Study cases	Generic skills
Fagerholm et ál.	2015	[8]	29	Industry	Study cases	Generic skills
Von Wangenheim et ál.	2013	[14]	67	Education	Learning based on games	Generic skills
Mesquida et ál.	2017	[19]	1	Industry	Learning based on games	Generic skills
Aldahmash et ál.	2017	[9]	1	Other	Other	Generic skills
Babb et ál.	2013	[10]	13	Education	Study cases	Generic skills
Dutra et ál.	2015	[25]	2	Other	Other	Generic skills
Pereira et ál.	2016	[16]	1	Industry	Learning based on games	Generic skills
Moe	2013	[26]	9	Industry	Study cases	Generic skills
Albuquerque et ál.	2016	[22]	0	Industry	Study cases	Generic skills
Martin et ál.	2017	[20]	1	Education	Study cases	Specific skills
Rodríguez et ál.	2012	[17]	6	Education	Learning based on games	Generic skills
Alqudah et ál.	2017	[27]	2	Industry	Other	Specific skills
Kropp et ál.	2014	[15]	19	Education	Study cases	Generic skills
Diel et ál.	2015	[28]	0	Industry	Other	Generic skills
Igaki et ál.	2014	[29]	14	Education	Study cases	Generic skills
Read et ál.	2014	[30]	8	Education	Problem based learning	Generic skills
Scharf et ál.	2013	[31]	20	Education	Learning based on games	Generic skills

3.1 Overview of Primary Studies

As mentioned above, systematic mapping of literature is a defined method for constructing classifications and conducting thematic analyzes [6]. In Fig. 3, the bubble chart of the systematic mapping made with the categories defined above is shown.

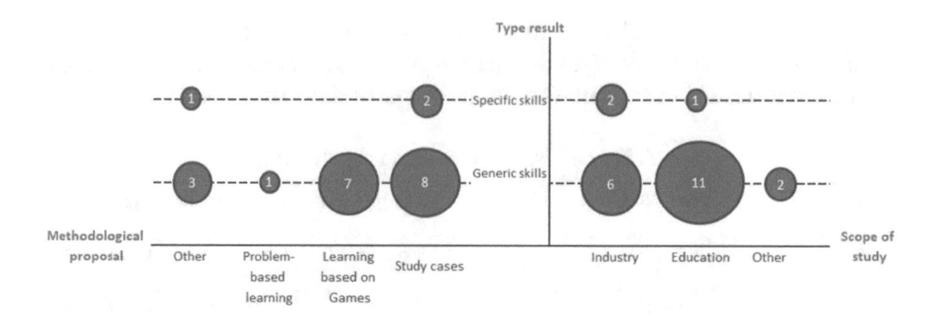

Fig. 3. Bubble chart of systematic mapping

In Table 2, shows the correspondence between the research questions defined and the primary studies selected.

Table 2. Correspondence between research questions and primary studies

Investigation questions	References to primary studies
RQ1: What skills are necessary to be part of an agile development team?	[3, 8–11, 13, 15, 22, 25–31]
RQ2: What methodological proposals exist to develop the teaching/learning process on agile methods?	[11, 12, 14–17, 19–21, 25]
RQ3: What evaluation methods will be adequate to measure achievement levels in the skills detected in RQ1?	[11, 14–17, 20, 21, 29]

3.2 RQ1: What Skills Are Necessary to Be Part of an Agile Development Team?

After reviewing the selected primary studies, it is remarkable the great diversity of approaches and contexts described, which allows having a broad view of factors and aspects considered in the different studies.

Most of the articles studied focus their research on generic skills and the factors that influence the performance of an agile team. This is due to the fact that the cases studied respond to situations applied to students belonging to the last years of the races, so that specific skills would not represent a problem. One of the reasons that can explain the above is that the software development industry itself has recently insisted on the importance of recruiting people with good levels of performance in generic skills.

For the success in the use of agile methodologies, the people who integrate a team must have an adequate level of generic skills to promote good results and a harmonious work environment.

In [11], the experience in teaching SCRUM using LEGO Workshops is presented. Through this type of method, it was possible to generate a favorable work environment to understand the functioning of the work methodology and the roles present in SCRUM. Both students and teachers identify communication and collaboration, as necessary skills to achieve a good performance at work with this type of methodologies. Likewise, the specific skills linked to the planning and estimation of work are recognized as weak aspects and that in some students were distracting elements within the methodological work that was intended to be achieved.

In [13], a survey is carried out to companies and professionals of the area, with respect to the appreciation they have regarding skills, in areas and environments of global software development for the design, implementation and testing phases. The survey was mainly attended by people from Finland, Canada and Germany, although there are also records of professionals from the US, Uruguay and Spain. It was demonstrated that the adaptability and cultural awareness represent the differences of competence requirements of requirements engineering, software design, implementation and testing. As a result of the above, it is not only important to recognize the skills necessary to work under agile methods, but also to determine the level of achievement necessary for each stage and for each role that the work methodology demands.

After making a detailed reading of the studies, it was possible to identify a set of skills necessary to work under agile methods. In Table 3, lists the identified skills that were classified into two groups: Specific and generic skills. Generic skills, in turn, are classified as interpersonal and social.

Table 3. Identified skills and reference articles

<table>
<tr><th colspan="2">Skills</th><th>Concepts</th><th>References</th></tr>
<tr><td colspan="2" rowspan="4">Specific</td><td>Knowledge of tools for work planning</td><td rowspan="4">[11, 12, 15, 20, 27]</td></tr>
<tr><td>Specify the software requirements</td></tr>
<tr><td>Knowledge of software development environments</td></tr>
<tr><td>Apply techniques for estimating effort</td></tr>
<tr><td rowspan="16">Generic</td><td rowspan="7">Interpersonal</td><td>Effective communication</td><td rowspan="7">[3, 8, 10, 16, 19, 26–30]</td></tr>
<tr><td>Collaboration</td></tr>
<tr><td>Adaptability</td></tr>
<tr><td>Self-management</td></tr>
<tr><td>Self learning</td></tr>
<tr><td>Creativity</td></tr>
<tr><td>Second language (English)</td></tr>
<tr><td rowspan="9">Social</td><td>Effective communication</td><td rowspan="9">[3, 8–11, 13–17, 19, 21, 22, 25, 27–29, 31]</td></tr>
<tr><td>Learning from error</td></tr>
<tr><td>Equipment configuration</td></tr>
<tr><td>Sense of collaboration</td></tr>
<tr><td>Team identity</td></tr>
<tr><td>Understand other perspectives needs and values of people</td></tr>
<tr><td>Manage diversity in the team</td></tr>
<tr><td>Improvement of work processes</td></tr>
<tr><td>Self-management</td></tr>
</table>

3.3 RQ2: What Methodological Proposals Exist to Develop the Teaching/ Learning Process on Agile Methods?

According to the selected primary studies, it was possible to recognize the use of the following methods for the teaching of agile methodologies. The recognized methods are:

- Study cases
- Games-based learning
- Problem-based learning

Game-based learning is the most widely used method in Software Engineering education. The possibilities offered by this method are so broad that they allow stimulating the methodological learning of students in agile methods. A good example of the above can be seen in [11], where the experience in teaching SCRUM using LEGO Workshops is presented. Through this type of games, it was possible to generate a favorable work environment to understand the functioning of the work methodology and the roles present in SCRUM. Similar situation occurs in [14], where from a simple game with pencil and paper, students showed a good level of knowledge of the Scrum framework.

Gamification has become an attractive tool to strengthen teaching in Software Engineering, because it focuses students' work on specific subjects that the teacher considers necessary, avoiding distracting elements that generate problems in student learning, just as it is presented in [11] with the topics related to work planning and effort estimation. Through gamification, it has been possible to demonstrate the improvement in the levels of commitment, motivation and performance of the agile teams [12, 15].

In [16], gamification is used to try to improve the commitment and performance of collaborators. Within the obtained results, it is indicated that the levels of commitment of the team were maintained while the performance of the collaborators increased by 30%, which contributed to the improvement of the agile process. In [12], the authors propose a framework for gamification composed of ontology, a methodology that guides the process and a tool.

The great difficulty facing education in Software Engineering, is the possibility of generating educational environments as close to reality. In [17], an interesting experience is described when using Virtual Reality to simulate an environment close to reality for the teaching of SCRUM. The results show the advantages of this type of tools that the students perceived as positive when planning the work, estimating efforts and to hold meetings.

The multiple benefits and applications of gamification in educational subjects in Software Engineering, is reflected in the systematic review focused on the use of gamification in work teams in Software Engineering [18]. In said study, they point out that gamification can be applied in equipment for:

1. Improve the skills of its members during the development of a software project.
2. Identify models of personalities to obtain the most effective combination in the conformation of a work team.

3. Motivate the members of a team to meet the objectives of their work. With the right motivation, the level of team cohesion could be improved.

3.4 RQ3: What Evaluation Methods Will Be Adequate to Measure Achievement Levels in the Skills Detected in RQ1?

The evaluation methods identified in the primary studies, respond more to mechanisms used to measure the perceived effects after the inclusion of the teaching method of agile methods. The evaluation of the teaching-learning process of agile methodologies was conducted by the following instruments: Surveys, Observation, Investigation - Action and expert judgment.

The evaluation of generic skills is a particularly complex issue, because in order to evaluate each of them, the instruments must necessarily be different, since there is no evaluation method that alone can provide all the necessary information for judge the skill of a professional. Therefore, it is necessary to combine different methods and tools to evaluate this type of skills.

In [23], the idea is reinforced that there is no single evaluation method or instrument that can provide all the information to judge the competence of a professional. On the other hand, he maintains that the ability can only be evaluated in the action. According to Miller's pyramid [24], there are four levels of training in order of complexity. At the base of the pyramid are the knowledge that a professional needs to know to develop their professional tasks effectively, at the top level will be the ability to know how to use this knowledge to analyze and interpret the data obtained. Not only is it necessary to know or know how to use it, but it is also necessary to demonstrate how they are used.

Therefore, in [23] he proposes the following strategy to perform the skills assessment:

1. Establishment of the skills or learning outcomes of the students.
2. Identification of relevant (professional) learning situations and their structure for the evaluation of professional competence.
3. Extraction of the criteria of merit.
4. Obtaining sufficient evidence of competence.
5. Determination of achievement levels of skills
6. Elaboration of competence scales.

4 Conclusions

This article presents a systematic mapping with the aim of being able to identify and analyze the existing scientific contributions related to the development of skills to form agile teams of high performance.

The findings are useful and will allow us to continue reflecting and developing our research in the design of a framework for the traceability of skills in the learning of agile methodologies.

In relation to the identification of skills, the appearance of concepts such as: Communication, collaboration and self-management is frequent. Undoubtedly, these skills represent compulsory entry behaviors so that students can perform efficiently in an environment of agile development. For this reason, it is considered that skill development levels must be diagnosed in a timely manner within the training process, in order to perform the necessary actions (individual and group) to achieve the expected levels of achievement.

The use of didactic elements, games, case studies and in particular, the gamification activities, allows us to adequately address the skills to be developed for agile methods. However, the scarce empirical evidence prevents having data that allows to analyze the experiences in more detail.

With the results obtained, our future works are:

- Validate empirically in the industry the skills found. The idea is to determine which are the most relevant skills for the software development industry.
- Conduct a study regarding the definitions of skills proposed by the ACM, IEEE, SFIA, Tuning LA and Swebook. With the purpose of designing an integral framework that allows to visualize both generic and specific skills.
- Propose teaching methods mapped to the skills with their respective assessment methods.

Acknowledgments. Héctor Cornide Reyes is a beneficiary of the INF-PUCV 2016 Scholarship in his Doctorate studies in Informatics Engineering at Pontificia Universidad Católica de Valparaíso.

References

1. Pineda E, Gonzalez C (2016) White paper: networking skills in Latin America. CISCO
2. Consejo Nacional de Educación. http://www.cned.cl/indices-educacion-superior. último acceso 15 June 2017
3. Melo CDO, Cruzes DS, Kon F, Conradi R (2013) Interpretative case studies on agile team productivity and management. Inf Softw Technol 55(2):412–427
4. Beck K, Beedle M, van Bennekum A, Cockburn A, Cunningham W, Fowler M, Thomas D (2001) Manifiesto por el desarrollo Ágil de Software. Obtenido de Agile Manifesto: http://www.agilemanifesto.org/iso/es/manifesto.Html
5. Anaya R (2006) Una visión de la enseñanza. Revista Universidad EAFIT 42(141):60–76
6. Petersen K, Feldt R, Mujtaba S, Mattsson M (2008) Systematic mapping studies in software engineering. EASE 8:68–77
7. Kitchenham BA, Budgen D, Pearl Brereton O (2011) Using mapping studies as the basis for further research a participant-observer case study. Inf Softw Technol 53(6):638–651
8. Fagerholm F, Ikonen M, Kettunen P, Münch J, Roto V, Abrahamsson P (2015) Performance Alignment Work: how software developers experience the continuous adaptation of team performance in Lean and Agile environments. Inf Softw Technol 64:132–147
9. Aldahmash A, Gravell AM, Howard Y (2017) A review on the critical success factors of agile software development. In: European conference on software process improvement, September 2017. Springer, Cham, pp 504–512
10. Babb JS, Hoda R, Nørbjerg, J (2013) Barriers to learning in agile software development projects. In: International conference on agile software development, June 2013. Springer Heidelberg, pp 1–15

11. Steghöfer JP, Burden H, Alahyari H, Haneberg D (2017) No silver brick: opportunities and limitations of teaching Scrum with Lego workshops. J Syst Softw 131:230–247
12. García F, Pedreira O, Piattini M, Cerdeira-Pena A, Penabad M (2017) A framework for gamification in software engineering. J Syst Softw 132:21–40
13. Holtkamp P, Jokinen JP, Pawlowski JM (2015) Soft competency requirements in requirements engineering, software design, implementation, and testing. J Syst Softw 101:136–146
14. Gresse C, von Wangenheim RS (2013) SCRUMIA—an educational game for teaching SCRUM in computing courses. J Syst Softw 86:2675–2687
15. Kropp M, Meier A, Mateescu M, Zahn C (2014) Teaching and learning agile collaboration. In: 2014 IEEE 27th conference on software engineering education and training (CSEE&T), April 2014. IEEE, pp 139–148
16. Pereira IM, Amorim VJ, Cota MA, Gonçalves GC (2016) Gamification use in agile project management: an experience report. In: Brazilian workshop on agile methods, November 2016. Springer, Cham, pp 28–38
17. Rodríguez G, Soria A, Campo M (2011) Teaching scrum to software engineering students with virtual reality support. In: International conference on advances in new technologies, interactive interfaces, and communicability, December 2011. Springer, Heidelberg, pp 140–150
18. Hernández L, Muñoz M, Mejía J, Peña A, Rangel N, Torres C (2017) Una Revisión Sistemática de la Literatura Enfocada en el uso de Gamificación en Equipos de Trabajo en la Ingeniería de Software. RISTI-Revista Ibérica de Sistemas e Tecnologias de Informação 21:33–50
19. Mesquida AL, Karać J, Jovanović M, Mas A (2017) A game toolbox for process improvement in agile teams. In: European conference on software process improvement, September 2017. Springer, Cham, pp 302–309
20. Martin A, Anslow C, Johnson D (2017) Teaching agile methods to software engineering professionals: 10 years, 1000 release plans. In: International conference on agile software development. Springer, Cham, pp 151–166
21. Scott E, Rodríguez G, Soria Á, Campo M (2016) Towards better Scrum learning using learning styles. J Syst Softw 111:242–253
22. Albuquerque R, Fernandes R, Fontana RM, Reinehr S, Malucelli A (2016) Motivating factors in agile and traditional software development methods: a comparative study. In: Brazilian workshop on agile methods, November 2016. Springer, Cham, pp 136–141
23. Tejada Fernández J, Ruiz Bueno C (2016) Evaluación de competencias profesionales en Educación Superior: Retos e implicaciones. [Evaluation of professional competences in Higher Education: Challenges and implications]. Educación XX1 **19**(1), 17–38, https://doi.org/10.5944/educxx1.12175
24. Miller GE (1990) The assessment of clinical skills/competence/performance. Acad Med 65(9):s63–s67. https://doi.org/10.1097/00001888-199009000-00045
25. Dutra AC, Prikladnicki R, Conte T (2015) Characteristics of high performance software development teams. In: International conference on enterprise information systems, April 2015. Springer, Cham, pp 345–363
26. Moe NB (2013) Key challenges of improving agile teamwork. In: International conference on agile software development, June 2013. Springer, Heidelberg, pp 76–90
27. Alqudah MK, Razali R (2017) Key factors for selecting an Agile method: a systematic literature review. Int J Adv Sci Eng Inf Technol 7(2):526–537

28. Diel E, Bergmann M, Marczak S, Luciano E (2015) What is agile, which practices are used, and which skills are necessary according to brazilian professionals: findings of an initial survey. In: 2015 6th Brazilian workshop on agile methods (WBMA), October 2015. IEEE, pp 18–24
29. Igaki H, Fukuyasu N, Saiki S, Matsumoto S, Kusumoto S (2014) Quantitative assessment with using ticket driven development for teaching scrum framework. In: Companion proceedings of the 36th international conference on software engineering, May 2014. ACM, pp. 372–381
30. Read A, Derrick DC, Ligon GS (2014) Developing entrepreneurial skills in IT courses: the role of agile software development practices in producing successful student initiated products. In: 2014 47th Hawaii international conference on system sciences (HICSS), January 2014. IEEE, pp 201–209
31. Scharf A, Koch A (2013) Scrum in a software engineering course: an in-depth praxis report. In: 2013 IEEE 26th conference on software engineering education and training (CSEE&T), May 2013. IEEE, pp 159–168

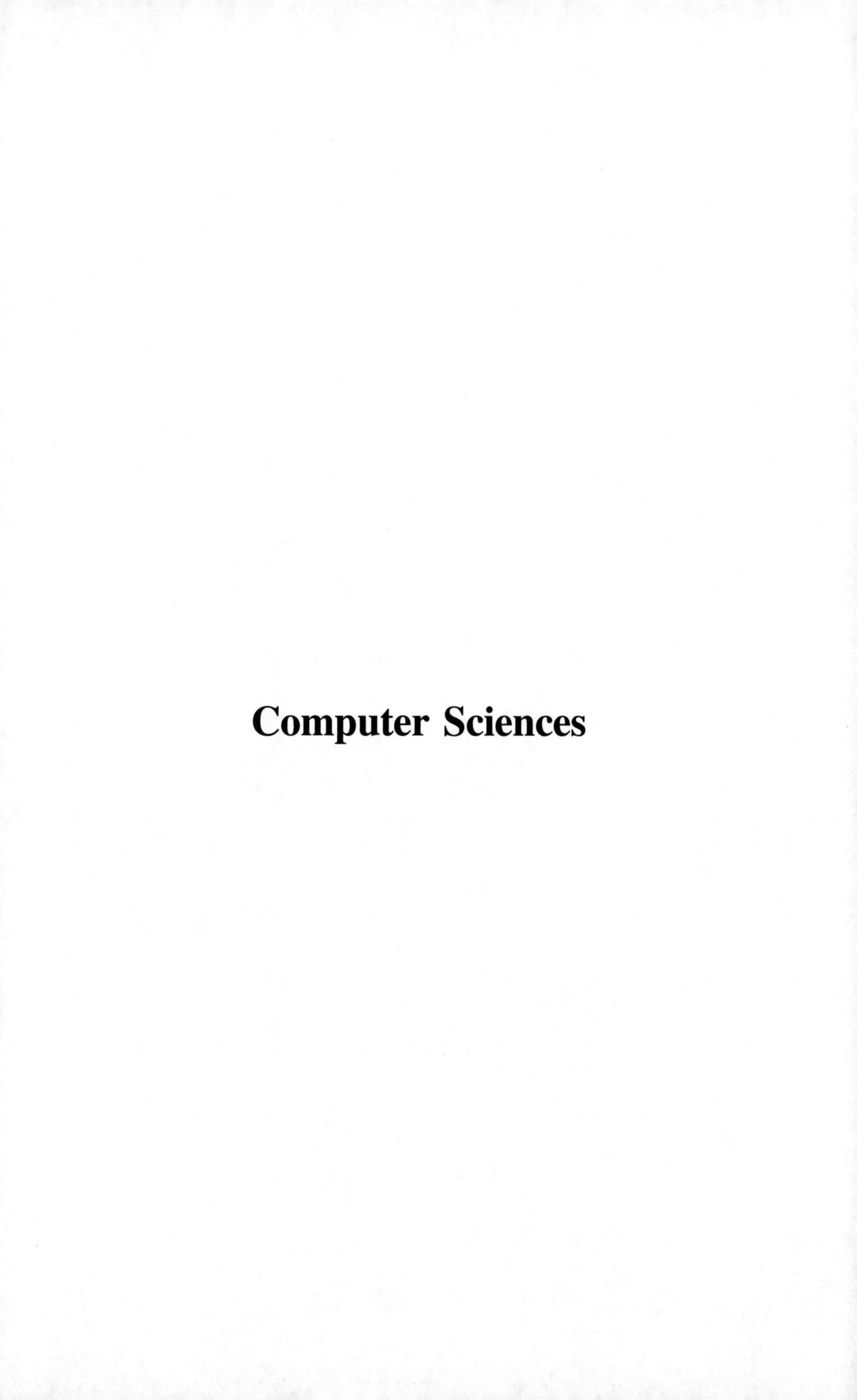

Computer Sciences

A Text Mining Approach to Discover Real-Time Transit Events from Twitter

Belén Arias Zhañay(✉), Gerardo Orellana Cordero, Marcos Orellana Cordero, and María-Inés Acosta Urigüen

Universidad del Azuay, Av. 24 de mayo 7-77, Cuenca, Ecuador
{barias,gorellana,marore,macosta}@uazuay.edu.ec

Abstract. The accelerated growth of the number of inhabitants in the cities brings with it the increase in the number of means of transport, generating new conflicts related to traffic and mobility. This growth and lack of alternative public transportation create a scenario where traffic becomes a serious problem. Such is the case in Cuenca, a city located in Ecuador with a population growth of 15% in the last 7 years, and so has the number of cars. Moreover, transit information is only delivered by traditional media which is not always accurate or in real-time. It is imperative to create a system to discover real-time events to help the population to acquire precise information. With the arising of social networks such as Twitter, new opportunities to solve the transit problem at its origin. Twitter users interact with the social network every day and inform their fellow users of different topics such as transit. We take Twitter as a source of information to feed a real-time system which infers transit data from tweets. We create a predictive model with the use of pre-processing techniques for data cleaning, Support Vector Machines for predictive modeling, dictionaries and Levenshtein distance for location discovery, and finally, association analysis for data pattern finding. Our results show that our approach outperforms the existing works in the field. Furthermore, we have achieved accuracy values greater than 90% in classification subroutines and more than 70% in location discovery. Thus, we have settled a successful prediction model to implement real-time transit discovery in Twitter.

Keywords: Text mining · Transit · Traffic · Twitter · Real-time analysis

1 Introduction

The accelerated growth of the number of inhabitants in the cities brings with it the increase in the number of means of transport, generating new conflicts related to traffic and mobility [26]. The growth in city population and lack of alternative public transportation create a scenario where traffic becomes a serious problem [26]. Such as it is the case of Cuenca, a city located in the mountains of Ecuador which population is approximately 603.269 inhabitants. Cuenca has grown 15% in the last seven years [20] and had 64.254 cars at 2016 [19]. The high amount of cars and the narrow streets in Ecuadorian cities create traffic jams [4]. Moreover, transit information is only delivered by traditional media which is not always accurate or in real-time. Thus, it is imperative to create a system to discover transit information in real-time to help the population get

M. Botto-Tobar et al. (Eds.): TICEC 2018, AISC 884, pp. 155–169, 2019.
https://doi.org/10.1007/978-3-030-02828-2_12

accurate and fast information. The growth in social networks such as Twitter brings several research opportunities [22]. Moreover, it is a source for people's opinion in different matters such as politics, economy, city problems and events, among others [24]. Therefore, this social network becomes important when analyzing people's daily activities such as mobility. People tweet about transit situations; they indicate the places, events, and sources of transit information in real time [9, 12, 13, 25]. Consequently, given the problem of high traffic in Cuenca and a rich source of transit information such as Twitter, we see an opportunity to settle a prediction system to give users rich information about the daily problems concerning this domain. A depth analysis of social networks involves the use of several different text mining strategies as it can be seen in [14]. Tweets are text which is an unstructured form of information, so, their analysis is not trivial. We propose an approach which uses text mining techniques and machine learning algorithms to extract knowledge and define predictive models to classify future tweets in real-time. Some previous examples of using predictive models in transit data can be seen in sentiment analysis applications which aim to determine the position of a person in a topic [1], and in traffic classification [9, 13]. The studies on traffic classification show an accurate predictive model. However, we see limitations on their location detection approaches and event classification. As a result, we get inspiration from these proposals and improve their location routines, enlarge their event classification, and add event-location pattern recognition analysis. Our work generates a predictive model to first classify whether a tweet belongs to transit or not, to identify the polarity of the tweet, to determine the event causing high transit, to find the location of the transit event, and finally to determine the main event-location patterns in transit data. We successfully performed the mentioned tasks for the tweets of Cuenca, Ecuador.

This study is organized as follows: Sect. 2 shows related work regarding transit using Twitter, Sect. 3, shows our approach to detect real-time transit events, Sect. 4, shows results and discussion, and finally Sect. 5, shows conclusions and future work.

2 Related Work

Researchers have used Twitter data as a rich source of information to study diverse topics such as crime and natural disasters [1], presidential elections [18, 31], finance [23], public transportation [32], among others. These have different approaches and uses; for instance, in [1], the authors created a method to help emergency teams to find disasters exact location and speed up the help systems. Other studies, such as [31] analyzed the United States people's sentiment about presidential elections to discover possible trends in the results. In [18], they did a similar work by implementing a system capable of predicting election results.

Our aim is to describe transit data in Twitter, thus, we discovered two main research interests in the literature review. The first one mainly relates to sentiment analysis of people's opinion about transit laws, gasoline prices, and related topics [5]. They aimed to find interesting patterns in Twitter data. On the other hand, other studies performed different classification tasks to discover traffic jams, their location, and the main originator causes [9, 13, 25]. They created predictive models to generate real-time systems to help finding very transited areas with potential transit-related problems.

In [25], the authors analyzed transit using a sentiment analysis approach with Twitter data. They collected tweets via Twitter API using transit-related keywords, then they used logistic regression for classification, and finally designed a system to report transit sentiment to the final users. This approach presented a valid method to inform about transit. However, the authors did not show the type of transit event and its location. In [9], the authors showed a more complete approach. They took into account not only the transit sentiment but also its location, event, and also they evaluated different classification algorithms. Also [13] performed a similar study in which they determined five different events which cause traffic and their location. An important matter to consider is the different approaches in the location detection. In [9], the authors used a dictionary to detect streets and in [13], the authors used the latitude and longitude metadata from the tweets. We see limitations in both approaches; the first one is dependent only on streets when the user may refer to other public places, while the second relies on coordinates that are not always available. Therefore, our approach aims to solve these limitations by using an enhanced version of the two approaches in order to obtain more accurate and relevant information.

The studies described in the paragraph above showed valid approaches which succeeded as real-time systems. However, it is not only important to inform transit in the shape of tweets, but also to analyze the patterns causing the transit in order to give users a richer form of information than plain text. In consequence, we get inspiration from [9, 13] and enhance their proposals by implementing improved location features, enlarging the events classification, and adding methods to do implement pattern analysis over the results.

3 Methodology

This study uses the CRISP-DM Methodology (Cross Industry Standard Process for Data Mining) [7] to mine Twitter data. Some works in text mining have successfully used this methodology when mining social networks data [8, 29]. We have adopted this methodology as a framework given that such work expresses a mining process from the data understanding to the evaluation and deployment of the solution. Therefore, we develop our domain specific approach from this.

Figure 1, presents an overview of our method which expresses the process to discover real-time transit events. The subroutines go from the data extraction to the pattern analysis. The following sub-sections describe each one in detail in a step-wise approach. It is important to note that every step depends on the results of the previous one.

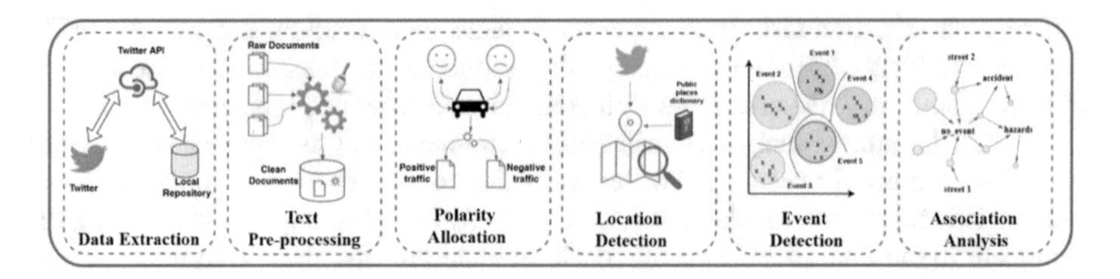

Fig. 1. A text mining approach to discover real-time transit events in Twitter

3.1 Tweets Extraction

There are two well-known methods for tweet extraction. One is based on a streaming process where a threat is extracting all tweets for undefined periods of time [30]; and the second, uses the Twitter API for tweet extraction in chunks of data based on a query. In order to extract domain specific tweets, many works have used the second method with the Twitter API [1, 3, 9, 10, 13, 31]. Therefore, we use a similar approach. We use the twitteR library to access the Twitter API developed for the R programming language. The twitter API bases its results in a search string. Therefore, we develop two search strings in order to obtain training and testing datasets. We focus on transit data in Cuenca, a city located in the mountains of Ecuador whose main language is Spanish. So, our search strings are presented in such language. We used the following search string for our training dataset.

- *(circulación OR circulacion OR tránsito OR transito OR flujo OR congestión OR congestion) AND vehicular*

The string was built by looking for synonyms of the word transit. Moreover, we look for most of the topic related tweets by using 'OR' operators and one 'AND' operator to narrow down the results to vehicle tweets. As it can be seen in [11], Twitter keywords are sensible to accents, punctuation, and diacritics. Therefore, we use our keywords with and without such characters in order to obtain the most relevant tweets from our search.

- *(circulación OR circulacion OR tránsito OR transito OR flujo OR congestión OR congestion) AND vehicular*
- *Latitude and Longitude: Parameters set to latitude and longitude of Cuenca's city center.*

The testing search string is presented above and differs from the first one with the location. This study focuses on the analysis in a specific city. Therefore, we added parameters related to the latitude and longitude to indicate the API to search only tweets located in a 10 miles radius from the specified coordinates.

3.2 Tweets Pre-processing

The tweets extraction results in a dataset of unstructured data with high complexity and dimensionality. Therefore, this dataset should be pre-processed in order to reduce its complexity. Moreover, text presented in a micro-blogging service such as Twitter, shows more typos, slang, ungrammatical structures, and emoticons than formal text representations [15]. We perform the following preprocessing techniques to simplify text:

- *Tokenization*: The text is split according to a separator, usually space or commas. The smallest unit in which they are divided is called token.
- *Lower Case*: Text transformation into a unified case to reduce the complexity and number of different terms in the dataset.
- *Stopwords*: This method assumes that there are words which are not useful or do not provide discriminative information. Thus, words such as articles, conjunctions, and others are filter out from the text.

Additionally, other general cleaning methods were used, such as URLs removal, punctuation removal, and isolated characters removal. We remove them taking into account that in our analysis we consider that they do not provide useful information to help discriminate events in the text that would improve predictive model accuracy. Moreover, the removal reduces data complexity by filtering out unnecessary characters in the text.

3.3 Polarity Allocation

At this point in our process, the tweets have been extracted and pre-processed. Ideally, these tweets should contain only transit data. However, this is not the case given the heterogeneity in language and the correspondent search string. Because of that, the tweets go over a second filter and are classified in transit or not transit. Moreover, this work focuses on transit tweets which causes discomfort among people. We perform a sentiment analysis to allocate the final polarity of the data.

- *Transit tweets*: In order to split tweets in transit and non-transit, we used a supervised approach. Therefore, we first manually classify the tweets into two categories. These categories are tweets which contain transit information and tweets which are unrelated to the study domain. We use Support Vector Machine (SVM) as our classification algorithm as it was done in [28].
- *Sentiment Analysis*: Social networks users express their feelings about topics. Therefore, once tweets are classified into transit related, we classify them into those expressing positive or negative polarity.

The classification categories described above need labeling data to be processed with a supervised algorithm such as SVM. Thus, we manually classified the tweets described in previous sections before each SVM classification. SVM is a classification algorithm which takes as input several parameters which modify the results performance. Some works which similarly optimized algorithm can be seen in [6, 17]. One parameter which can be taken into consideration for the optimization process is the Kernel. However, it is shown in [2, 16, 21], that the Radial Basis Function (RBF) Kernel performs the best in text mining tasks. This kernel is focused on non-linear, non-separable problems as it is the case with text. The following are the optimization parameters corresponding to this kernel:

- *Gamma* (γ): This parameter takes into account the shape of the data. A high value of γ assumes that the texts are far away from each other, a low-value assumes texts closer to each other in a high dimensional space.
- *C*: This parameter is the cost function associated with miss-classification, it sets the classification penalty. High values of C can produce better classification results. However, such high values can also produce overfitting.

As it was done in [16], we use the following γ values: 2^{-15}, 2^{-13},..., 2^{3} and C values: 2^{-5}, 2^{-3},..., 2^{15} to optimize our model with a Grid Search approach. It is important to note that the optimization can be based on different metrics such as precision, recall, f-measure, and accuracy. Such metrics are shown in Table 1 and are calculated with the confusion matrix values in which every classified sample is assigned one of the following categories: true positive (TP), true negative (TN), false positive (FP), and false negative (FN). We use these metrics to evaluate our classifiers as done in [9] and perform optimization based on f-measure. We do this, given that this metric provides a combined recall and precision value which is more expressive than the others.

Table 1. Model evaluation metrics

Metric	Equation
Recall	$\frac{TP}{TP+FN}$
Precision	$\frac{TP}{TP+FP}$
F-score	$\frac{2*(Recall*Precision)}{Recall+Precision}$
Accuracy	$\frac{TPP+TN}{TP+FP+FN+TN}$

3.4 Location Detection

It can be seen in [9, 13] that there are two different approaches to obtain a tweet location. On one side, the coordinates of the tweet can be taken as metadata coming from the tweet, and on the other side, the location can be inferred by the text in the tweet. The first approach is more accurate, but a large percentage of the studied tweets do not have this information available. The second approach takes the text of the tweet and infers the location, it is less accurate than the first one because it is based on dictionaries and can be sensible to typos and missing information.

The present study takes the two methods into consideration in order to build a more robust system to locate as many tweets as possible. The first method is self-explanatory given that this basic information is taken from the Twitter API. However, the second method uses a more complex process which is explained below:

Dictionary Creation: This method is dependent on a set of dictionaries based on a specific city. We build such dictionaries based on open sources. To obtain streets information we use OpenStreetMap[1], additionally, we obtain public places and sectors in the city by manually crawling the web.

Location Detection: We use several natural language processing tools and algorithms to identify the location of the tweets with the previously mentioned dictionary. This process is split into the following two sub-processes.

1. Tweet split into n-grams: An n-gram is a sequence of consecutive words from a text according to [27]. Thereby, we divide the tweet into 1, 2, 3, and 4 n-grams in order to compare them later with our dictionary of locations. We use this number of grams given that our dictionary analysis tells us that most of the places have at most four words. We use the *NLTK* library from the Python language for this purpose.
2. Distances measurement: We compare the n-grams with every entity in our dictionary. For this matter, we use the *Levenshtein* distance algorithm which uses dynamic programming to assess the likelihood between two sets of strings.

These two processes produce a set of terms and its percentage likelihood to a part of the tweet text. Thus, we define a threshold of 0.8 to select those which are more likely to be a street or sector from the dictionary.

3.5 Event Detection

To determine the event which causes high traffic, we use the negative transit tweets as it was done in Subsect. 3.4 and manually classify them in the several classes determined by [13]. We consider that these classes meet the reality of Cuenca. However, by a manual inspection of the tweets, we determine that many users do not explain the event causing high traffic. Therefore, we added a "no event" to the following:

- *No event*: There are tweets indicating the presence of high traffic. However, the reason that originated the traffic is not mentioned.
- *Accidents*: The tweet shows that there is high traffic mainly produced by crashes involving vehicles and/or pedestrians.
- *Road Works*: The tweet indicates that there are works in the road, which are delaying vehicles and therefore causing high traffic.
- *Weather - Hazards*: The high traffic is produced by bad weather or hazards on the road which causes vehicles to go slower or closed streets.
- *Public events*: The high traffic is produced by public events, which cause crowds of people to mobilize. Some examples of these events can be concerts, parades, sports events, among others.
- *Hindering Objects*: When vehicles, animals or other objects standing in the road produced the high traffic and interfere the normal transit flow.

[1] https://www.openstreetmap.org/

We build a predictive model in a base of our manually classified tweets with the SVM algorithm with an RBF kernel and optimized Gamma(γ) y C parameters as done in [16].

3.6 Association Analysis

As it is suggested in the CRISP-DM methodology [7], a key stage in the mining process is to evaluate that our model accomplished its goal. As an evaluation tool and a source of real-time extra information, we use an algorithm to find association rules in the data. We use an a-priori algorithm for such task. This type of algorithm finds patters by employing level-wise search. In our case, we aim to find location-event transit information between the location of the event described in Subsect. 3.4 and the tweet event described in Subsect. 3.5. Moreover, our association analysis proposes the following two granularity levels.

As it is suggested in the CRISP-DM methodology [7], a key stage in the mining process is to evaluate that our model accomplished its goal. As an evaluation tool and a source of real-time extra information, we use an algorithm to find association rules in the data.

1. Narrow place - event granularity: This analysis comprehends the located places as they are. This means that located streets and public place names are not replaced with any tag and are associated with the event. As a result, this part of the method obtains specific transit problematic locations.
2. Wide place - event granularity: This analysis changes the location granularity. Those tweets which were detected to have a location are marked with the tag lbl:lugar replacing the specific street or public place, while those which do not are placed with the tag lbl:no lugar. This analysis helps to comprehend the main transit problems happening at a certain time and if they show a location related to them.

The a-priori algorithm takes a support and confidence value as input parameters. The support is a threshold that indicates the number of values that should be present to create a rule. The confidence, on the other hand, shows the probability that a rule may be true. These parameters, however, might be difficult to define and are dataset specific. We define them with a manual trial-error approach.

4 Results and Discussion

As a result of the data extraction, we obtained 2500 tweets for our training and testing dataset. From the total amount of tweets, 700 (28%) belong to the target city Cuenca and the rest belong to any other part of the world. A small overview of how the dataset looks can be seen in Table 2.

Table 2. Twitter extraction sample data

Tweet Id	Tweet text	Coordinates	
		Latitude	Longitude
1	si te encuentras en una zona de alto flujo vehicular, conduzca con paciencia y evite enfrascarse en discusiones. #selmejor		
2	ecu reporta mayo max uhle llamada indica accidente trnsito unidades circulacin vehicular habilitada	−2.8974039	−79.004483400000
3	ro de los remedios, presenta carga vehicular moderada	−2.8974039	−79.004483400000
4	hermilio mena y ro de los remedios, presenta buen avance vehicular		

As it can be seen in Table 2, we obtained the tweets' text and its coordinates as metadata.

4.1 Polarity Allocation

Polarity allocation is a two-step process in which we first filter out non-transit related tweets and after we perform a sentiment analysis. We show two different classification results below:

- Transit tweets: Table 3, shows that we obtained an f-measure of 0.8510. However, when looking at the precision and recall, we observed a lower value of recall and a very high value for precision. We assume that this lower recall might be produced by a very unbalanced dataset. The dataset contained 2311 tweets which correspond to the transit category while there are 289 which were non-transit. This step to categorize tweets in transit or non-transit is not trivial given that by the analysis of our dataset we discover that 11.56% of the tweets do not correspond to transit. As a result of this process, we can say that the classification in transit or non-transit has an accuracy of 95.62 which is superior to the obtained in [13] and equivalent to [9]. However, to obtain a better recall, a larger dataset would be advisable.

Table 3. Processes' optimization parameters and results

Parameters				Results		
Latitude	Gamma	C	Recall	Precision	F-measure	Accuracy
Transit tweets	0.03125	4	0.8079	0.9450	0.8510	0.9562
Sentiment analysis	0.0625	2	0.9213	0.9398	0.9303	0.9259
Location detection	NA	NA	0.8043	0.8078	0.8061	0.7129
Events detection	0.0156	16	0.6906	0.8979	0.7539	0.9083

NA: Not available for the classification.

- Sentiment Analysis: Table 3 shows the classification results. These results present more stable values of precision, recall, and f-measure. We assume this is because the polarity among the training set is more balanced. We compare our work with a previous study such as [9] in which the authors obtained a 95.75% accuracy which is equivalent to our results.

4.2 Location Detection

Location processes took the 348 tweets from Cuenca which had a negative polarity. Additionally, the data crawler resulted in a dictionary with 1905 different terms. Finally, in order to check the validity of our approach, we defined a ground of true based on a manual inspection and control of the data. The manual data control and algorithm execution resulted in the confusion matrix shown in Table 4 in which two categories are shown. We classify as TP when our algorithm detected a location that was in the tweet, TN when we did not discover any location in the tweet and there were indeed no location associated to it, FP when we detected a location which was not present, and FN when we did not find a location in the tweet when there was actually one or more locations in it.

Table 4. Location confusion matrix

	Correct	Incorrect	Totals
Precision	370	90	460
Recall	88	72	160
Totals	458	162	620

Table 3, shows the location detection accuracy, precision, recall, and f-measure obtained from the computation of the confusion matrix in Table 4 accordingly to the equations expressed in Table 1. The values suggest that our algorithm performs well with this dataset giving values above 70% in all measures. We attribute this success to the meticulous process to create the locations dictionary from Cuenca.

4.3 Event Detection

The polarity allocation received 2500 tweets as dataset. However, this process filtered out all non-transit related tweets. Thus, the event detection had 1603 labeled relevant records as consequence. We have an unbalanced dataset in which most of the events are "no event", Table 5 shows the data distribution. This distribution explains that when Twitter users talk about problematic situations with the transit in Cuenca, they mostly do not specify the cause causing the transit. So, we mainly focus in the 24.95% of tweets which explain further transit incidents.

Table 5. Event data distribution

Category	Number of tweets	Percentage
Accident	121	7.55%
Road works	96	5.99%
Public events	52	3.24%
Weather – Hazards	86	5.36%
Hindering objects	45	2.81%
Not event	1203	75.05%

Table 3, shows the event detection results. We obtain a classification accuracy of 90.83% that compared with other studies such as [9] is higher even though we also use the SVM as classifier. Furthermore, we also obtain better results in recall and precision which we emphasize over accuracy given that we have an unbalanced dataset. We assume the main reason for our improvement is the parameter optimization we performed. The values of precision, recall, and f-measure were greatly influenced by the optimization; precision increased from 0.82 to 0.90 showing an increment of 9.86%, recall from 0.66 to 0.69 showing an increment of 4.57%, and f-measure from 0.69 to 0.75 showing an increment of 8.73%. At the end, we consider that parameter fine-tuning can improve classification results considerably.

4.4 Association Analysis

Figure 2 shows the results of applying the a-priori algorithm and its visualization over our dataset.

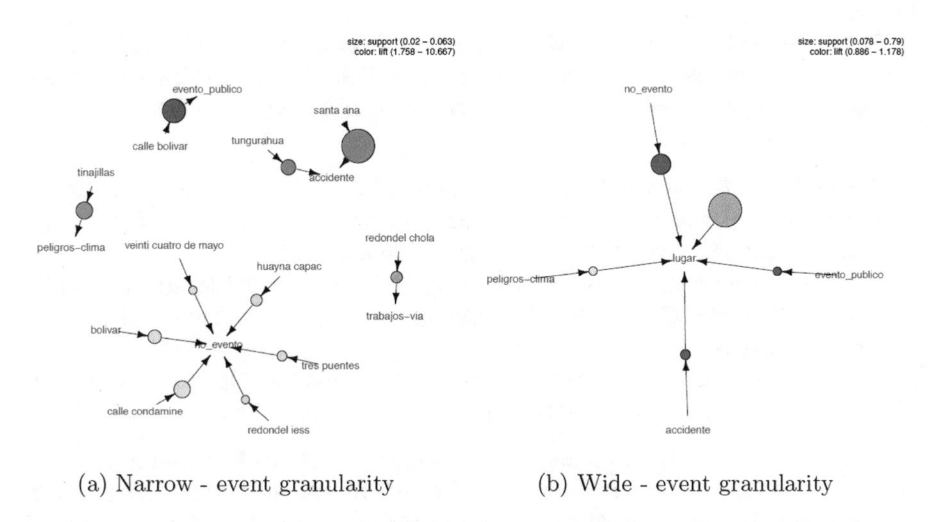

(a) Narrow - event granularity (b) Wide - event granularity

Fig. 2. Association analysis results over two levels of granularity

The visualization analysis shows different patterns represented with a connected graph. The resulting graphs show the set of different rules found in the data which are remarked as circles with different size and tonality. These rule properties show interest measures of the data. Thus, we can draw conclusions about how the transit problems appear in the data with two different granularity levels from such rules. Figure 2a, clearly suggest that most of the transit problems do not show an associated event to them. Also, some rules do show a strong predominance which shows higher transit here than in other rules, an example of this can be seen in the rule "calle_condamine to no_ event". This is explained by the higher amount of tweets which do talk about a specific transit even. Other rules do show that there is a presence of other events too, however, they do not have a predominant effect given a lower number of tweets in comparison with the "no_evento" event. Figure 2b, on the other hand, shows the rules strength in the presence or absence of a place. It is seen that the "no_evento" event is stronger than other which confirms the rules shown in Fig. 2a.

The association analysis has shown to be an interesting tool to analyze real-time transit events. Furthermore, it is also important to note that picking the right parameters is fundamental to achieve an accurate analysis. We determined a support and confidence of 0.01 and 0.6 respectively by a manual inspection of the parameters, which showed the known reality of Cuenca. These values are low, however, we attribute this to the high variability in the data given and that text is an unstructured source of information.

5 Conclusions

This study builds an approach to discover real-time transit events from a social network such as Twitter. Moreover, our model finds not only event related information but also the incident location and proposes as well an association analysis to discover problematic locations.

We performed three different classification processes: transit detection, sentiment analysis, and event detection. Moreover, we used the SVM algorithm which the literature reveals is the most accurate for text classification tasks. The classification insights are further described below:

- We obtained a transit detection accuracy of 95.62%. However, this might be biased by a very unbalanced dataset. We analyzed also precision and recall measures and obtained measures above 80% showing an efficient classifier.
- Our transit sentiment analysis showed very high values in all measures (92.13% recall, 93.98% precision, and 92.59% accuracy), proving that our model outperforms other works done in the field.
- The dataset showed a very unbalanced event distribution with high predominance of "no event". Thus, we put more emphasis on the f-measure with a value of 75.39%.

Our location detection algorithm uses geographic data and text data to discover streets and public places where the transit is located. We obtained values above 70% in all metrics which suggests an efficient location detection method. For this study, we used the *Levenshtein* distance. However, we consider interesting to analyze different distance algorithms that could improve the metrics in the future.

We found no predominance of an specific event in traffic jams. However, by our association analysis, we found specific problematic places which do affect peoples' activities.

The association analysis was particularly sensitive to the support and confidence values. Therefore, picking a right value is not trivial. We performed a trial and error approach to find the values that show our dataset at its best and found low values of support and confidence.

This study has deepen the potential of text mining techniques and machine learning algorithms to extract and create prediction models from Twitter. In the domain of transit information, we created association models, which show the problematic locations and events. Such interesting discoveries are shown in self-explanatory graphs that can influence real-time user decisions and improve life quality by avoiding high transit situations. Even though this study was designed for a specific city such as Cuenca, we consider it would be easier to apply and adapt to others. In this study, we created a complete model for real-time prediction. However, our future work will focus on the implementation of a platform which will allow us to report this information in real-time to people. We consider that the benefits of such implementation can change the transit by giving people the chance to pick alternative routes, which we believe it could create an auto-regulated system worth studying.

Acknowledgments. This research was supported by the Vice-rectorate of Investigations of the Universidad del Azuay. We thank our colleagues from Laboratorio de Investigación y Desarrollo en Informática (LIDI) de la Universidad del Azuay who provided insight and expertise that greatly assisted this research.

References

1. Arias M, Arratia A, Xuriguera R (2014) Forecasting with twitter data. ACM Trans Intell Syst Technol 5(1):8:1–8:24
2. Banados JA, Espinosa KJ (2014) Optimizing support vector machine in classifying sentiments on product brands from twitter. In: The 5th international conference on information, intelligence, systems and applications, IISA 2014, pp 75–80. IEEE
3. Bisio F, Meda C, Zunino R, Surlinelli R, Scillia E, Ottaviano A (2015) Realtime monitoring of twitter traffic by using semantic networks. In: Proceedings of the 2015 IEEE/ACM international conference on advances in social networks analysis and mining 2015, ASONAM 2015, New York, NY, USA, pp 966–969. ACM
4. Borsdorf A, Stadel C (2015) The Andes: a geographical portrait. Springer, Switzerland
5. Cao J, Zeng K, Wang H, Cheng J, Qiao F, Wen D, Gao Y (2014) Web-based traffic sentiment analysis: methods and applications. IEEE Trans Intell Transp Syst 15(2):844–853
6. Chapelle O, Vapnik V, Bousquet O, Mukherjee S (2002) Choosing multiple parameters for support vector machines. Mach Learn 46(1–3):131–159
7. Chapman P, Clinton J, Kerber R, Khabaza T, Reinartz T, Shearer C, Wirth R (2000) Crisp-dm 1.0 step-by-step data mining guide
8. Contreras Chinchilla L, Rosales Ferreira K (2016) Análisis del comportamiento de los clientes en las redes sociales mediante técnicas de minería de datos

9. D'Andrea E, Ducange P, Lazzerini B, Marcelloni F (2015) Real-time detection of traffic from twitter stream analysis. IEEE Trans Intell Transp Syst 16(4):2269–2283
10. Das TK, Acharjya D, Patra M (2014) Opinion mining about a product by analyzing public tweets in twitter. In: 2014 international conference on computer communication and informatics (ICCCI), pp 1–4. IEEE
11. Twitter Developers: Rules and filtering
12. Dhavase N, Bagade A (2014) Location identification for crime & disaster events by geoparsing twitter. In: 2014 international conference for convergence of technology (I2CT), pp 1–3. IEEE
13. Gu Y, Qian ZS, Chen F (2016) From twitter to detector: Real-time traffic incident detection using social media data. Transp Res Part C Emerg Technol 67:321–342
14. Gupta V, Lehal GS et al (2009) A survey of text mining techniques and applications. J Emerg Technol Web Intell 1(1):60–76
15. Han B, Baldwin T (2011) Lexical normalisation of short text messages: Makn Sens a #twitter. In: Proceedings of the 49th annual meeting of the association for computational linguistics: human language technologies, vol 1, pp 368–378. Association for Computational Linguistics
16. Hsu CW, Chang CC, Lin CJ et al. (2003) A practical guide to support vector classification
17. Hsu CW, Lin CJ (2002) A comparison of methods for multiclass support vector machines. IEEE Trans Neural Netw 13(2):415–425
18. Ibrahim M, Abdillah O, Wicaksono AF, Adriani M (2015) Buzzer detection and sentiment analysis for predicting presidential election results in a twitter nation. In: 2015 IEEE international conference on data mining workshop (ICDMW), pp 1348–1353
19. Instituto Ecuatoriano de Estadísticas y Censos INEC: Anuario de transporte (2016). http://www.ecuadorencifras.gob.ec/documentos/web-inec/Estadisticas_Economicas/EstadisticadeTransporte/2016/2016_AnuarioTransportes_PrincipalesResultados.pdf. Accessed 26 Apr 2018
20. Instituto Ecuatoriano de Estadísticas y Censos INEC: Conozcamos Cuenca a través de sus cifras. http://www.ecuadorencifras.gob.ec/conozcamos-cuenca-a-traves-de-sus-cifras/. Accessed 26 Apr 2018
21. Jadav BM, Vaghela VB (2016) Sentiment analysis using support vector machine based on feature selection and semantic analysis. Int J Comput Appl 146(13):26–30
22. Kim HJ, Kim YH (2016) Recent researches on application and influence of twitter. Int J Appl Eng Res 11(11):7501–7504
23. Luong TT, Houston D (2015) Public opinions of light rail service in Los Angeles, an analysis using twitter data. In: iConference 2015 proceedings
24. Madani A, Boussaid O, Zegour DE (2015) Real-time trending topics detection and description from twitter content. Soc Netw Anal Min 5(1):59
25. Pandhare KR, Shah MA (2017) Real time road traffic event detection using twitter and spark. In: 2017 International conference on inventive communication and computational technologies (ICICCT), pp 445–449. IEEE
26. Pattanaik V, Singh M, Gupta P, Singh S (2016) Smart real-time traffic congestion estimation and clustering technique for urban vehicular roads. In: 2016 IEEE region 10 conference (TENCON), pp 3420–3423. IEEE
27. Radovanovíc M, Ivanovíc M (2008) Text mining: Approaches and applications. Novi Sad J. Math 38(3):227–234
28. Suárez EJC (2014) Tutorial sobre máquinas de vectores soporte (SVM). Tutorial sobre Máquinas de Vectores Soporte (SVM)

29. Treboux J, Cretton F, Evéquoz F, Le Calvé A, Genoud D (2016) Mining and visualizing social data to inform marketing decisions. In: 2016 IEEE 30th international conference on advanced information networking and applications (AINA), pp 66–73. IEEE (2016)
30. Twitter: Overview - Twitter developers, May 2018
31. Wang H, Can D, Kazemzadeh A, Bar F, Narayanan S (2012) A system for real-time twitter sentiment analysis of 2012 U.S. presidential election cycle. In: Proceedings of the ACL 2012 system demonstrations, ACL 2012, Stroudsburg, PA, USA, pp 115–120. Association for Computational Linguistics
32. Wang X, Wei F, Liu X, Zhou M, Zhang M (2011) Topic sentiment analysis in twitter: a graph-based hashtag sentiment classification approach. In: Proceedings of the 20th ACM international conference on Information and knowledge management, pp 1031–1040. ACM

Automatic Microstructural Classification with Convolutional Neural Network

Guachi Lorena[1(✉)], Guachi Robinson[2], Perri Stefania[3], Corsonello Pasquale[3], Bini Fabiano[2], and Marinozzi Franco[2]

[1] Yachay Tech University, Hacienda San José s/n, 100119 Urcuquí, Ecuador
lguachi@yachaytech.edu.ec
[2] Sapienza University of Rome, via Eudossiana 18, 00184 Rome, RM, Italy
[3] University of Calabria, via P. Bucci, 87036 Arcavacata di Rende, Italy

Abstract. Microstructural characterization allows knowing the components of a microstructure in order to determine the influence on mechanical properties, such as the maximum load that a body can support before breaking out. In almost all real solutions, microstructures are characterized by human experts, and its automatic identification is still a challenge. In fact, a microstructure typically is a combination of different constituents, also called phases, which produce complex substructures that store information related to origin and formation mode of a material defining all its physical and chemical properties. Convolutional neural networks (CNNs) are a category of deep artificial neural networks that show great success in computer vision applications, such as image and video recognition. In this work we explore and compare four outstanding CNNs architectures with increasing depth to analyze their capability of classifying correctly microstructural images into seven classes. Experiments are done referring to ultrahigh carbon steel microstructural images. As the main result, this paper provides a point-of-view to choose CNN architectures for microstructural image identification considering accuracy, training time, and the number of multiply and accumulate operations performed by convolutional layers. The comparison demonstrates that the addition of two convolutional layers in the LeNet network leads to a higher accuracy without considerably lengthening the training.

Keywords: Microstructure characterization · Image processing · CNN

1 Introduction

In the recent years, Convolutional Neural Networks (CNNs) have gained the ability of extensively supporting applications in Artificial Intelligence, Pattern Recognition, and in Material Science fields due to its considerable precision in patterns recognition and classification from pixel images. Particularly, in Material Science the ability to analyze microstructures relies on finding the reliable description and classification of the structures, which determine the microstructural image.

Some image processing algorithms suitable for microstructures analysis and classification are known in Literature [1]. However, they often provide processing, segmentation, and analysis methods that have to be recombined for every analysis task, thus

M. Botto-Tobar et al. (Eds.): TICEC 2018, AISC 884, pp. 170–181, 2019.
https://doi.org/10.1007/978-3-030-02828-2_13

requiring time and expert knowledge [2]. Recently, innovative automatic methods based on CNNs that belong to deep learning technique have been presented in [3–5]. CNNs perform pixel-wise analysis each with its own architecture that influences the achieved accuracy (i.e. classification, recognition, segmentation, etc.), the number of operations, the training time, the model size, among others.

In order to demonstrate how the network depth influences the accuracy achieved in microstructural classification, in this work, we explore and compare four outstanding and standard CNN architectures [6–9], which outperform others methods in several visual recognition tasks. As a case study, we used Ultrahigh carbon steels (UHCS) microstructural images dataset [10] since the metallic materials and its alloys have become a fundamental part of industrial development, such as lighter sheet metal structures [11] and automotive applications, for which UHCS are considered a promising technique to create "ultrahigh strength" sheet materials [12]. Moreover, in some cases, to obtain a thick-layer, bonding steps are required that commonly involve mechanical working by pressing, forging, rolling [13], or extrusion.

The rest of the paper is organized as follows. Section 2 describes most relevant related works. The explored CNN architectures are described in Sect. 3. Implementation details are presented in Sect. 4. Experimental results, obtained processing UHCS dataset, are presented and discussed in Sect. 5. Finally, Sect. 6 deals with the concluding remarks.

2 Related Works

The mechanical properties of the bodies are particularly influenced by their microstructures that are determined by size of phases, shape and distribution. Microstructures have different appearance features determined by several parameters, such as treatment, post-treatments, granulometry law, alloying elements, rolling setup, among others. Thus, techniques that assure robustness to classify microstructures and that can be easily configured to analyze different microstructures are essential to perform efficient and reliable microstructural image analysis.

Some specialized algorithms can automatically analyze microstructures with a great effort in understanding particular materials or classes of microstructures [14–19]. For example, in [14] a microstructural and morphological characterization approach is presented. It uses CHIMERA software and performs estimation by means of morphological properties as porosity, volume fraction, and distribution of solid and pore phases. The methods presented in [15] and [16] exploit a bag of visual features representation as a general microstructure descriptor inspired by the bag of words representation for document classification (it models a class of documents as a probability distribution of word occurrences over the set of words in a vocabulary) to train a support vector machine (SVM) classifier, where SVM automatically classifies a microstructure into one of several classes. The method presented in [17] employs the diameter ratio and the area ratio as geometric shape descriptors. Then, a fuzzy rule based classifier is built based on known feature values, in order to classify graphite inclusions into one of the three classes. Also, Artificial Neural Network has captured special attention to segment and quantify the microstructures from images [19].

Recent research has begun to explore deep CNNs in computer vision methods applied to Material Science for microstructural image analysis. As an example, the work presented in [5] examines and compares Mid-level image patch descriptors with VGG-16 [20] and VGG-19 [8] CNN architectures in two ways: (i) by classifying microstructures on the basis of their primary components; (ii) by classifying microstructures depending on their annealing conditions. As demonstrated in [4], fully CNNs provided with max-voting schemes can exploit pixel-wise segmentation, thus achieving higher effectiveness over the prior state-of-the-art for steel microstructural classification without requiring separate segmentation and feature extraction stages. Motivated by robust accuracy achieved by CNNs and knowing that materials classification is a demanding task, in [21], the mean average precision achieved by pre-trained CNN architectures [22–24] has been evaluated. The pipeline is compounded by the CNN and the SVM, where the output of each CNN is used to train SVM classifier.

3 Methods

In order to automatically classify microstructural images into one of seven classes, we apply and compare four computer vision approaches for microstructural image classification by using CNNs, which is a kind of multi-layer network focused on recognizing visual patterns from pixel images. CNNs are attractive for image recognition and classification due to their high accuracy. The generic CNN is trained with datasets built from thousands of gray-scale images. Only in the training dataset, each microstructural image provides labeled information to depict the class to belonging the microstructure. Each training image passes through a sequence of convolutional layers, layers of activation functions, pooling layers, and fully connected layers to output the classification label. The general workflow of a CNN is schematized in Fig. 1 for several architectures. It can be seen that a CNN is compounded by several repeating kinds of layers, each named on the basis of the performed operations:

- Convolutional layer: it convolves the input data with an established number of (k) linear convolution filters with a defined size. This layer slides the filters over the input, computing dot products between the input data and the entries of the filters to detect and learn patterns from the previous layer, as shown in Eq. (1), where $k = 1, \ldots, K$ is the index of the k-th feature map; (i, j) is the index of neuron in the k-th feature map, and x is the input data; W_k and b_k are trainable parameter weights and biases from the visible units to the hidden units; and $(f_k)_{ij}$ is the learning k-th feature map.

$$(f_k)_{ij} = (W_k * x)_{ij} + b_k \quad (1)$$

- Pooling layer: it is a non-linear down-sampling layer for immediately reducing the spatial size after obtaining features using convolutional layers to diminish the amount of parameters and computations in the network. Its purpose is to merge semantically similar features into one. Each pooling layer pools the convolved features by a filter of size $F \times F$, then it divides convolved features into disjoint $F \times F$ regions, and takes the mean or maximum feature activation over disjoint regions to obtain pooled convolved features, which can then be used for classification. Summary statistics based on mean "mean pooling" or max "max pooling" values of a particular feature over a region of the input data are likely computed to obtain much lower dimension with respect to using all the extracted features, which can also control overfitting (the production of an analysis that corresponds too closely or exactly to a particular set of data).

- Fully connected layer: it is a linear combination, where each node is connected to all nodes of the previous layer. Its output is computed by matrix multiplication followed by bias offset, as shown in Eq. (2), where y_k is the k-th output neuron and W_{kl} is the kl-th weight between x_l and y_k.

$$y_k = \sum_l^L W_{kl} x_l + b_k \tag{2}$$

- Activation function: it often follows a pooling layer or a fully-connected layer and applies a non-linear activation operation to encode complex patterns of the input, through transformations of the weighted sums of inputs transferred to the artificial neurons. Deep neural networks typically use the ReLU activation function given in Eq. (3), since it is not continuously differentiable and does not vanish for high activations.

$$f(z) = \max(0, x) \tag{3}$$

- Output layer: it is the last layer in the network and holds the name of the loss function used for training the network for multiclass classification, the size of the output (number of output units), and the class labels.
- Classifier layer: it computes a class label probability of the input data. Its goal is to transform all the net activations in the final output layer to a series of values that can be interpreted as probability vector values between 0 and 1 (which have to be positive, smaller than 1, and sum up to 1). Softmax classification layer, which is the most extensively used in CNNs, applies a categorical probability distribution based on exponential function, as denotes Eq. (4) for the k-th class and an input X.

$$P(y = j|X; W, b) = e^{X^T W_j} / \sum_{k=1}^K e^{X^T W_j} \tag{4}$$

- Loss layer: it computes the difference between true class labels and the predicted class labels to determine the goal of learning by matching parameter settings as current network weights to a scalar value establishing the "badness" of the settings. Thus, the learning purpose is to find weights settings that minimize the loss (called also error, cost, or objective) function. The cross-entropy loss function given by Eq. (5) is one of the loss layers most widely used in deep CNNs.

$$E = -\sum_x P'(x) \log P(x) \tag{5}$$

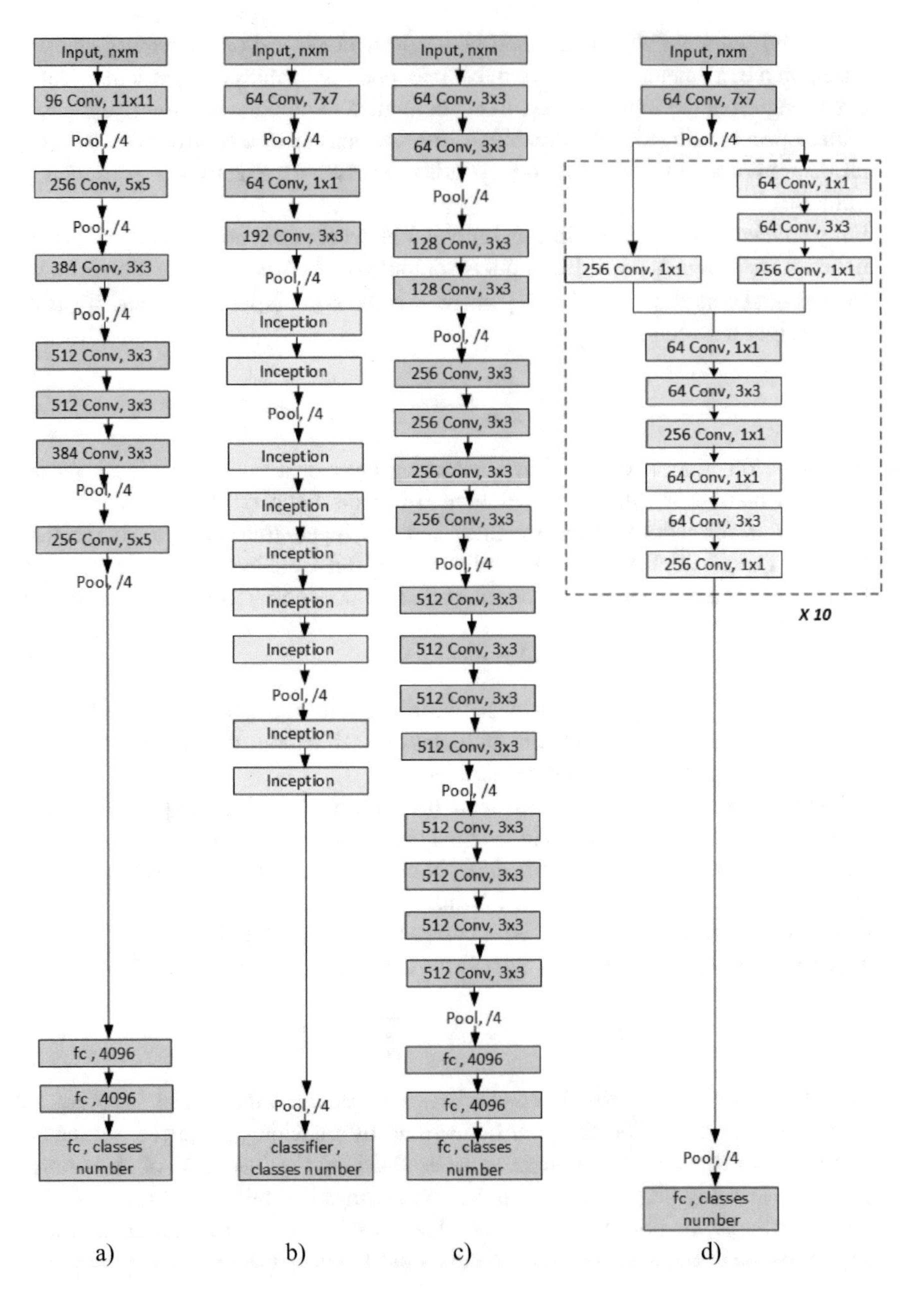

Fig. 1. State-of-the-art CNN architectures: (a) LeNet [6]; (b) GoogleNet [7]; (c) VGG-19 [8]; (d) ResNet-101 [9].

There $P(x)$ is the predicted probability value for class x and $P'(x)$ is the true probability for that class.

- Inception block: it is used in GoogleNet architecture for improving the performance of the network. It consists of multiple convolutional layers (multiple features) from multiple filters and its goal is to cover a bigger area, keeping a fine resolution for small information on the images. Therefore, this block convolves in parallel different filter sizes from 1 × 1 to a bigger 5 × 5.

In the following subsections, the four architectures of CNNs that we apply for microstructural image classification are described.

3.1 LeNet

This model introduced the convolutional network approach in 1998 for digits classification, which takes 32 × 32 gray scale images as input data. In this approach, increasing the size and the number of convolutional layers is required to process higher resolution images.

Based on the original model, LeNet network was implemented starting with the first convolutional layer, where each input microstructural image is convolved with 96 learnable filters of size 11 × 11. After that, a down-sampling is performed, through the pooling layer of size 3 × 3 with a stride of 2, to reduce the spatial size of representation to a ¼ of its previous size, and to also reduce the amount of parameters and computations (256, 384, 512, 512, 384, 256 learnable filters have been used for the subsequent convolutional networks respectively, while the same settings were applied for all pooling layers). At the end, three fully connected layers act as the classifiers and contain 4096, 4096 and 7 output units, respectively.

Differently from the original LeNet [6], which uses a layer with sigmoid activation function after each fully connected layer, in the revisited version of the LeNet CNN here adopted all layers are followed by a ReLU activation layer that causes sparsity in the features. Additionally, in this work, one fully connected layer, two convolutional and two pooling layers have been added, in order to process images with resolution higher than 32 × 32.

3.2 GoogleNet

It was inspired by LeNet but includes a novel inception module able to perform several very small convolutions in order to increase accuracy and significantly reducing the number of parameters. In 2014, it won the ImageNet Large Scale Visual Recognition (ILSVRC 2014) competition, showing a performance very close to human level with an error rate of 6.67%.

GoogleNet [7] was implemented with a total of 66 convolutional layers, where feature kernels range from 1 × 1 to 5 × 5 in inception blocks; while 7 × 7, 1 × 1, and 3 × 3 sizes have been applied for the first one as depicted in Fig. 1b.

3.3 Vgg-19

It consists of 16 convolutional layers, and 3 fully connected layers. This uniform approach has been widely used for extracting features from images. For implementation purposes, 64, 128, 256, and 512 features maps were used with a kernel size of 3×3 in all convolutional layers as depicted Fig. 1c.

3.4 ResNet101

This approach was introduced in 2015, with residual connections ("skip connections") also called gated recurrent units, which are highly similar to successful unit applied in Recurrent Neural Networks (RNNs). At ILSVRC 2015, it achieved an error rate of 3.57%, being able to train a network with 12 layers maintaining a complexity lower than VGG. It was implemented with 104 convolutional layers. ResNet [9] is focused on solve the saturated accuracy problem with the network depth increasing. It learns establishing the residual function based on stacked non-linear layers as shown in Eq. (6), where $F(x)$ and x are the stacked non-linear layers and the identity function (input = output) respectively.

$$F(x) = H(x) - x => H(x) = F(x) + x \tag{6}$$

In all applied models, the purpose of the fully connected layers is to use the learnt features for classifying the input image into seven classes based on the training dataset. The sum of the probabilities outputted by the last fully connected layer is 1. This is ensured by using the softmax as the activation function that takes the output of the last fully connected layer as the input and provides the image probability distribution over the labels. The softmax function takes a vector of arbitrary real-valued scores and forces it into a vector of values between zero and one that sum to one.

4 Implementation Details

4.1 Data Preparation

Prior to training stage, 75% of tif microstructural images [10] (which correspond to scanning electron microscopy micrographs of Ultrahigh Carbon Steel subjected to a variety of heat treatments and taken at several different magnifications), were preprocessed to generate a LMDB dataset then used to train each CNN. LMDB is a compressed format for working with large dataset. The preprocessing operation is focused on data augmentation to head the problem of unbalanced classes in the dataset. This operation includes several transformations, such as rotations, zooming, and transpose (horizontal and vertical flip), that are useful to obtain much larger dataset with slightly altered images. Therefore a LMDB dataset with approximately 35×10^3 tif images of size of 645×522 was generated for training purposes. In our experiments it was possible to use the original image resolution thanks to the availability of a GPU memory (11441 MiB provided by Tesla K80 NVIDIA GPU). Figure 2 depicts some examples of different input training microstructural images that were manually labeled by human in order to assign the corresponding class value.

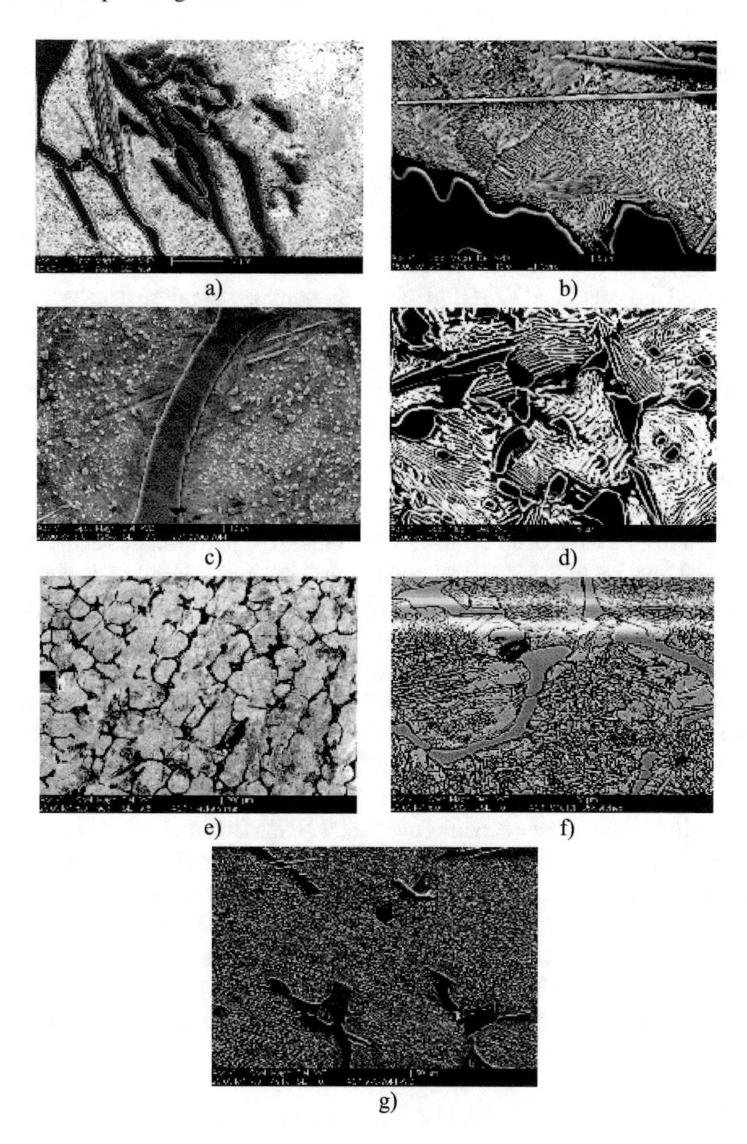

Fig. 2. Some examples of microstructure classes. (a) Pearlite+widmanstatten; (b) Pearlite; (c) Spheroidite; (d) Pearlite+spheroidite; (e) Network; (f) Martensite, (g) Spheroidite +widmanstatten.

4.2 Training

CAFFE framework and a Tesla K80 NVIDIA GPU were used for training all the selected CNNs, in order to learn the weights of convolutional and fully connected layers by back-propagation with the cross-entropy loss function given by Eq. (5). The loss function E measures how far the predicted output of the CNN is from the correct output; the back-propagation process propagates the loss function gradient to previous layers using stochastic gradient descent as learning algorithm for computing the gradient descent of the cost function for every mini-batch of a certain number of training examples. The gradient descent is focused on minimize the loss function repeatedly by updating the layers parameters in the opposite direction of the gradient of the loss function.

CNNs used the generated LMDB dataset compounded of microstructural tif images, where all images were sampled randomly from the available samples of training. We used a learning rate of 0.001, a momentum of 0.9, and a stepsize of 5×10^3. The training stage in all cases was performed in 10×10^4 iterations for evaluation purposes.

5 Experimental Results

Experimental results have been evaluated to validate the performance achieved by each trained model on a set of 3.5×10^2 microstructural images. Therefore, accuracy metric, based on the percentage of images correctly classified, was computed to measure the ability of classifying microstructural images into corresponding class. Table 1 presents maximum (MAX), minimum (MIN), and average (AVG) accuracy values. Experimental results clearly show that the proposed variation of LeNet model takes advantages of the used training settings (i.e. learning rate of 0.001, and momentum equal to 0.9) achieving its highest accuracy already at the iteration 10^4, which indeed leads to the lowest accuracy values for deeper networks. Performed tests also demonstrated that GoogleNet, VGG-19 and ResNet-101 architectures can improve their MIN, MAX and AVG accuracies by up to 39%, 17% and 25% only if more than 20000 training iterations are performed. Table 1 also shows that deeper CNNs have MIN accuracy values significantly lower than the LeNet architecture. This behavior is due to high unbalanced classes within the training dataset. As an example, the VGG-19 architecture achieves its lowest MIN accuracy value in the presence of input images belonging to the pearlite+spheroidite class. This happens because only 10.5% of the images within the training dataset belong to this class.

Table 1. Accuracy values achieved at the iteration 10^4.

	LeNet variation Depth = 7	GoogleNet [7] Depth = 66	VGG-19 [8] Depth = 16	ResNet-101 [9] Depth = 104
MAX	97.1%	96.6%	71.4%	62.1%
MIN	42.9%	7.1%	4.7%	8.6%
AVG	78.5%	56.2%	17%	36.2%

As it is well known, the CNN depth has a fundamental impact on the overall classification, as well as on the overall achieved performance. For this reason, the above examined CNNs approaches were characterized also in terms of the number of multiply and accumulate operations (MACs) required in convolutional layers, foreward-backward time, and foreward time. Results presented in Fig. 3 show that the proposed variation of LeNet network reaches the highest AVG accuracy score but it is characterized by an amount of MACs, a time for foreward-backward and a time for the foreward process much lower than the deeper examined counterparts. From scores achieved by VGG-19 [8] (MACs = 132.3G; Foreward-Backward = 128.2 ms) and ResNet-101 [9] (MACs = 52.1G; Foreward-Backward = 346.2 ms) it is clear that a higher number of operations not always leads to higher processing time or higher number of convolutional layers. For example, VGG-19, with the highest MACs value, gets relatively low foreward-backward time. The efficiency of the proposed LeNet variation is confirmed also in Fig. 3e, which compares the examined CNNs in terms of normalized average error rates, number of MACs, foreward-backward and foreward times.

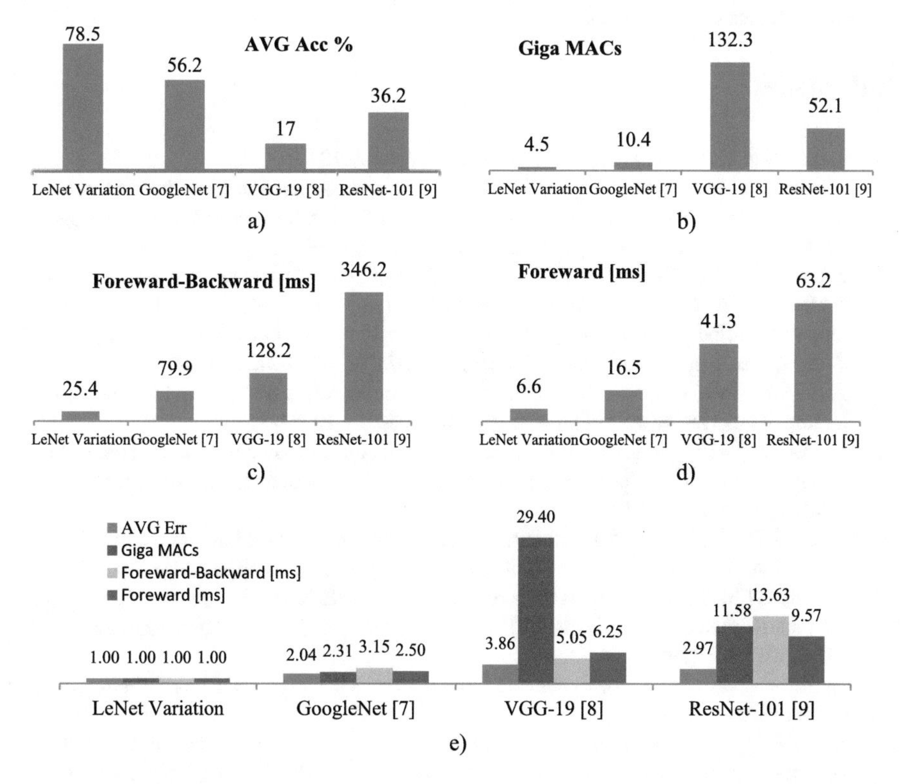

Fig. 3. Comparison results: (a) average accuracy; (b) number of MACs; (c) foreward-backward time; (d) foreward time; (e) normalized parameters.

6 Conclusion

This paper has empirically compared the performance and the suitability of outstanding CNN architectures for microstructural classification. They are different in the number of convolutional layers and in the way that they are organized. As well as, a lower number of multiplying and accumulate operations not always lead to a lower foreward-backward time. As well as, the strong difference between maximum and minimum accuracy by classifying in one of seven classes is given by considerable difference of quantity of data for each class (a relation from 10 to 0.23), that was not greatly solve with data augmented technique.

Based on evaluation of trained networks, as it was expected deeper networks requires more training for increasing accuracy. These approaches could be efficiently introduced for microstructural images classification.

Acknowledgements. This work used the supercomputer of the National Supercomputing Service Yachay EP of Ecuador (Quinde I).

References

1. ImageJ. http://rsb.info.nih.gov/ij/. Accessed 10 June 2018
2. Boschetto A, Campana F (2012) Morphological characterisation of cellular materials by image analysis. In: Computational modelling of objects represented in images III: fundamentals, methods and applications, Rome, pp 391–396
3. Ronneberger O, Fischer P, Brox T (2015) U-Net: convolutional networks for biomedical image segmentation. In: Navab N, Hornegger J, Wells W, Frangi A (eds) Medical image computing and computer-assisted intervention – MICCAI 2015. Lecture notes in computer science, vol 9351. Springer, Cham, pp 234–241. https://doi.org/10.1007/978-3-319-24574-4_28
4. Azimi S-M (2018) Advanced steel microstructural classification by deep learning methods. Sci Rep 8(1):2128. https://doi.org/10.1038/s41598-018-20037-5
5. DeCost B (2017) Exploring the microstructure manifold: image texture representations applied to ultrahigh carbon steel microstructures. Acta Mater 133:30–40. https://doi.org/10.1016/j.actamat.2017.05.014
6. LeCun Y (1995) Learning algorithms for classification: a comparison on handwritten digit recognition. In: Neural Networks: The Statistical Mechanics Perspective pp 261, 276
7. Szegedy C (2015) Going deeper with convolutions. In: IEEE conference on computer vision and pattern recognition (CVPR). IEEE Xplore, Boston, pp 1–9. https://doi.org/10.1109/CVPR.2015.7298594
8. Simonyan K (2015) Very deep convolutional networks for large-scale image recognition. In: ICLR. arXiv:1409.1556
9. He K (2016) Deep residual learning for image recognition. In: Proceedings of the IEEE conference on computer vision and pattern recognition. IEEE Xplore, Las Vegas, pp 770–778. https://doi.org/10.1109/CVPR.2016.90
10. UHCSDB microstructure explorer. http://uhcsdb.materials.cmu.edu/. Accessed 12 June 2018
11. Placidi F (2008) An efficient approach to springback compensation for Ultra High Strength Steel structural components for the automotive field. In: New developments on metallurgy and applications of high strength steels, pp 193–206

12. Wadsworth J (2010) The evolution of ultrahigh carbon steels - from the Great Pyramids, to Alexander the Great, to Y2K. In: TMS annual meeting the minerals, metals, & materials society, United States
13. Mancini E (2011) Surface defect generation and recovery in cold rolling of stainless steel strips. J Tribol 133(1):012202. https://doi.org/10.1115/1.4002218
14. Lanzini A (2009) Microstructural characterization of solid oxide fuel cell electrodes by image analysis technique. J Power Sources 194(1):408–422. https://doi.org/10.1016/j.jpowsour.2009.04.062
15. DeCost B (2015) A computer vision approach for automated analysis and classification of microstructural image data. Comput Mater Sci 110:126–133. https://doi.org/10.1016/j.commatsci.2015.08.011
16. Sundararaghavan V (2005) Classification and reconstruction of three-dimensional microstructures using support vector machines. Comput Mater Sci 32(2):223–239. https://doi.org/10.1016/j.commatsci.2004.07.004
17. Prakash P (2011) Fuzzy rule based classification and quantification of graphite inclusions from microstructure images of cast iron. Microsc Microanal 17(6):896–902. https://doi.org/10.1017/S1431927611011986
18. Saheli G (2004) Microstructure design of a two phase composite using two-point correlation functions. J Comput-Aided Mater Des 11(2–3):103–115. https://doi.org/10.1007/s10820-005-3164-3
19. de-Albuquerque V (2008) A new solution for automatic microstructures analysis from images based on a backpropagation artificial neural network. Nondestruct Test Eval 23(4):273–283. https://doi.org/10.1080/10589750802258986
20. Ballas N (2015) Delving deeper into convolutional networks for learning video representations. arXiv:1511.06432
21. Kalliatakis G (2017) Evaluating deep convolutional neural networks for material classification. arXiv:1703.04101
22. Krizhevsky A (2012) ImageNet classification with deep convolutional neural networks. In: Advances in neural information processing systems, pp 1097–1105
23. Zeiler M, Fergus R (2014) Visualizing and understanding convolutional networks. In: Fleet D, Pajdla T (eds) Computer vision – ECCV 2014. ECCV 2014. Lecture notes in computer science, vol 8689. Springer, Cham, pp 818–833. https://doi.org/10.1007/978-3-319-10590-1_53
24. Sermanet P (2013) Overfeat: integrated recognition, localization and detection using convolutional networks. arXiv:1312.6229

Clustering Algorithm Optimization Applied to Metagenomics Using Big Data

Julián Vanegas and Isis Bonet(✉)

EIA University, km 2 + 200 Vía al Aeropuerto José María Córdova, Envigado, Antioquia, Colombia
julianvanegas357@gmail.com, ibonetc@gmail.com

Abstract. In metagenomics, the amino acid sequences, due to the extraction process, are separated in DNA fragments of variable sizes. These fragments are used afterwards to determine which of the already recognized species are present in the samples and what portion of these amino acid sequences have not been previously categorized. Seeking for this method for identification to produce better results, clustering algorithms will be used as enablers in the identification process for the different species. These algorithms group amino acid sequences with a certain similarity rate, producing DNA fragments clusters, so these can be compared in group and be analyzed faster. One of the problems when analyzing metagenomic databases is that they are very large, which makes the algorithms have a high computational time. New technologies already provide platforms to develop and run algorithms achieving better temporal performance. Platforms like Apache Spark and TensorFlow were used with the objective of reducing the execution times, as they include native implementations of these clustering algorithms in their libraries. With these libraries as a base, an implementation of Iterative k-means was implemented and then used as a comparison point. In the results iterative k-means reduce the execution time with respect to the traditional implementation. The use of TensorFlow improved the execution times in general, with a more significative difference in the case of the Iterative k-means, with the disadvantage that it requires much more processing power.

Keywords: Metagenomics · TensorFlow · Spark · K-means · Clustering

1 Introduction

Microorganisms, even though they are the most abundant living beings and they dominate our planet in almost every environment, are still a big mystery in their investigation. In a 2016 study it was predicted that Earth could house approximately 1 trillion of these microbial species, of which only 0.01% have been identified [1]. A considerable number of these microorganisms are relevant in areas of human studies, particularly in agriculture and medicine, so the interest for increasing this identification percentage of the microbial species has grown in the latest years, as it could provide solutions for problems of daily life.

M. Botto-Tobar et al. (Eds.): TICEC 2018, AISC 884, pp. 182–192, 2019.
https://doi.org/10.1007/978-3-030-02828-2_14

The popularity growth in the study of microorganisms has led to the search for more effective alternatives useful in their identification. Nonetheless, a great obstacle in this study is the limitation for these lifeforms to survive in artificial environments, which is more aligned to the traditional method used for these studies, which was cultivating them isolated in a laboratory. Some of these species can't even survive outside their natural habitat. The discipline of metagenomics was created to fill this gap, which allows sequencing the genome of microbes directly from environmental samples, without the need to culture them artificially [2, 3]. The use of metagenomics also aids to control the bias introduced to the samples and experiments when these are artificially produced in a laboratory and allows the discovery of new lines of bacterial life [4].

Analysis of the results provided by the metagenomic process entails big challenges, as while it can be guaranteed in laboratory cultures that the genomic sequences belong to a single species, in a metagenomic process the environmental sample contains homogeneous groups of microorganisms, and the result will produce the genome sequences of any species in the sample, without distinction to the group it belongs to, and it could contain up to 10000 species [2].

A possible solution to this obstacle is using the existing databases to filter subsequences of microorganisms that are already registered and identified, separating the partial genomes of species that have yet to be discovered. However, this also leads to another challenge that must be overcome, the speed of the comparison considering not only the number of subsequences that are returned by the metagenomic process, but also the massive volume of the databases of identified species, making difficult a quick and efficient analysis of the results.

Metagenomics is being used currently to aid in the identification of amino acid sequences from microorganisms present in several environmental samples. These sequences are then compared to a database of all the known amino acid sequences, to test against any coincidence of them, or to check if the original sequence is not indexed yet.

For this specific case, this comparison can take from days to weeks due to the size of the database used as a reference, and due to that initially the sequences were compared one by one. Using clustering algorithms and artificial intelligence, it's possible to make groups of sequences that have similar characteristics, and that possibly belong to the same organism. Using algorithms like k-means with several distances and considering several characteristics a clustering of the sequences was made, looking that in each of the clusters different microorganisms could be identified. These clusters are then analyzed as a group, instead of taking each sequence separately.

Using distributed computing and machine learning platforms, like Apache Spark and TensorFlow, it's possible to implement an optimization of the clustering algorithms used in this process. These implementations manage to accelerate the processing of the great volume of data that need to be analyzed, also taking advantage of cloud computing, and micro optimizations present in the used algorithms.

2 Related Works

The process in metagenomics that allows us to identify which group of fragments belong to a single organism is called binning. The most complicated part of this process is that we have many mixed organisms and we do not know how many or which ones we have. According to the previously developed investigations, the research of the binning process can be divided on two methodologies: composition-based (supervised or unsupervised method) and similarity-based methods (supervised method) [5]. Similarity-based binning uses a database of known genes or proteins such as BLAST with similarity techniques in order to find to which organism each fragment belongs [6].

On the other hand, composition-based binning is focused on representing the fragments with features that allow them to be standardized at the same size and represent information that helps to separate them taxonomically. Some features used to represent the fragments are GC content, codon usage or oligonucleotide frequencies. Examples of supervised implementations are NBC [7], TACOA [8] and Phymm [9].

Some known unsupervised binning methods, are TETRA [10] and MetaCAA [11]. The main differences between the methods are based on the clustering algorithms such as Self-Organizing Maps (SOM) method [12, 13], k-means [14], fuzzy *k*-means algorithm [15], k-median [16, 17] and expectation maximization (EM) [18, 19].

PhymmBL [9] and new versions of MetaCluster [17] are examples of hybrid algorithms that combine the composition-based methodology along with alignment-based methods.

One of the advantages of the methods of composition, above the methods based on similarity is related to their computational complexity and are time-consuming. Considering that, we focused on composition-based methodologies. They also are used as a preliminary step to minimize complexity of the similarity-based methodologies, creating the groups that belong to each organism before searching them in the databases.

Considering the need and use of these methods, something important is, in addition to being accurate, they have to be computationally fast. This paper is based on a previous work [14], which use a variant of k-means (iterative *k*-means) with *k*-mers as features to represent the fragments, to improve the computational time.

3 Methods and Data

The purpose of this section is to describe the data and methods that were used in our experiments, making a description of the composition of our database considering the size of the metagenomic sequences present in the sample, and the diversity of the organisms within.

We present the features used to represent the metagenomic sequences, using the composition as a base. We also introduce the iterative clustering method based in k-means, and the distance functions used in the algorithm. Finally, we present the quality measures used to compare the results.

3.1 Data and Features

A sample containing the assembled genomic sequences was downloaded from the FTP site of the Sanger institute (ftp.sanger.ac.uk). This sample contains sequences at the contig level of different types of organisms, including viruses, bacteria, and eukaryotes. With the purpose of having representations of different domain groups, and variety within the groups, the database consists of 9 eukaryotes, 2 bacteria, and 5 viruses.

The description of each organism, including its domain, number of contigs in the sample for each species, and the minimum and maximum length of their contig is included in the Table 1. This shows the heterogeneity of the database used in the experiments.

Table 1. Organisms present in the database.

Organism	Domain	Contigs	Min length	Max length
Ascaris suum	Eukaryote	137650	50	30000
Aspergillus fumigatus	Eukaryote	295	1001	29660
Bacteroides dorei	Bacteria	1928	500	29906
Bifidobacterium longum	Bacteria	18	540	26797
Bos taurus	Eukaryote	315841	101	5000
Candida parasilopsis	Eukaryote	1540	1003	29956
Chikungunya	Virus	1	11826	11826
Dengue	Virus	64	10392	10785
Ebola	Virus	1	18957	18957
Glossina morsitans	Eukaryote	20334	101	29996
HIV	Virus	1	9181	9181
Influenza	Virus	8	853	2309
Malus domestica	Eukaryote	66739	102	5000
Manihot esculenta	Eukaryote	7192	1998	4998
Pantholops hodgsonii	Eukaryote	159729	50	5000
Zea mays	Eukaryote	161235	102	5000
		872576	50	30000

The variation of the number of contigs is also very large. While some organisms like Bos Tauru have 315841 contigs, with lengths between 101 and 5000 bases, other like Ebola have a single contig with size of 18957 bases.

We used a composition-based feature to represent the DNA fragments, due to the spread not only in the number of contigs between species but also the variation in size between them.

K-mers is one of the most used features to represent DNA fragments. This features is a combination of nucleotides (A, C, G, T) of size k, resulting in a subset of pieces of DNA of length *k*. Tetranucleotide ($k = 4$) is the most common used [10, 20], this means there are 256 features representing possible 4-mers.

GC- content is another feature which represent the percent of G and C in the fragment.

The features considered were the *k*-mers of both length 3 and 4, and the gc-content. For the *k*-mers all combinations of the nucleobases, that is, guanine, adenine, cytosine and thymine of the lengths described were considered.

3.2 Clustering Method

K-Means

K-means algorithm is a non-supervised clustering method, meaning it doesn't need to know the classification of each of the points prior to its training. Its objective is to assign the different points to a determined k clusters, by approximation of the centers using the mean of the distance of all of the points assigned to their center during each iteration of the algorithm. K-means++ is a variant to the initialization method of k-means, which improves the selection of centroids for the clusters. This algorithm finds the initial k centroids using a weighted probability distribution where a point x from the initial data is chosen based on the probability proportional to a distance function. This will ensure that the centroids are distant from each other. With these initial centroids chosen, the algorithm proceeds as the standard k-means clustering.

For this clustering model the value of k must be known prior to the execution, which makes difficult its use in cases where it isn't clear the number of groups that exist in the data.

Two different distance functions were used in conjunction with the K-means algorithm, with the following formulas:

$$\text{Euclidean distance } dE(x, y) = \sqrt{\sum\nolimits_{i=1}^{n} (x_i - y_i)^2}$$

$$\text{Cosine distance } dC(x, y) = \frac{\sum\nolimits_{i=1}^{n} (x_i \times y_i)}{\sqrt{\sum\nolimits_{i=1}^{n} x_i} \times \sqrt{\sum\nolimits_{i=1}^{n} x_i}}$$

Iterative K-means

Iterative K-means is a variation on the standard K-means algorithm. In each iteration of Iterative K-means, a full run of K-means is executed, the best clusters are chosen according to some margin, and the data in the rest of the clusters that weren't chosen is used as the input data for another run of K-means. This will make it so in each iteration of the K-means algorithm the best clusters for that run are chosen, and the rest can be reprocessed into possibly better clusters.

Each iteration of K-means can use different k, and different distance functions can also be sued to evaluate different combinations of parameters that produce the best results.

For our use of Iterative K-means in these experiments, the margin was calculated using the average of the distance of each centroid with the rest of the clusters, averaging these calculations for all clusters in the data, and adding the standard deviation. We interpreted that the clusters that have a higher average distance than this margin, is because they are more separate from the rest of the clusters, while the ones below this margin are still too grouped up and could be reprocessed with the clustering algorithm.

3.3 Clustering in Apache Spark

Apache Spark provides a machine learning package, called spark.mllib. [20] This package contains several models available for clustering, including K-means, beside others like Gaussian Mixture Model and Power Iteration Clustering.

Spark has APIs for several programming languages, like Scala, Java, and Python. We decided to use Scala in our experiments, as it is the API native language and also is the more updated out of all of them.

A great disadvantage of the use of K-means in Spark is that only the Euclidean distance is available, while other platforms like TensorFlow also allow the use of cosine distance.

3.4 Clustering in TensorFlow

TensorFlow has a native API based in Python and provides several libraries of machine learning. The *tf.contrib.factorization* module contains several models and operations related to factorization, or clustering, of data. Its two main models are Gaussian Mixture Model, or GMM, and K-means [21].

Each of these models has two implementations, using two different APIs, with Graphs and Estimators. The main difference between the use of these two is that with Graphs one must first create a graph of operations that will be executed in order in a TensorFlow session, while the Estimator-based API uses eager execution, which brings it closer to an imperative programming model where operations and instructions are executed immediately.

For our experiments we used the K-means implementation based in Estimators, using the KMeansClustering module.

3.5 Result Validation

As for the measures used in the validation of the resulting clusters, both purity and dispersion were considered. A cluster purity represents the existence of a single species inside of it, and it's calculated as the highest percent of the classes contained within it. Dispersion is a measure of all the contents of the cluster, represented as a matrix of the percentages of all of the classes assigned to the cluster.

These measures allowed us to make a visual and quick validation of the effectiveness of the clustering algorithms related to the separation of the different species. The ideal result cluster would have a purity of 100%, meaning that only a single species was assigned to it, also meaning that it wouldn't have any dispersion.

4 Results and Discussion

Using the implementation of iterative K-means in Apache Spark and TensorFlow, we proceeded to run several executions of the algorithm, with different parameters, comparing the results and registering the execution times.

As a first comparation point we executed the non-iterative K-means algorithm with both distances, using k = 30. With these observations as a base we concluded that for our base the cosine distance produced the best results, shifting our focus primarily on TensorFlow, considering Apache Spark doesn't have the cosine distance as an alternative in its implementation of K-means.

We also tested the features of the data, with experiments for different combinations of them. First, we used all features, and then using exclusively the k-mer of length 4, or 4mer. This is thanks that previous results showed that the best results were obtained using these features [22, 23]. These results were confirmed with our experiments, a considerable difference in the purity of the clusters was registered based on the features used during the algorithm execution.

The results obtained from executing the non-iterative K-means, with cosine distance, 4mer as features and k = 30 can be appreciated in Fig. 1.

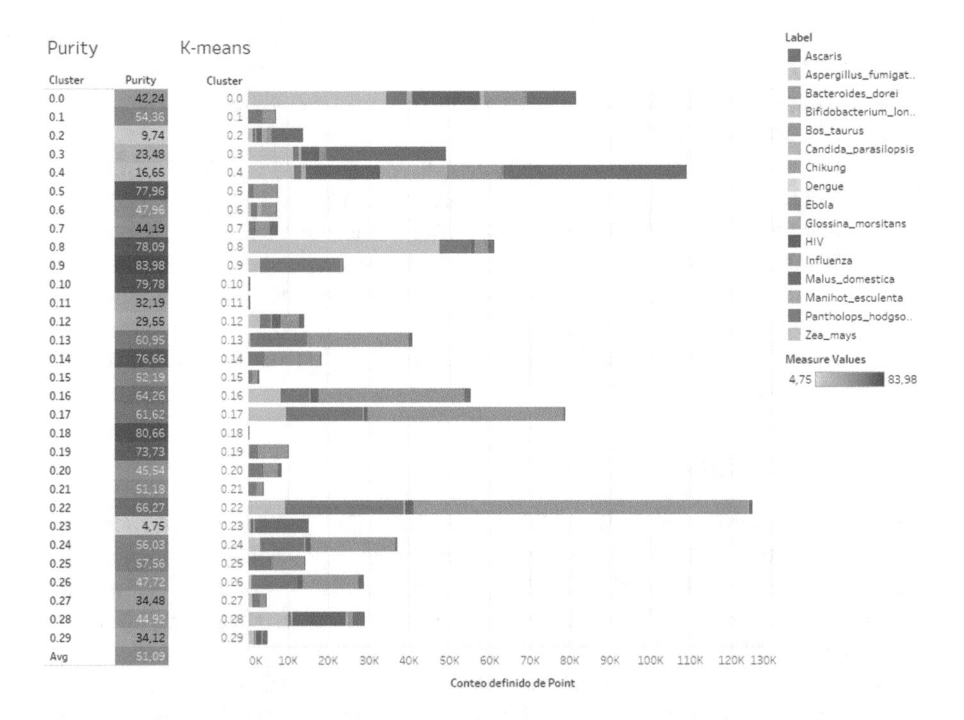

Fig. 1. Clusters using non-iterative K-means, cosine distance, and k = 30

We can appreciate in the figure that most of the clusters are very dispersed, and the clusters with less dispersion are also the smaller ones, as is the case for clusters 0.10, 0.11, and 0.18.

In the left side of the figure the purity of each of the clusters can be seen, with a higher percentage representing a better measurement for the clusters.

For the next experiment, we decided to run a second iteration over these same results, to make a comparison of the variation by maintaining the best clusters of this first iteration and regrouping the rest with a second K-means run.

For this, we executed the iterative K-means, using the same cosine distance and 4mer as features, but this time with two iterations, using k = 30 for the first one and k = 20 for the second one. The results of this experiment can be seen in Fig. 2.

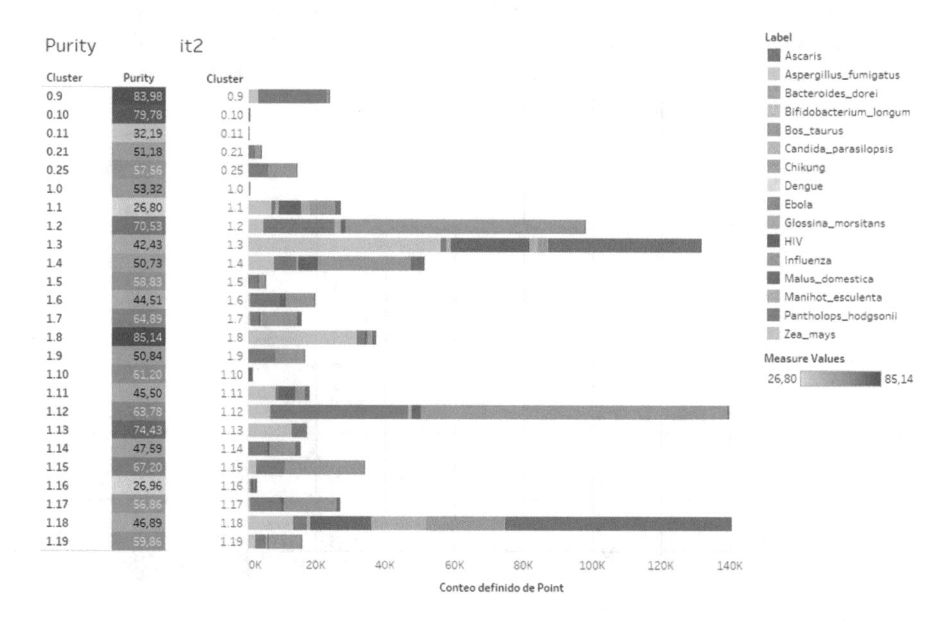

Fig. 2. Clusters created with iterative K-means, cosine distance, and k = 30, 20

After adding a second iteration with k = 20, the total amount of clusters was reduced, as from the first 30 clusters only 5 were selected as the best, and the rest was reprocessed. This brings the total for this experiment to 25 clusters. Introducing more iterations will create a lot more variation in the number of clusters, which could help overcome the limitation of having to know the value of k beforehand.

The clusters that remained from the first iteration are 0.9, 0.10, 0.11, 0.21, and 0.25. Out of these, the cluster 0.9 has an especially good purity, with 83.98% of its composition belonging to *Pantholops hodgosonii*, while the rest, although they have a purity between 30% and 80%, are much smaller in comparison.

The *Ascaris* species, while in the alternative of the non-iterative K-means is a lot more disperse, in these results it's almost completely contained in the clusters 1.3 and 1.18, showing a better separation for this organism in the second experiment.

It's possible that after merging these two clusters and their reprocessing it could be better separated from the rest, giving indication of a possible improvement to the algorithm, taking account similar clusters that could be merged prior to their regrouping from another iteration of K-means.

While in the results for the non-iterative K-means, the minimum purity was 4.75%, with a second iteration the minimum purity reaches 29.80%, showing a considerable improvement in the composition of the new clusters. The average of these new purities is 56.12%, that compared to the 51.09% of the first result, shows an improvement across the board for the purity of all the clusters.

Looking to reduce the noise that existing in the data sample, we made an analysis by filtering by the length of the contigs for each data point, only taking into account lengths larger than a specific number of bases and measuring the change in the average purity for these new clusters, also using the results from the iterative K-means of Fig. 2. The resulting graphic of the averages can be seen in Fig. 3.

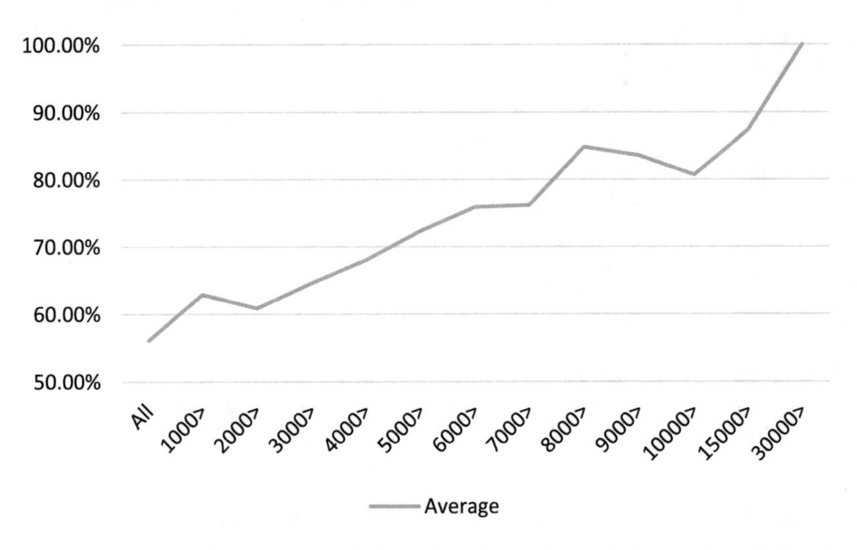

Fig. 3. Evolution of the average purity based on filtering by a minimum of contig length.

Increasing the minimum length considered by the filter also shows an increment in the average purity of the clusters, reducing the amount of data that is being considered. For the filter of length greater than 30000, there exists only a single data point, belonging to a single cluster, making the purity 100%. This also shows another problem, by filtering very aggressively, the amount of data is reduced to the point of not providing any useful information. For this, a balance would need to be found between filtering the noise of the smaller organisms but also not eliminating the majority of data that needs to be analyzed.

To measure the execution times of the algorithms, we used a high memory machine with 52 GB of RAM, taken from Google's cloud computing service, Google Cloud Platform. In this machine, we ran several executions of the K-means algorithm simultaneously. In results from previous experiments, with the non-iterative K-means, using a Weka implementation and executing it also in a high memory machine, the minimum execution time was of 16 min [22]. With our alternative developed in TensorFlow and the same algorithm parameters, we measured an execution time of approximately 13 min. For this algorithm specifically the difference is negligible, considering the execution time was quite short from the start.

As for the iterative K-means, the execution times can fluctuate due to all the related variables, with the number of iterations taking a considerable effect on the execution time. Nonetheless, in previous experiments with the Weka-based implementation, the minimum execution time for cases with two iterations was approximately 1 h, and for more complex cases it could increase to 4 to 8 h [22]. For our alternative using TensorFlow, the minimum time for cases with two iterations was of 25 min, and for the more complex cases it reached 1 h. This is a measurable improvement, considering that depending on the experiments this algorithm could be executed several times per day, saving hours in each run.

Remarkable as a result of work done is decreased runtime Iterative k-means for implementation in previous work, with reductions of up to 7 h in more complex executions.

Conclusions

In this paper we proposed an alternative of iterative K-means developed using TensorFlow, taking advantage of its machine learning platform, that reduces the execution time of the clustering algorithm. Apache Spark was also considered to reduce the execution times, but due to its limitation in its API to use the cosine distance, its use is a lot more restricted.

An existing alternative is to use the distributed computing capabilities of Apache Spark, while using the TensorFlow API to develop the clustering algorithms, including iterative K-means. Apache Spark provides support for executing Python in its platform, which means the execution of TensorFlow in Spark is possible.

In the results we can confirm that the quality of the clusters by using iterative K-means is improved compared to the alternative of the non-iterative K-means, and we also reaffirm that the cosine distance with 4mer as features is providing the best results in the clusters regarding their purity and dispersion.

We also were able to reduce the execution times of the iterative K-means by several hours with our implementation using TensorFlow, compared to a more traditional, Weka-based approach. These improvements will be more apparent the greater the number of iterations is used, as we could see that while in a single iteration the reduced time was only minutes, by the second iteration the reduction was of several hours, leading to the conclusion that with even more iterations the gap will increase even more.

The results look promising in the fact that they were tested in a single machine, with even a greater reduction being possible by extending the algorithm so it runs over Apache Spark in a cluster.

References

1. Locey KJ, Lennon JT (2016) Scaling laws predict global microbial diversity. Natl Acad Sci
2. Wooley JC, Godzik A, Friedberg I (2010) A Primer on Metagenomics. PLoS Comput Biol 6(2):e10006672010
3. Thomas T, Gilbert J, Meyer F (2012) Metagenomics-a guide from sampling to data analysis. Microb Inform Exp
4. Handelsman J (2004) Metagenomics: application of genomics to uncultured microorganisms. Microbiol Mol Biol Rev
5. Kislyuk A, Bhatnagar S, Dushoff J, Weitz J (2009) Unsupervised statistical clustering of environmental shotgun sequences. BMC Bioinform 10(1):316
6. Camacho C et al (2009) BLAST + : architecture and applications. BMC Bioinform 10(1):421

7. Rosen GL, Reichenberger E, Rosenfeld A (2010) NBC: The Naïve Bayes classification tool webserver for taxonomic classification of metagenomic reads. Bioinformatics
8. Diaz NN, Krause L, Goesmann A, Niehaus K, Nattkemper TW (2009) TACOA–Taxonomic classification of environmental genomic fragments using a kernelized nearest neighbor approach. BMC Bioinf 10:56–56
9. Brady A, Salzberg SL (2009) Phymm and PhymmBL: metagenomic phylogenetic classification with interpolated markov models. Nat Methods 6(9):673–676
10. Teeling H, Waldmann J, Lombardot T, Bauer M, Glockner F (2004) TETRA: a web-service and a stand-alone program for the analysis and comparison of tetranucleotide usage patterns in DNA sequences. BMC Bioinf 5(1):163
11. Reddy RM, Mohammed MH, Mande SS (2014) MetaCAA: A clustering-aided methodology for efficient assembly of metagenomic datasets. Genomics 103(2–3):161–168
12. Abe T, Kanaya S, Kinouchi M, Ichiba Y, Kozuki T, Ikemura T (2003) Informatics for unveiling hidden genome signatures. Genome Res 13(4): 693–702
13. Zouari H, Heutte L, Lecourtier Y (2005) Controlling the diversity in classifier ensembles through a measure of agreement (in English). Pattern Recognit 38(11):2195–2199
14. Bonet I, Escobar A, Mesa-Múnera A, Alzate JF (2017) Clustering of metagenomic data by combining different distance functions. Acta Polytech Hung 14(3)
15. Woods K, Kegelmeyer WP, Bowyer K (1997) Combination of multiple classifiers using local accuracy estimates (in English). IEEE Trans Pattern Anal Mach Intell 19(4):405–410
16. Leung HC et al (2011) A robust and accurate binning algorithm for metagenomic sequences with arbitrary species abundance ratio (in eng). Bioinformatics 27(11):1489–1495
17. Wang Y, Leung H, Yiu S, Chin F (2014) MetaCluster-TA: taxonomic annotation for metagenomic data based on assembly-assisted binning (in English). BMC Genomics 15(1), 1–9. Article no. S12
18. Partalas I, Tsoumakas G, Katakis I, Vlahavas I (2006) Ensemble pruning using reinforcement learning. In: Advances in artificial intelligence, proceedings, Lecture Notes in Computer Science, vol 3955. Springer, Berlin, pp 301–310
19. Nanni L, Lumini A (2006) FuzzyBagging: a novel ensemble of classifiers. Pattern Recognit 39(3):488–490
20. MLlib Clustering (2018) In: Apache Spark Docs ed
21. Module (2018) tf.contrib.factorization. In: Tensorflow Python API Docs ed
22. Bonet I, Escobar A, Mesa-Múnera A, Alzate JF (2017) Clustering of metagenomic data by combining different distance functions. Acta Polythecnica Hung 14(3)
23. Bonet I, Montoya W, Mesa Múnera A, Alzate JF (2014) Iterative Clustering Method for Metagenomic Sequences
24. Apache Software Foundation (2018) MLlib Clustering. https://spark.apache.org/docs/2.3.0/mllib-clustering.html
25. Google, Module: tf.contrib.factorization(2018). https://www.tensorflow.org/api_docs/python/tf/contrib/factorization

Intelligent System of Squat Analysis Exercise to Prevent Back Injuries

Paul D. Rosero-Montalvo[1,2,3(✉)], Anderson Dibujes[1], Carlos Vásquez-Ayala[1], Ana Umaquinga-Criollo[1], Jaime R. Michilena[1], Luis Suaréz[1], Stefany Flores[1], and Daniel Jaramillo[1]

[1] Universidad Técnica del Norte, Ibarra, Ecuador
pdrosero@utn.edu.ec
[2] Instituto Tecnológico Superior 17 de Julio, Ibarra, Ecuador
[3] Departamento Informática y Automática, Universidad de Salamanca, Salamanca, Spain

Abstract. The sports ergonomics study allows a bio-mechanical analysis in order to evaluate the impact produced by different muscle conditioning exercises such as the squat. This exercise, if carried out in an erroneous way, it can cause lumbar injuries. The present electronic system acquire the data of the Smith bar and the back by means of accelerometer sensors. This is done in order to implement an intelligent algorithm that allows to recognize if the athlete performs the exercise properly. For this, a stage of prototypes selection and a comparison of classification algorithms (CA) is carried out. Finally, a quantitative measure of equilibrium between both criteria is established for its proper selection. As a result, the k-Nearest Neighbors algorithm with k = 5 achieves a 96% performance and a 50% training matrix reduction.

Keywords: Squat analysis · Intelligent systems · Embedded systems Back injuries

1 Introduction

Ergonomics refers to the study of man when performing physical activities in his work environment in order to know the muscular abilities and limitations of the human being [1]. In this way, we want to reduce the risks of bad postures that can cause injuries and to improve the techniques in the tasks performed. Sports ergonomics focuses on providing a safe environment for the athlete's muscular development [2]. Therefore, bio-mechanics is of great importance, since it allows to perform an analysis in the different training sessions and to explain how injuries happened. As a result, appropriate techniques can be determined in the development of exercises [3].

The Higher Council of Sports and other sports associations worldwide, defined the types of legal ergonomic aids and doping substances. These can be: (i) material aids, such as the design of sports equipment such as insoles, shoes,

M. Botto-Tobar et al. (Eds.): TICEC 2018, AISC 884, pp. 193–205, 2019.
https://doi.org/10.1007/978-3-030-02828-2_15

among others. (ii) physiological aids, which helps in the increase of red blood cells in the blood. (iii) psychological aids, which are related to the mood of the people and (iv) pharmacological that are supplementary to improve the physical activity of the athlete [4,5]. The above ergonomic techniques are intended to improve the performance of the athlete without affecting their health in the short or long term. One of the areas of the body that is more likely to suffer an injury, are the lumbar [6].

There are exercises where maximum effort is required, which can increase the load on the lumbar spine and predispose the athlete to physical problems, such as injuries for excessive strength and muscle fatigue. These can cause a rupture (damage) of one or more of their vertebral bodies, inter-vertebral discs and ligaments. The main affections can be the sciatica and low back pain, can be caused by the performance of a squat with a bad technique when lifting dead weights [7]. This injuries occur in the lower part of the back with different degrees of severity [8,9]. For this reason, 30% of high performance athletes of both genders have injuries in this area of the body [8].

One of the common exercises in most sports disciplines that exposes athletes to the problems mentioned above are squats. Since performing this exercise without proper supervision and technique there is a 60% chance of suffering low back pain. This exercise can be performed in two ways, half squat (the lower thighs are parallel to the ground at an angle of 90 degrees) and full squat (the athlete performs a deep flexion of the legs, the hip moves slightly downward and forward) [9]. This exercise can be done with dumbbells, steel bar or Smith machine. The greater incidence in injuries is with the steel bar because the balance of the body is needed.

In order to evaluate the impact produced by the squat in the lumbar spine, a bio-mechanical analysis is performed (treating the body as a mechanical system formed by several separated parts) and verify the correct functioning of the system. This analysis shows that when performing the squat, the spine is flexed and the muscles that extend the spine are from the pelvis (gluteus major and minor) [10]. In addition, it was observed that, from the 40° of flexion of the spine, the lumbar muscles stop acting [11]. At that time, an imbalance of the weight of the Smith bar can cause damage. In order to avoid muscle problems, the athlete must perform the exercise in a correct way. That is, keep the back straight and the back muscles contracted isometrically. In this way, the weight of the bar does not generate pressure on the inter-vertebral discs [12]. The Fig. 1 shows the bio-mechanics of the trunk when performing good and bad posture.

In order to study of the bio-mechanics of the back, many embedded electronic systems (formed by a processing unit, memory blocks, data input and output blocks, a communication medium and a battery) seek to represent the operation of the body to alert bad postures and avoid injuries [13]. For this, the use of sensors becomes fundamental in the development of the system, since it allows the data collection process. This is developed by a transducer, which transforms from a physical quantity to an electric quantity. It is also necessary to have a signal coupling circuit. The union of the transducer and its coupling circuit is

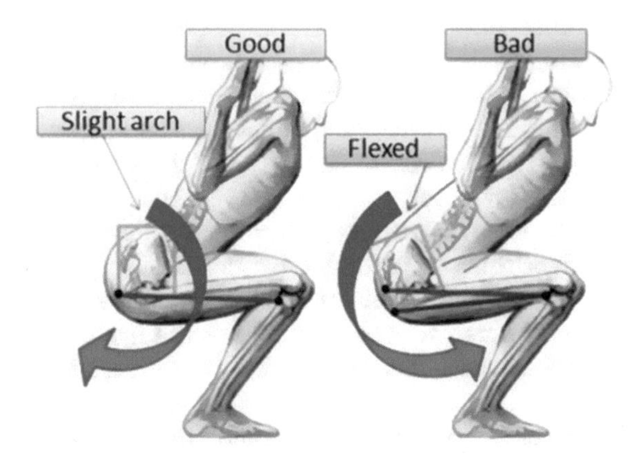

Fig. 1. Back bio-mechanical analysis

called a sensor. During the data collection process, the data can be influenced by conditions that cannot be controlled, as well as environmental factors that affects data validity and usability The physical components that integrate the sensors are far from been ideal and introduce sources of uncertainty in the measurement process, consequently, it is possible to represent only an estimate of the physical phenomenon to be studied. Also, the speed of acquisition of these systems is considerably high in relation to the actions of the human being. As a result, a large amount of information is collected, and this information cannot be valid for the system. An embedded system must adapt to the different conditions that were exposed [14]. This implies in some way, the ability to make decisions using intelligent mathematical algorithms. The same ones that need a set of data entry to train themselves. Under this concept, it is expected that the sensors and data collected are the most appropriate [15]. To do this, it is necessary to determine that the set of information to be stored within the system allows to maintain the intrinsic knowledge of the large volume of information found by the sensors [16,17]. In addition, the reduction of the training set allows optimizing the limited resources of an embedded system, considering that the more data is processed, the battery life will be reduced proportionally [18].

The algorithms of machine learning when implemented within an embedded system, separates the human being in the decision-making process. For this reason, it is necessary to provide an interface that allows knowing the decision made by the system to the user. In this way, the user can know the operation of the system and assess the accuracy of the response when exposed to real conditions [19].

Some Works like [20,21], explain kinematic parameters that characterize good and poor leg squat performance. This diagnostic is testing with two-dimensional video and three-dimensional motion analysis data were collected. However, the reliability was inadequate by excessive frontal plane motion of the knee and hip. For these reasons, the squat analysis still remains an open problems.

The present embedded system detection of postures in the squat exercise works in the following way. It has 3 nodes, located in the waist, in the exercise bar and in a computer. The computer node acts as a receiver of information and sends it for visualization. The remaining nodes have an accelerometer-gyroscope sensor. Those sensors allow to collect the angles of the movement at the moment of the exercise of the low back and the exercise bar. To have the best data set, data filters are used to collect sensor variables. Subsequently, an analysis of machine learning algorithms is performed by means of a quantification measure of the data representation algorithms with the different classification algorithms. This in order to determine the most suitable of the system. Finally, they are implemented and tested with different performance measures of classification algorithms.

The rest of the work is structured as follows: Sect. 2 shows, the design and materials of the electronic system and the proposed methodology. Section 2.2 indicates data analysis with machine learning techniques. The results of the system are shown in Sect. 3. Finally, Sect. 4 presents the conclusions and future work.

2 Materials and Methods

This section, for one hand shows the electronic design and diagram connection between sensors, embedded system and principal node. For the other hand, shows the data acquisition and the propose methodology.

2.1 Electronic Design

The electronic system has been designed according to the IEEE 29148 standard. It explains how the system can obtain the hardware and software requirements [22]. Some works use an electronic diagram connection to show the embedded system functionality [15]. In Table 1, the most system relevant requirements are shown. The same ones that were determined by the user and developers needs.

Later, with the established requirements. The electronic elements are selected for the system under the criteria like: usability, speed, price, programming easily, among others. In Fig. 1 its connection diagram is shown. The embedded system works as follows: (i) accelerometer sensor acquires data of Y axis angle of the back and the exercise bar. (ii) The button is responsible for activating and deactivating the system. The same one that gives the user one minute until it starts working. The buzzer has the functionality of emitting sounds to give attention to the activation of the system (2 beeps) and bad back position (1 beeps). (iii) The communication means chosen is radio-frequency because of its coverage. (iv) As a microprocessor system is Arduino, this it considered for its ease of programming and work environment with sensors. (v) The battery provides all the proper voltage for the system (5 volts for all elements).

The development of the system is carried out by nodes: (i) back node, the same one that has the activation button and synchronizes the communication

Table 1. System requirements

System requirements (SHr)					
#	Requirements	Priority			Relationship
		High	Medium	Low	
SHr1	Ease of system use	X			SHr3
SHr2	Error bad posture system alarm	X			SHr6
SHr3	The system must has a visualization interface	X			SHr1
SHr4	The data analysis must be in real time	X			SHr8
SHr5	The system must has a light weight	X			SHr1
SHr6	The system must measure the angle of back and bar	X			SHr2
SHr7	The system must have a wireless communication	X			SHr1
SHr8	The load computational must be the less as possible		X		SHr4

with the following nodes. (ii) The bar node is activated by the back node, this system is responsible for acquiring the data of the bar angle in Y axis. (iii) Central node, only has the communication module between the other node and the computer. The Fig. 2 show the electronic diagram connection of back node.

2.2 Data Analysis

The data acquisition speed of the electronic system is 8MHz. Therefore, the system memory is not adequate to store all the information are reading by sensors to train the classification algorithm. Due to this, a data analysis is done to choose the best training set to store at the system and still to keep a high performance of the classifier [23]. As a first step, the data collection is done with the supervision of a trained professional, 20 athletes performs the exercise in a correct and erroneous way. As result, there are 208 repetitions of the exercise (104 good repetitions and 104 bad repetitions). The Fig. 3 shows the data acquisition representation of the exercise [17].

After that, the performance comparison of the best classification algorithms with optimization approaches is performed, which are: (nearest neighbor rule (k-NN), support vector machines (SVM), Bayesian classifier and decision tree. Five tests were carried out with different training sets to validate the performance of each classifier. With this result, we can reduce the training matrix and determine the best data set with its classification algorithms. Table 2 shows the summary of the performance of each classifier.

The data analysis result was only about confusion matrix. For this reason, it is necessary to know the border decision line by each training set label. To do this, a matrix was made of less data separation that comprises from the

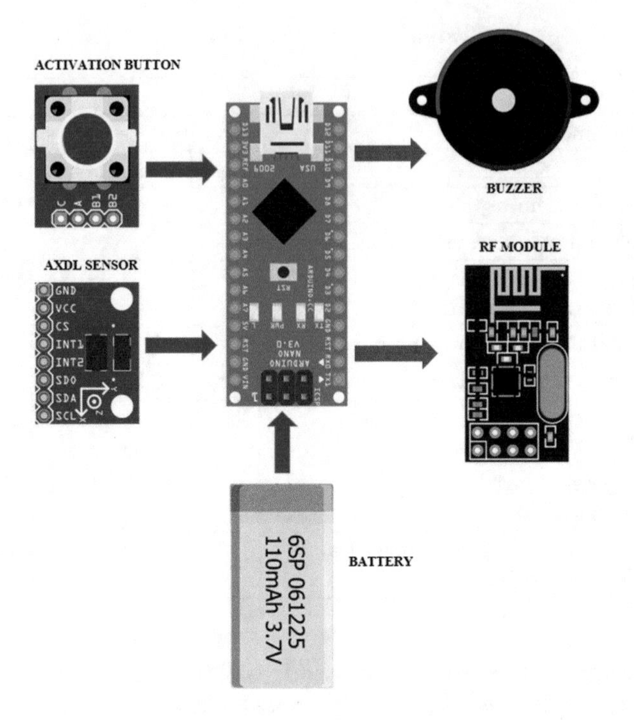

Fig. 2. Electronic diagram connection

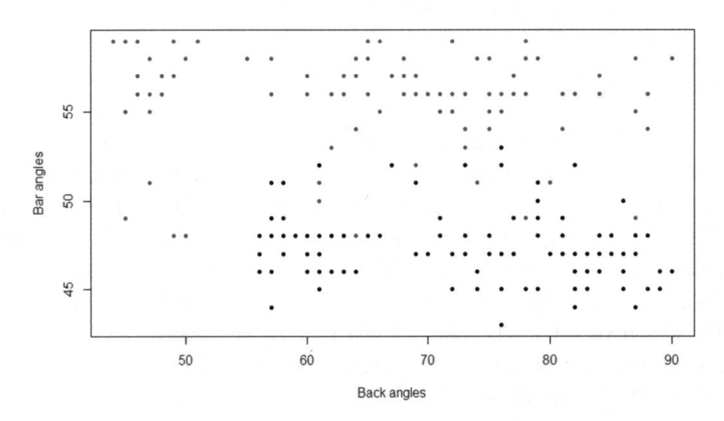

Fig. 3. Data acquisition by squat exercise

maximum and minimum points of each training set characteristic. In relation to points, a new matrix is created with 0.01 distance between each point of the back and the bar angles. As a result, Fig. 4 shows how the decision algorithms have learned from the training set data to establish their decision edge. On the one hand, the green color establishes a suitable squat exercise. On the other hand, the color tomato, defines a bad bio-mechanical action of the person.

Table 2. Classifier performance

Tests with aleatory date set						
	Test 1	Test 2	Test 3	Test 4	Test 5	Average Error
k-NN						
k = 1	0,058	0,096	0,058	0,038	0,115	0,073
k = 3	0,077	0,077	0,038	0,058	0,077	0,065
k = 5	0,058	0,077	0,058	0,077	0,077	0,069
SVM						
kernel = sigmoid	0,038	0,058	0,115	0,115	0,058	0,077
kernel = polynomial	0,077	0,077	0,096	0,096	0,115	0,092
kernel = linear	0,058	0,096	0,115	0,135	0,058	0,092
kernel = radial	0,058	0,096	0,096	0,135	0,058	0,088
BAYES						
Normal equation	0,058	0,058	0,115	0,077	0,115	0,085
DECISION TREE						
Normal equation	0,058	0,077	0,077	0,115	0,096	0,085

In Fig. 4 it can be seen that k-NN represents in a better way the distribution of the data in relation to its label and agrees with the confusion matrix performance. For this reason and the easily software implementation, k-NN is chosen as algorithm for this system. As a next step, to determine the best training set, the criterion of CNN (Condensed Nearest Neighborhood) is used. CNN is used as a data editing technique, since it seeks to eliminate prototypes that are in spaces of representation different from the majority of the cases that belong to its same class. The application of this technique, in addition to eliminating the erroneously classified data from the original training set, leads to a better grouping by label in a new set of the training matrix [18]. Specifically, CNN acts in the sense of eliminating data at the borders of each cluster with the aim of smoothing the decision lines. To achieve this goal, it is necessary to normalize the data (scale between 0 to 1). Subsequently, a selection criterion must be defined to choose the appropriate one. To verify the best training set, It tested by a quantitative assessment between the percentage of instances removed with different CNN criteria and the classifier performance. In the Table 3 show the removed instances (RM), classifier performance (CP) and the quantitative equilibrium measure (QEM), between RM and CP.

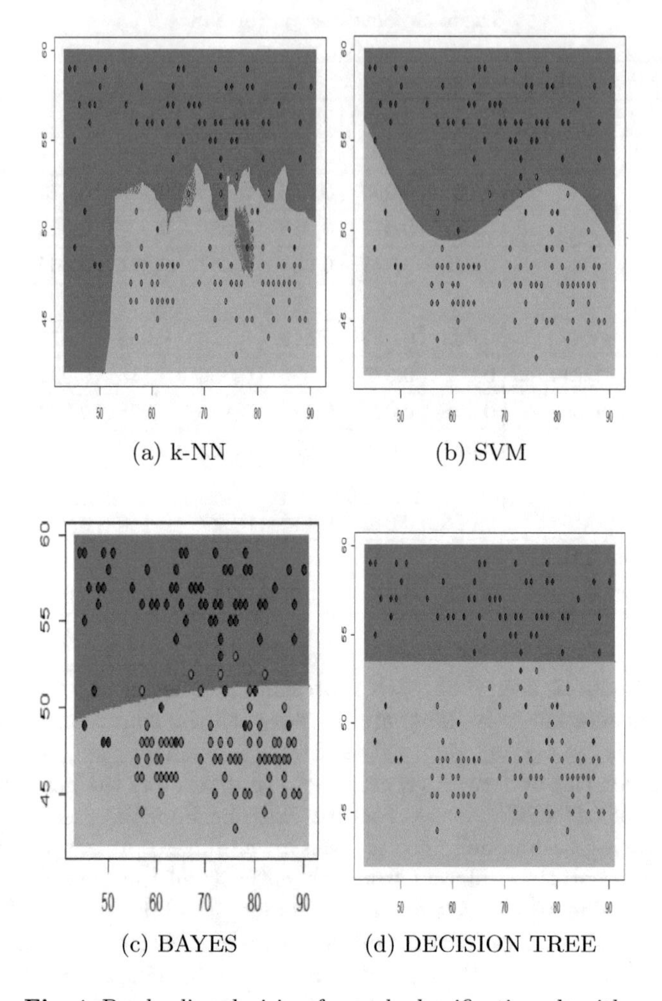

(a) k-NN (b) SVM

(c) BAYES (d) DECISION TREE

Fig. 4. Border line decision for each classification algorithm

3 Results

Once the stage of the development of the electronic system and its data analysis is completed. For the one hand, it proceed to implement the optimal training set for the classifier. It must be taken into consideration that the initial matrix, considered as **A** with ($\mathbf{m}x\mathbf{n}$) dimensions, where $\boldsymbol{m} = 208$ and $\boldsymbol{n} = 2$, and label vector called **C**. The reduced matrix **B** with ($\mathbf{p}x\mathbf{n}$), were $\boldsymbol{p} = 110$ and $\boldsymbol{n}$. This data set was defined by (>0.6) CNN criteria. For the other hand, the classifier chosen was k-NN with $k = 5$ is given by the performance analysis. The k-NN pseudo-code is:

$Start$
$Input : B = \{(B_1, C_1), ..., (B_N, C_N)\}$
$X = (x_1, ...x_n) New \ \ cases \ \ for \ \ clasification$
$For \ \ all \ \ new \ \ objet \ \ classified(x_i, c_i)$
$calculate \ \ d_i = d(X_i, X) \ Order \ \ to \ \ d_i \ (i = 1, ..., N)$
$in \ \ ascending \ \ order$
$Keep \ \ to \ \ K = 5 \ \ cases \ \ B_i^5$
$To \ \ assign \ \ to \ \ x \ \ the \ \ most \ \ frecuently \ \ case \ \ in \ \ B_i^5$
End

Table 3. QEM summary

Label 1: Right Exercise				
Criteria	Removed instances	% RM	C. P.	QEM
<0.2	68	0.80	0.5	0.4
<0.3	60	0.71	0.50%	0.35
<0.4	53	0.63	0.50	0.31
>0.6	31	0.36	0.85	0.31
>0.7	38	0.45%	0.85%	0.38
>.8	23	0.27	0.80	0.21
Label 2: Bad exercise				
<0.2	56	0.66	0.80	0.53
<0.3	48	0.57	0.80	0.45
<0.4	36	0.42	0.80	0.34
>0.6	67	0.79	0.80	0.63
>0.7	62	0.73	0.70	0.51
>0.8	73	0.86	0.70	0.60

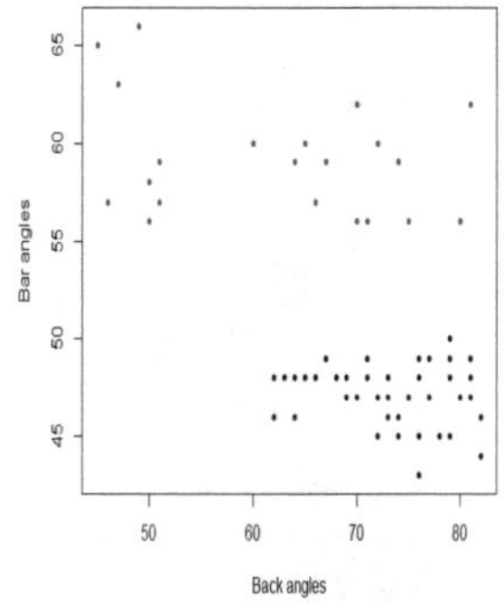

(a) Training Set

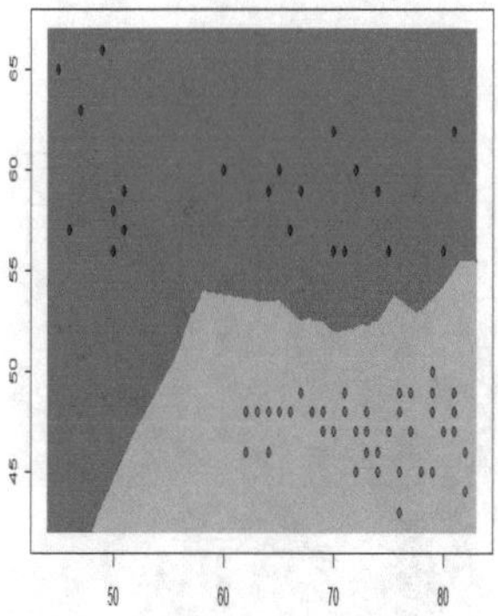

(b) k-NN, k=5

Fig. 5. Border line decision for finally classification algorithm

Figure 5 shows the set of data used in order to enter them into the system memory. In addition, it indicates how the classifier has the decision edge. The same system was put to the test with 52 data between good and bad repetitions, achieving a 96% performance. That is, the system selected has chosen 50 times correctly.

Once the last software tests were done, the algorithm was implemented and the embedded system developed to be used in a gym, the same as shown in Fig. 6.

As a final result, an interface was developed for visualization of embedded system classification. The same one that receives the decision of the node of

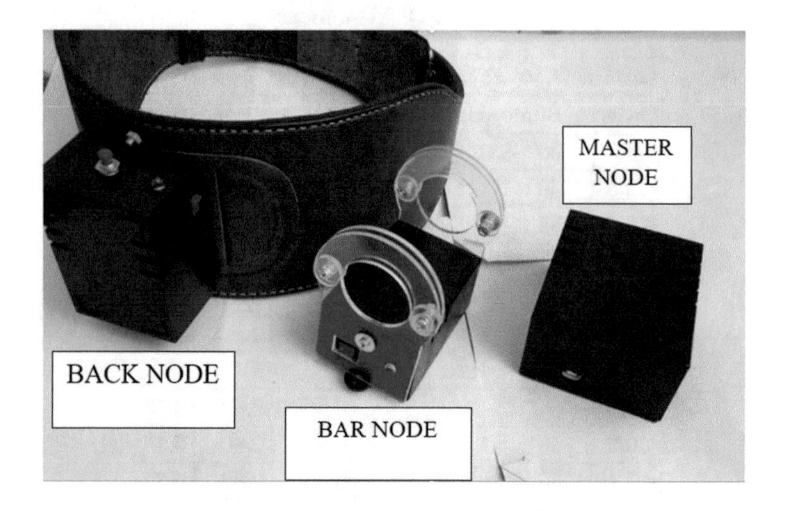

Fig. 6. Embedded system result

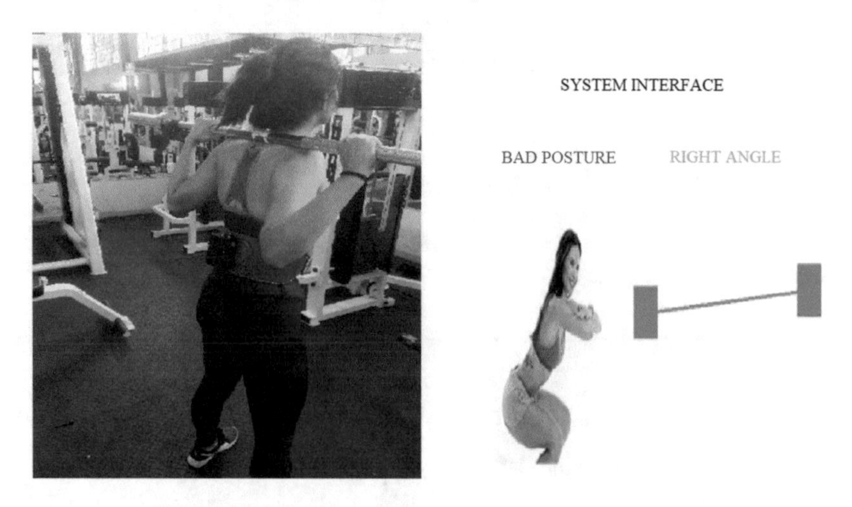

Fig. 7. System test in real conditions by athletic woman

the back that is sent to the node connected to the computer. The interface is developed in Processing and indicates if the position is correct or not and stores the repetitions performed by each user. In Fig. 7 we can observe a athletic woman using the system together with the visualization of the exercise in the computer.

4 Conclusions and Future Work

In bio-mechanic analysis of the back in exercise as the squat allows to have a control of the activity of the athlete to avoid injuries. Because of this, the perform corresponding data analysis, allows an electronic system to make decisions automatically. For this reason, the CNN prototype selection algorithm with the criterion (> 0.6) allowed to reduce the training matrix in order to improve the response time of the system and alert in real time of bad athlete movement. In addition, selecting the classification algorithm appropriately allows us to know which of them can have an adequate decision edge for its operation under normal conditions. In such a way, that the k-NN algorithm could better represent the data set and be the selected one.

The visualization interface allows the athlete and trained to have an agile monitoring tool. With a way to store data and recreate these movements in order to improve the athlete's technique.

As future work, it is expected to increase the performance of the system, improve the interface and perform more exhaustive tests on the operation in real conditions.

Aknowledgements. This work is supported by the "Smart Data Analysis Systems - SDAS" group (http://sdas-group.com).

References

1. Kianifar R, Lee A, Raina S, Kulic D (2017) Automated assessment of dynamic knee valgus and risk of knee injury during the single leg squat. IEEE J Transl Eng Health Med 5:1–13. http://ieeexplore.ieee.org/document/8089739/
2. DiMattia MA, Livengood AL, Uhl TL, Mattacola CG, Malone TR (2005) What are the validity of the single-leg-squat test and its relationship to hip-abduction strength? J Sport Rehabil 14(2):108–123. https://doi.org/10.1123/jsr.14.2.108
3. Crossley KM, Zhang W-J, Schache AG, Bryant A, Cowan SM (2011) Performance on the single-leg squat task indicates hip abductor muscle function. Am J Sports Med 39(4):866–873. https://doi.org/10.1177/0363546510395456
4. Weeks BK, Carty CP, Horan SA (2012) Kinematic predictors of single-leg squat performance: a comparison of experienced physiotherapists and student physiotherapists. BMC Musculoskelet Disord 13(1):207. https://doi.org/10.1186/1471-2474-13-207
5. Ageberg E, Bennell KL, Hunt MA, Simic M, Roos EM, Creaby MW (2010) Validity and inter-rater reliability of medio-lateral knee motion observed during a single-limb mini squat. Musculoskelet Disord 11(1):265. https://doi.org/10.1186/1471-2474-11-265

6. Poulsen DR, James CR (2011) Concurrent validity and reliability of clinical evaluation of the single leg squat. Physiother. Theory Pract 27(8):586–594. https://doi.org/10.3109/09593985.2011.552539
7. Penafiel BF (2015) Biomechanical analysis for different techniques of the full squat. In: 2015 CHILEAN conference on electrical, electronics engineering, information and communication technologies (CHILECON), pp 225–228. IEEE, October 2015. http://ieeexplore.ieee.org/document/7400380/
8. Weeks BK, Carty CP, Horan SA (2015) Effect of sex and fatigue on single leg squat kinematics in healthy young adults. BMC Musculoskelet Disord 16(1):271. https://doi.org/10.1186/s12891-015-0739-3
9. Nakagawa TH, Moriya ÉT, Maciel CD, SerrãO FV (2012) Trunk, pelvis, hip, and knee kinematics, hip strength, and gluteal muscle activation during a single-leg squat in males and females with and without patellofemoral pain syndrome. J Orthop Sport Phys Ther 42(6):491–501. https://doi.org/10.2519/jospt.2012.3987
10. Levinger P, Gilleard W, Coleman C (2007) Femoral medial deviation angle during a one-leg squat test in individuals with patellofemoral pain syndrome. Phys Ther Sport 8(4):163–168. http://linkinghub.elsevier.com/retrieve/pii/S1466853X07000466
11. Padua DA, Marshall SW, Boling MC, Thigpen CA, Garrett WE, Beutler AI (2009) The Landing Error Scoring System (LESS) is a valid and reliable clinical assessment tool of jump-landing biomechanics. Am J Sports Med 37(10):1996–2002. https://doi.org/10.1177/0363546509343200
12. Graci V, Van Dillen LR, Salsich GB (2012) Gender differences in trunk, pelvis and lower limb kinematics during a single leg squat. Gait Posture 36(3):461–466. http://linkinghub.elsevier.com/retrieve/pii/S0966636212001324
13. Kulas AS, Hortobágyi T, DeVita P (2012) Trunk position modulates anterior cruciate ligament forces and strains during a single-leg squat. Clin Biomech 27(1):16–21. http://linkinghub.elsevier.com/retrieve/pii/S0268003311001902
14. Hallgren KA (2012) Computing inter-rater reliability for observational data: an overview and tutorial. Tutor Quant Methods Psychol 8(1):23–34. http://www.tqmp.org/RegularArticles/vol08-1/p023
15. Nunez-Godoy S (2016) Human-sitting-pose detection using data classification and dimensionality reduction. In: 2016 IEEE Ecuador technical chapters meeting (ETCM), pp 1–5. IEEE, October 2016. http://ieeexplore.ieee.org/document/7750822/
16. Pohjalainen J, Räsänen O, Kadioglu S (2015) Feature selection methods and their combinations in high-dimensional classification of speaker likability, intelligibility and personality traits. Comput Speech Lang 29(1):145–171 http://linkinghub.elsevier.com/retrieve/pii/S0885230813001113
17. Rosero-Montalvo P (2017) Prototype reduction algorithms comparison in nearest neighbor classification for sensor data: empirical study. In: 2017 IEEE Second Ecuador Technical Chapters Meeting (ETCM), pp 1–5. IEEE, October 2017. http://ieeexplore.ieee.org/document/8247530/
18. Rosero-Montalvo P, Umaquinga-Criollo A, Flores S, Suarez L, Pijal J, Ponce-Guevara K, Nejer D, Guzman A, Lugo D, Moncayo K (2017) Neighborhood criterion analysis for prototype selection applied in WSN data. In: 2017 International conference on information systems and computer science (INCISCOS). IEEE, November 2017, pp 128–132. http://ieeexplore.ieee.org/document/8328096/
19. Kuncheva LI (1995) Editing for the k-nearest neighbors rule by a genetic algorithm. Pattern Recogn Lett 16(8):809–814. http://linkinghub.elsevier.com/retrieve/pii/016786559500047K

20. Horan SA, Watson, SL, Carty, CP, Sartori M, Weeks BK (2014) Lower-limb kinematics of single-leg squat performance in young adults. Physiotherapy Canada
21. Raïsänen A, Pasanen K, Krosshaug T, Avela J, Perttunen J, Parkkari J (2016) Single-leg squat as a tool to evaluate young athletes' frontal plane knee control. Clin J Sport Med 26(6):478-482
22. IEEE (2011) IEEE 29148-2011 - ISO/IEC/IEEE International Standard - Ingeniería de sistemas y software - Procesos del ciclo de vida - Ingeniería de requisitos. https://standards.ieee.org/findstds/standard/29148-2011.html
23. Rosero-Montalvo P, Peluo-Ordonez D, Godoy P, Ponce K, Rosero E, Vasquez C, Cuzme F, Flores S, Mera ZA (2017) Elderly fall detection using data classification on a portable embedded system. In 2017 IEEE Second Ecuador Technical Chapters Meeting (ETCM). IEEE, October 2017, pp 1–4. http://ieeexplore.ieee.org/document/8247529/

Architecture

Multifunctional Exoskeletal Orthosis for Hand Rehabilitation Based on Virtual Reality

Patricio D. Cartagena[1], José E. Naranjo[1], Lenin F. Saltos[1], Carlos A. Garcia[1], and Marcelo V. Garcia[1,2(✉)]

[1] Universidad Tecnica de Ambato, UTA, 180103 Ambato, Ecuador
{pcartagena6205,jnaranjo0463,lf.saltos,ca.garcia,mv.garcia}@uta.edu.ec
[2] University of Basque Country, UPV/EHU, 48013 Bilbao, Spain
mgarcia294@ehu.eus

Abstract. Within the field of physical rehabilitation in patients with fine motor deficits due to tendon injuries, this article is a novel proposal for the treatment and recovery of hand mobility. The mechanism works within virtual environments designed according to the needs of the beneficiary, through a mechatronic prototype controlled by algorithms based on fuzzy logic, the data sent by the Unity3D graphics engine and bending sensors is verified. The results of this system are focused on the process of digital signals for activation of the force feedback mechanism, this is done through the use of a flexible orthosis that allows the flexion and contraction of the fingers of the hand, thanks to this, excellent control results and an adequate performance in the development and execution of the proposed tasks in the virtual environment are obtained in a way that significantly promotes and improves the quality of life of the user.

Keywords: Virtual reality · Exoskeletal orthosis · Rehabilitation

1 Introduction

The technology of virtual reality has had a significant advance in the last decade, its use extends from the fields of medicine to the training of industrial machinery, among other applications, where specialists in this field are trained by VR systems [1]. The application of Virtual Reality (VR) in the rehabilitation of several deficiencies resulting from injuries belonging to the nervous system [1,2], has assumed an important role in the development of this field. According to WHO data, 15% of the world population live with some type of disability [3]; mostly in the upper limbs, where hands are compromised due to their function in daily life [1,2].

Rehabilitation techniques that allow the improvement of fine motor function allow to compensate or recover the loss of the skill, in addition to preventing or diminishing the wear of these functions [4]. In the process of rehabilitation

M. Botto-Tobar et al. (Eds.): TICEC 2018, AISC 884, pp. 209–221, 2019.
https://doi.org/10.1007/978-3-030-02828-2_16

three important concepts are known: (i) Deterioration, known as the loss of physiological functions or anatomical anomalies [5]; (ii) Disability, impediment to perform an activity considered normal for human beings [6] y (iii) Disadvantage, is the result of deterioration or disability, which constitutes an impediment to the development of the usual activities of each individual taking into account the age, sex and cultural factors [7].

The technical and scientific foundations that support the relationship between virtual reality and physical rehabilitation based on motor learning, are characterized by the use of an orthosis or prosthesis that allows the proper development of inaccurate movements to obtain accurate results in terms of patient rehabilitation. The rehabilitation process requires practice through the use of virtual tools where digitalized environments are developed with exercises and activities for the functioning of physical abilities, perception, attention, reasoning, abstraction, memory and orientation.

The rehabilitation tasks generated in virtual environments unlike conventional motor recovery activities, become more efficient, since they are innovative, personalized and attract the patient's attention. Regarding VR studies in patients with cerebrovascular damage, therapies with an average of three to four intensive sessions a day for a period of two weeks show an accelerated process of fine motor recovery, improving the strength and elasticity [8].

As described above, this article proposes the development of a bilateral fine motor rehabilitation system based on fuzzy logic. For a better control of the patient's progress, an active orthosis is presented as movement support, which generates a feedback of forces while allowing the patient's interaction with a virtual environment. The orthosis makes use of servo motors such as force actuators, flexibility sensors and the Leap Motion^{TM} optoelectronic device that generates virtualization of the hand in the 3D environment. In addition, an application developed in virtual reality using Unity 3D^{TM} graphics engine is presented [9]; The interface allows the patient to visualize the predetermined movements focused on the affected area of the lesion.

This article is divided into 6 sections including the introduction. In Sect. 2, the state of technology is analyzed. Section 3 presents the case study while Sect. 4 explains the implementation proposal. Section 4 details the results obtained and finally in Sect. 5 the conclusions and future research are presented.

2 State of Technology

2.1 Virtual Rehabilitation

Motor rehabilitation in virtual reality environments allows the user to interact in real time in computerized three-dimensional environments through multisensory devices, in order to recover motor mobility; being a functional, useful and motivating choice. The process of designing virtual environments in the field of motor rehabilitation is carried out under the following aspects:

- Selection of the rehabilitation protocol to examine the motion kinematics, according to the needs of the patient.

- Collect information for compliance with protocol objectives.
- Based on numerical descriptions, build geometric figures that give shape to the mechanism.
- Locate the objects in the 3D environment and focus the cameras.
- Convert mathematical information into screen pixels [10].

2.2 Graphical Environment

The graphic environment defines the workspace where environments modeled by collision regions are manipulated in a real environment controlling the necessary procedures for graphic display. The objective is to interact with the surfaces and environments of the interface to generate the rehabilitation of the patient in the affected area. To improve and give more reliability to the proposed rehabilitation system, the needs of the patient are first identified in order to specify an adequate environment that fits the severity of the motor injury [11]. Fig. 1, exemplifies the programmed virtual environment for hand rehabilitation.

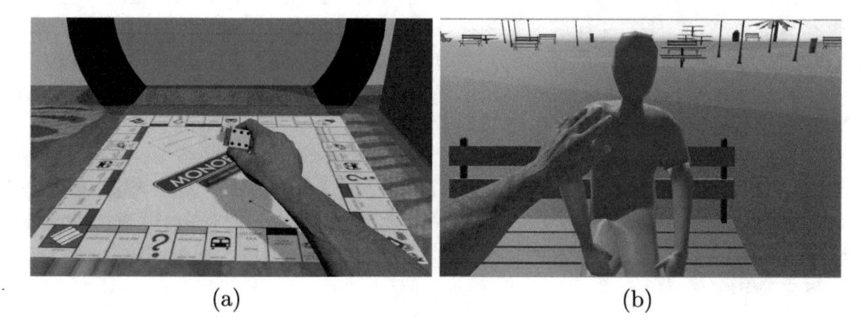

(a) (b)

Fig. 1. Unity 3D™ virtual environment (a) For hand moving rehabilitation (b) For wrist rehabilitation.

2.3 Physical Rehabilitation

It contains aspects of recovery of balance and stabilization of movement. The recovery of balance consists in the acquisition of minor movements through light flexions of joints of the hand, and the stabilization that entails muscle strengthening through activation exercises [12,13]. In a general way of rehabilitation, the exercise that the patient must execute basically will be the stretching and contraction of the fingers as shown in Fig. 2.

The flexion angles for these exercises are gradual, initially the angles of the metacarpo-phalangeal joint (MCF) must be between $0°$ to $30°$, later they reach $50°$ and finally there must be a difference greater than $50°$ until reaching the appropriate $90°$ mobility value. The movements that are executed in the process of rehabilitation of the hand are active, that is to say when the patient executes the movement in an individual way; and passive, when the movement is assisted by external elements [14].

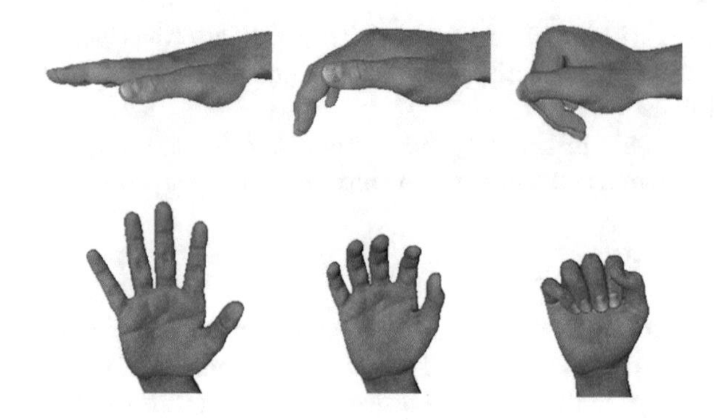

Fig. 2. Fingers mobility.

2.4 Fuzzy Control

It is expressed as control through words, interpreted with common logic and sentences, however, the processes are measured numerically. That is why the variables must be adapted before introducing their status to the controller, this adaptation process is called fusion and consists of giving a degree of membership within possible expressions.

Subsequent to the fusion, the result are linguistic variables, in which typical logical relations are applied, such as: (IF-THEN). Finally, once the controller interprets the relationships, these are translated from linguistic to numerical expressions and through a digital-analog converter the signal to be used by the plant is produced [15].

2.5 Hand Orthosis

To meet the objectives of stabilization, limitation of amplitude, and improvement of movements with the suppression of pain within the rehabilitation, a therapeutic device called orthosis is applied, which also allows the control of possible medical complications. The limb to be treated should be fully covered, however, when it comes to the hand, the palm areas should not be completely covered. The use of flexible thermoplastics is recommended, these provide comfort and facilitate movement, depending on the case to be treated [16]. Figure 3, shows the orthosis used in this research.

3 Case Study

The joint work of virtual rehabilitation and medicine has opened up in the last five years. The use of technological tools that allow greater flexibility has been introduced through devices with a high degree of immersion for its users. The use of the patient's senses plays a fundamental role in the rehabilitation since a feedback of the tasks is performed.

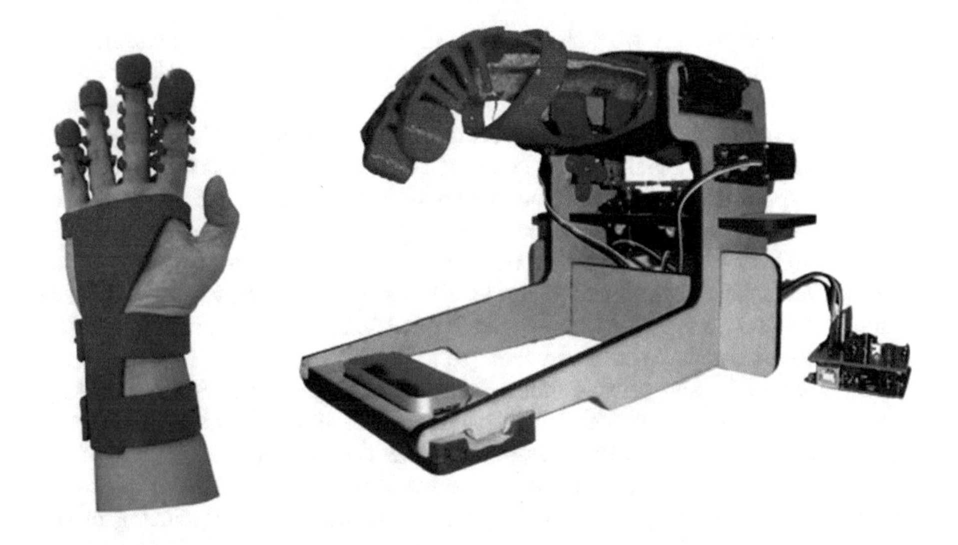

Fig. 3. Flexible material orthosis and armrest.

In fact, if a comparison is made between conventional physical rehabilitation and rehabilitation with the use of virtual reality, it can be seen that VR not only emulates the progress that physical rehabilitation offers but surpasses it. This novel therapy has been used to optimize learning processes or relearning movement patterns in people with cerebral stroke, perceptive-motor deformity, acquired brain injury, Parkinson's disease, orthopedic rehabilitation, balance training, functional activities of training for daily life and tele-rehabilitation.

The case study is applied in Ecuador, where approximately 202,216 people have some type of physical disability; of this total, in the Province of Tungurahua, there are 4,616 people who have a reduction in their physical capacities [17] in which the proposed system or future variations can be applied. That is why this province has been chosen for the application of rehabilitation with virtual reality.

The objective is to use a flexible hand orthosis as a means of fine motor rehabilitation in people with musculoskeletal disorders. Here the Leap Motion sensor is used for detection and virtualization of hand movements. The purpose of this control system is to integrate, through the use of fuzzy logic, personalized virtual systems with external devices that modify the functional-structural aspects of the neuromusculoskeletal system.

4 Implementation Proposal

4.1 Rehabilitation Tasks

The function of the orthosis is similar to a physiotherapist, in which the patient performs rehabilitation exercises through guided manipulation and daily movements of catching and leaving elements of various sizes. The purpose of the orthosis is also to comply with the requirements demanded by the rehabilitation exercises and the ability to adapt to the state of recovery of the patient.

The device must adapt to the degree of deficiency and use passive, assistive and active modes of support. In passive mode the system follows the desired path without considering the user's activity, i.e., the orthosis acts independently with the purpose of allowing the patient to learn how the device works. In the assistive mode, the user tries to move the fingers of the hand and the device helps him to carry out the task by performing assistive forces in the direction of the movement. In the assistive mode, the user tries to move the fingers of the hand and the device helps him to carry out the task by performing assistive forces in the direction of the movement [18].

The protocol for the passive mode considers that the patients for this stage do not have any degree of mobility in their injured limb, 6-minute sessions of programmable movements are necessary to allow the opening and closing of the hand (a rest period of 2 min between sessions is necessary). In assisted mode, the study subject must perform two types of exercises: opening-closing of the hand; and mobility of fingers for 5 min with the purpose that the affected part of the patient acquires both sensitivity and flexibility in their articulations. Finally, in active mode the patient selects the required rehabilitation exercise, for this case opening-closing of hand; once this exercise has been completed (10 min), the patient takes a rest period (3 min) and repeats the rehabilitation process three times or, in turn, can begin to perform exercises with movement of individual fingers.

4.2 Proposed Architecture

The proposed system allows feedback of hand force in patients with no or little mobility in their upper limbs. This system allows a safe, adaptable and relatively economical rehabilitation. The system is related to the user through a bilateral communication:

(i) Initially in the graphic interface designed using the Unity 3D™ framework, the user knows the objectives of the rehabilitation exercise, as well as the position and angle reference of the injured limb.

$$(h_{ref}, \Theta_{ref}) \tag{1}$$

(ii) The patient generates an initial movement in the fulfillment of the preset exercise, in this step the position (h_d) and angle (Θ_d) of the hand is measured.

(iii) Due to the poor motility of the patient's fingers, it is possible that in a certain instance the rehabilitation objective cannot be completed, in this step the flexion of the fingers is measured using the flexion sensor, then the position and angular errors are calculated.

$$(F_e, \Theta_e) \tag{2}$$

These together with the size of the virtual object allow the fuzzy controller to give the control action to the orthosis to progressively fulfill the proposed exercise. The orthosis plays an important role in the system since the forces emitted by the servomotors must be controlled in magnitude and direction to contribute to the direct rehabilitation without collateral damage in the moving limbs. In Fig. 4 the interaction between the patient and the proposed system is described.

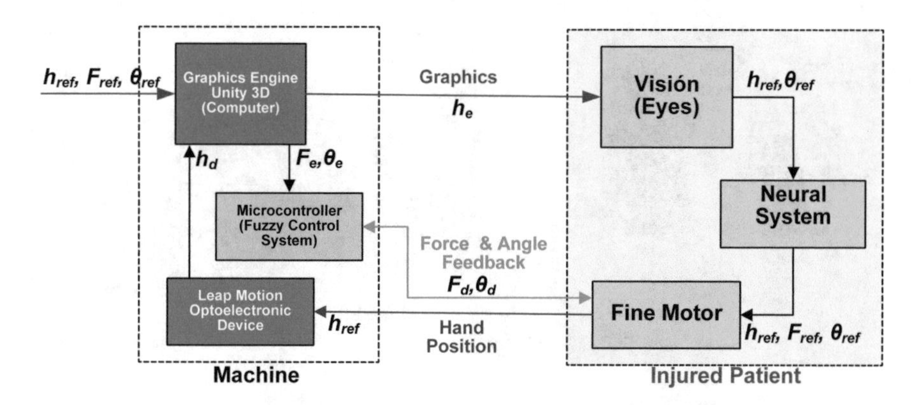

Fig. 4. Rehabilitation Architecture Block diagram.

4.3 Proposed Hardware Platform

The proposed hardware platform is an active hand orthosis controlled by four servo-motors that generate a pulling force on each finger for greater assisted maneuverability in each rehabilitation session. The semi-assisted rehabilitation is effective because the structure is controlled by four actuators responsible for the generation of traction force in each finger. The system also includes an armrest that, in conjunction with the orthosis, provides support for the actuators, which share fluorocarbon filaments, invisible and resistant, useful in the transmission of force. The architecture has the following elements:

- Virtual Interface, through the use of a computer, the exercises created in the Unity3D graphics engine are displayed.
- Control system, composed of the Raspberry Pi 3^{TM} microcontroller, which processes the information sent by Unity3D^{TM} and the bending sensors. This in order to perform the comparison process with the fuzzy sets and thus send the appropriate control signal to the servomotors.
- Leap Motion^{TM}, is an optoelectronic device that captures the movement of the hand, in inertial reference to the three coordinate axes X, Y, Z.
- Actuators, are a group of servomotors HS-311 the same that support a current of 180 mA and can generate a torque of up to 3.0 kg/cm. In addition, a driver for servo motors is incorporated to manipulate the actuators without voltage drops that could compromise the controller.

- Power supply, an external source is used that converts 110 [v] of alternating current (AC) to 5 [v] of direct current (DC) in order to supply the voltage and current necessary for the control system.

For a better understanding Fig. 5 shows the diagram that constitutes the hardware of the system.

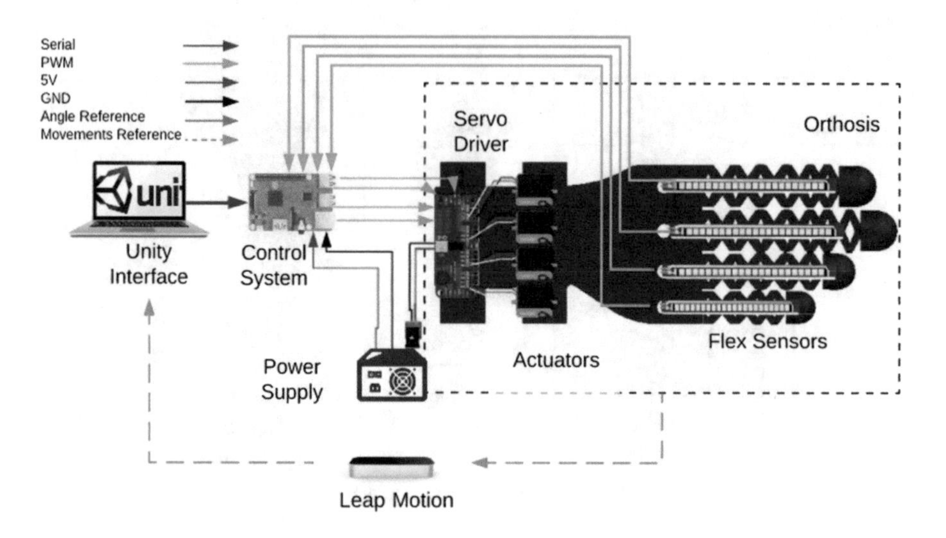

Fig. 5. System hardware construction diagram.

4.4 Fuzzy Control

In this section we present the methodology to obtain the fuzzy control design which will be implemented and programmed in the RPI card. The control of the system is based on knowing the transfer function of the actuators. A unipolar rectangular signal is applied to the servomotor and recorded by means of a transducer, as well as a data acquisition device. The information obtained by the acquisition system is interpreted by the Matlab^{TM} software (Control System Toolbox) in order to generate the third-degree transference model. See Eq. (3).

$$F_{ref}(s) = \frac{\Theta(s)}{V_a(s)} = \frac{19630}{s^3 + 198s^2 + 6280s} \tag{3}$$

Once the transfer function (3) has been obtained, a fuzzy logic controller (FLC) is designed for a higher order system. Prior to this, the Fuzzy Inference System (FIS) is built using Matlab's Logic Fuzzy Design software^{TM}, which shows the system's response stability through graphics. The basic structure of an FLC consists of three stages that are fusion, fuzzy inference and defuzzification

The fusion stage consists in assigning data with a certain degree of membership to fuzzy sets, it is essential to know the membership functions of the system for which the application of two input membership functions and one exit function is necessary:

- *"Size"* entry membership function, generated by values emitted due to the interaction among virtual objects, it obtains its place when the digital image of the hand comes in contact with virtual objects of different sizes, sending this information to the controller.
- *"Angle"* entry membership function, is generated by the values resulting from the analog-to-digital conversion of the bending transducer which fulfills the function of indicating the angular position of the fingers.
- *Output function "degrees"* consisting of the values included in the servo motor operating range (0° to 200°).

The final stage occurs in the decision making of the system. These are obtained by the centroid method, which is simplified by solving Eq. (4).

$$Cog = \frac{\int_a^b F(x)xdx}{\int_a^b F(x)x} \tag{4}$$

Through the Center of gravity (Cog) of the figures formed by the fuzzy sets, the system begins to make decisions in compliance with the control rules.

4.5 Virtual Environment

The design of the rehabilitation environment is developed by the Unity3D™ graphic engine, where the control scripts for each virtual device are programmed, so that they respond in accordance with the rehabilitation movements performed by the patient.

The proposed stages for the virtual environment are: Input Peripherals allow control of the closed loop between the patient and the virtual environment by means of movement in each exercise, the Leap Motion device is used to sense the vectorized position of the hand either with or without the use of the orthosis. Furthermore, there is an option to use the Head Mounted Display (HMD), Oculus Rift, device that allows the full immersion of the patient in a virtual reality environment. The next stage is Interface module that develops virtual environments so that the patient can interact with objects of different sizes and so the system can adapt to various levels of rehabilitation. It is important to maintain motivation in the patient, so virtual environments based on the performance of daily activities and entertainment are proposed. Finally, the Output Peripherals module requires communication ports and advanced programming to generate a software link between the actuators of the orthosis and the virtual environment, as well as the graphic response obtained with the application of an HMD.

The Leap Motion device allows interaction with the virtual environment, here the strategic points of the hand are digitized and verified when carrying out activities of rehabilitation. In Fig. 6 the capture of the strategic points of the hand by the Leap Motion sensor is observed in full execution of rehabilitation tasks.

Through this research, the development of virtual environments Rehabilitation Gaming System (RGS) is proposed, these provide the rehabilitation of

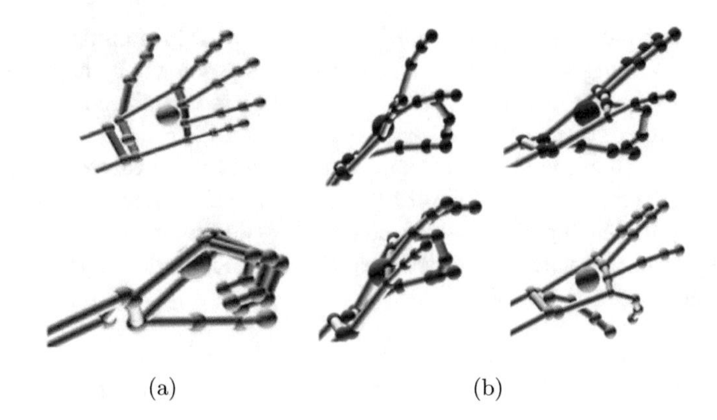

Fig. 6. Hand movements virtualization in execution of rehabilitation exercises. (a) Opening-closing of the hand (b) Individual movement of the fingers.

fine motor function and combine the execution of physical movements and the observation of the actions taken by virtual members. Environments have been designed with different levels of rehabilitation according to the needs of the patient. Figure 7, shows examples of RGS environments.

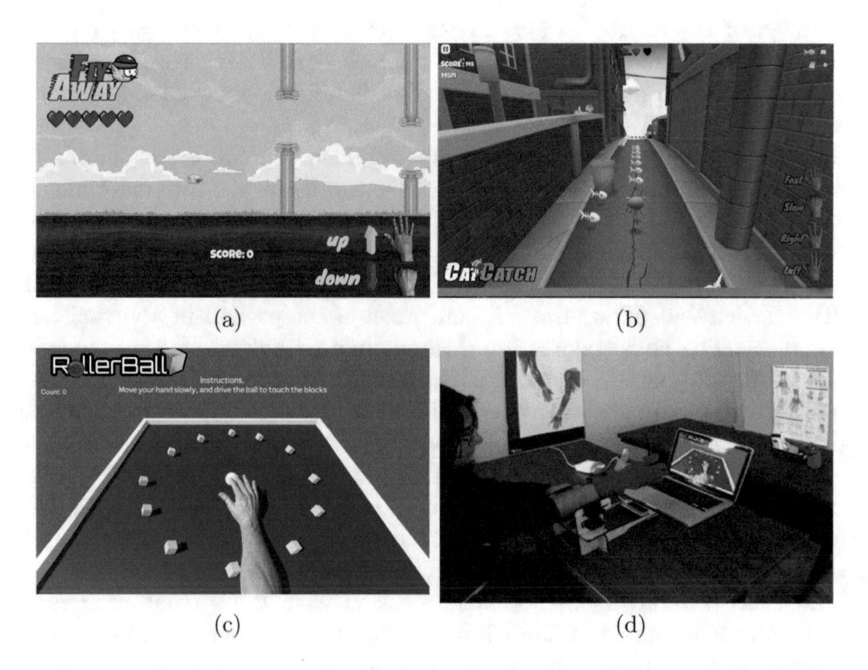

Fig. 7. RGS environments for individual and collective rehabilitation of the fingers. (a) Opening-closing hand Rehabilitation Task 1. (b) Opening-closing hand Rehabilitation Task 2. (c) Individual movement of the fingers Task 1. (d) Hand Orthosis real use

4.6 Experimental Analysis and Results

The proposed tasks to improve the fine motor skills of the subjects under study cover a wide range of options. This is due to the motor difficulty of each patient that can be divided into three groups: mild, medium or chronic. The rehabilitation task that most closely matches the injury can be selected.

The prototype is a significant support for the recovery of muscle memory. To verify the functioning of the system, the orthosis is applied without the help of the control system in order to visualize the mobility of a patient with mild injuries of the hand. In the graphic engine, when a small object is visualized, the patient needs to execute a 90° movement to complete the exercise, however, due to the low mobility of the injured limb, the task cannot be completed, generating an error between the desired position and the rehabilitation objective as shown in Fig. 8.

Subsequently, the patient performs the same rehabilitation exercise, but this time supported by the control system. As shown in Fig. 9. there is a point at

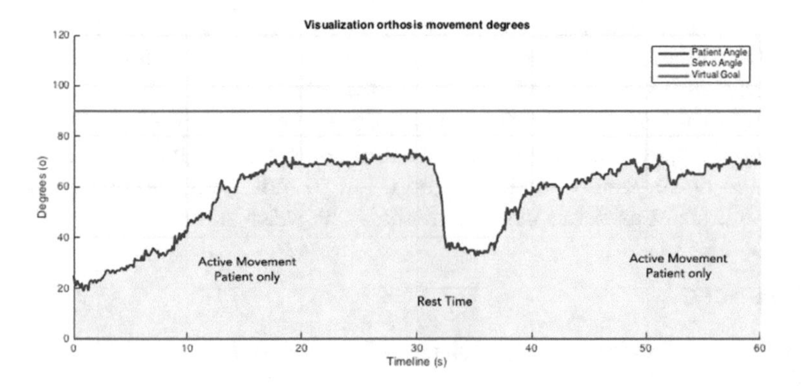

Fig. 8. Result of the exercise without control system.

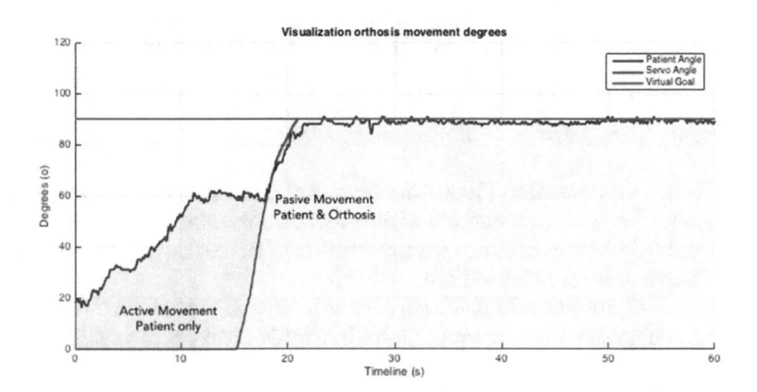

Fig. 9. Result of the exercise with fuzzy control system.

which the patient fulfills his goal through the actuators implemented in the orthosis, i.e., the control system detects when the user is unable to move forward with the exercise, activating the actuators with the necessary force feedback to complete the established protocol.

5 Conclusions and Ongoing Work

Fine motor rehabilitation in virtual environments through force feedback significantly improves both the physical and emotional function of patients, it motivates them to continue with the therapy and improve their motor skills. It is important to mention that depending on the degree of disability that the patient presents, at least two daily rehabilitation processes will be required for three weeks. Unlike the conventional method of motor rehabilitation, force feedback is a procedure that generates several possibilities in the field of physical rehabilitation, since it immerses the patient in a virtual environment that adapts to the user, as well as being dynamic and effective in the rehabilitation process.

Through an exoskeleton, the user is allowed to perform several exercises that improve the active mobility of the hand. The orthosis has a flexible structural design that adapts to the size of the affected extremity of the patient. Fuzzy logic works with systems with several inputs and outputs, it also adapts to the variation of the angles of the fingers and the size of the virtual objects.

For future research, it is proposed to improve the visual interface through the application of augmented reality, in addition to modifying the orthosis so that the rehabilitation exercises can also include the wrist and arm of the patient.

References

1. Holden MK (2005) Virtual environments for motor rehabilitation: review. Cyberpsychol Behav 8:187–211
2. Rose FD, Brooks BM, Rizzo AA (2005) Virtual reality in brain damage rehabilitation: review. Cyberpsychol Behav 8:241–262
3. Mundial B (2011) Informe mundial sobre la discapacidad 2011
4. Song Z, Guo S, Yazid M (2011) Development of a potential system for upper limb rehabilitation training based on virtual reality. Presented at the May 2011
5. Petersen RC, Smith GE, Waring SC, Ivnik RJ, Tangalos EG, Kokmen E (1999) Mild cognitive impairment: clinical characterization and outcome. Arch Neurol 56:303
6. World Health Organization (1980) International classification of impairments, disabilities, and handicaps: a manual of classification relating to the consequences of disease, published in accordance with resolution WHA29. In: 35 of the twenty-ninth World Health Assembly, May 1976
7. van Swieten JC, Koudstaal PJ, Visser MC, Schouten HJ, van Gijn J (1988) Interobserver agreement for the assessment of handicap in stroke patients. Stroke 19:604–607
8. Jack D, Boian R, Merians AS, Tremaine M, Burdea GC, Adamovich SV, Recce M, Poizner H (2001) Virtual reality-enhanced stroke rehabilitation. IEEE Trans Neural Syst Rehabil Eng 9:308–318

9. Kim SL, Suk HJ, Kang JH, Jung JM, Laine TH, Westlin J (2014) Using unity 3D to facilitate mobile augmented reality game development. Presented at the 2014 IEEE World Forum on Internet of Things (WF-IoT)
10. Vallejos D, García J, Muñoz E, Flórez J. Entorno Gráfico de un Entrenador Virtual de Prótesis de Mano. Artíc. Investig. Univ. Cauca Recuperado. www.unicauca.edu.co/ai/publicaciones/VallejosGarcia2009.pdf
11. Matos N, Santos A, Vasconcelos A (2015) A virtual rehabilitation solution using multiple sensors. In: Fardoun HM, Penichet VMR, Alghazzawi DM (eds) ICTs for improving patients rehabilitation research techniques. Springer, Heidelberg, pp 143–154
12. Brunon-Martinez A, Romain M, Roux J-L (2006) Rehabilitación de las lesiones tendinosas traumáticas de la mano. EMC - Kinesiterapia - Med Física 27:1–21
13. Terrade P, Ovieve J-M, Chapin-Bouscarat B (2010) Rehabilitación de las lesiones osteoligamentosas de los dedos de la mano. EMC - Kinesiterapia - Med Física 31:1–17
14. Florez CAC, Montanez JAM, Moreno RJ (2013) Design and construction of a prototype rehabilitation machine to hand and wrist. Presented at the October 2013
15. Kouro S, Musalem R (2002) Control mediante lógica difusa. Téc Mod Autom 7:1–7
16. Paysant J, Foisneau-Lottin A, Gable C, Gavillot-Boulangé C, Galas J-M, Hullar M, Kwiatek H, Lechaudel C, Pétry D, André J-M (2007) Ortesis de la mano. EMC - Kinesiterapia - Med Física 28:1–15
17. Vélez DA, Coello HA (2017) Discapacidad: Un reto para la inclusión participativa y la igualdad. Dominio Las Cienc 4:16
18. Mancisidor A, Zubizarreta A, Cabanes I, Bengoa P, Hyung Jung J (2018) Dispositivo Robótico Multifuncional para la Rehabilitación de las Extremidades Superiores. Rev Iberoam Automática E Informática Ind 15:180

Intelligent Territory Management

Subregion Districting to Optimize the Municipal Solid Waste Collection Network: A Case Study

Israel D. Herrera-Granda[1], Juan C. León-Jácome[1], Leandro L. Lorente-Leyva[1(✉)], Fausto Lucano-Chávez[1], Yakcleem Montero-Santos[1], Winston G. Oviedo-Pantoja[1], and Christian S. Díaz-Cajas[2]

[1] Facultad de Ingeniería en Ciencias Aplicadas, Universidad Técnica del Norte, Av. 17 de Julio, 5-21, y Gral. José María Cordova, Ibarra, Ecuador
{idherrera, jcleonj, llorente, fblucanoc, ymontero, wgoviedo}@utn.edu.ec

[2] Department of Mechanical and Aerospace Engineering (DIMEAS), Politecnico di Torino, Corso Duca degli Abruzzi 24, 10129 Turin, Italy
christiansantiago.diazcajas@studenti.polito.it

Abstract. This paper proposes the implementation of a subregion district model to optimize the collection of solid waste generated in the residential and commercial zones of the Ibarra city in Ecuador. The work begins with a review of cases and methodologies previously applied. Later, the initial condition was determined and the optimum number of districts or sub-areas in which the zones should be divided, starting from the model proposed by the Pan American Health Organization, The City Development and Zoning Plan, and some guidelines given by the Department of Social Development of Mexico in the year 1997. Furthermore, the homogeneous distribution of the districts on the city was also carried out taking into account of factors such as: the components of the collection network, area of each district, current and available road network, surface slopes, waste production rates, collection frequency, and optimal vehicle fleet dedicated to harvesting operations. As support tools for the analyses carried out QGIS 2.18.2, MS-Excel, Global Mapper and LingoV17 were used. Finally, it proposes an optimal distribution of the districts in which the waste collection crews would operate in the city.

Keywords: Optimization · Subregion districting · GIS · Municipal solid waste Collection network · Logistics

1 Introduction

Globally, there is a great interest in developing a holistic approach to the integral management of municipal solid waste (MSW), which considering its entire life cycle can prevent and minimize its waste, in addition to collaborating with the sustainable populations development and contribute to climate change mitigation. According to this approach, MSW's harvesting operations are extremely important to ensure the

M. Botto-Tobar et al. (Eds.): TICEC 2018, AISC 884, pp. 225–237, 2019.
https://doi.org/10.1007/978-3-030-02828-2_17

proper sanitary management of a population and therefore in an optimal and sustainable manner should be carried out [1, 2].

Several studies have found that the use of Municipal Solid Waste Transfer Stations minimizes the total distances traveled by the fleet of collecting trucks, especially in cities with more than one million inhabitants. The countries of Latin America and the Caribbean have begun to implement the transfer stations, especially in cities with a larger quantity of inhabitants; however, this resource still needs to be exploited mostly in developing countries [1, 3, 4].

Among the most notable techniques to improve, the management of the collection networks is the subregion districting [5]. One of the first research was carried out in the United States [6], which using a heuristic algorithm, the optimal districts and circuits allocated for truck rides in a network of streets without consider their addresses was simultaneously determined [7, 8]. That allowed to minimize the costs caused by traveling a street more times than it would need to collect the MSW associated with the street, in addition, the sum of the loads of RSM within the area allocated to each truck should not be exceed the truck load capacity.

By the year 1980, the World Health Organization (WHO) published a manual for the collection routes of MSW design [9], in which the parameters of the model were: the area, population density, and waste generation per capita (PPC) in each zone to be served. It is also considered characteristics of the collection equipment, waste density in the collector truck, frequency of harvesting desired in the area, number of feasible trips to be carried out by truck during the normal working day. As well as the total population, total area, population density, available collection equipment, collection frequency, journeys in working hours and working days per week. As model output of the model was obtained: the optimal number of districts or sub-sectors in which the commercial and residential zones should be divided and the number of districts per truck.

By the year 1997, the Secretariat of Social Development of Mexico (SEDESOL) published a handbook with the guidelines for designing optimal routes for MSW collection. Which includes guidelines for the subregion districting design, and some mathematical models based on linear programming for the determination of the optimal vehicle fleet of collection in a village, taking into account parameters such as collection frequencies, daily travel numbers available per truck, costs per hour of collection, price, efficiency and collecting truck capacity. It is also considered waste generation per capita, total population, daily collection budget, and total budget for the acquisition of new trucks [5].

Yueh et al. [10] proposes a mixed-integer programming model (MIP) which integrates factors such as: compactness, integrity of the road network, cost of collection and regional proximity to treat the subregion districting problem in the MSW collection system [10]. This model puersues to balance the route density index (RDI) between the districts proposed, which is equivalent to the relation between the total amount of MSW to the total length of the route within the district. The proposed model is applied in two areas of the Taiwan city in which, in the first area is produced 51 ton/day while in the second one is produced 130 ton/day; therefore, the model divides into two and four districts respectively. Similarly, the author's showed important data that may be relevant in the elaboration of the subregion districting. In addition, the application is usually carried out manually as in USA cities, it means they are drawn by hand; however, such implementations have been successful in cities such as Phoenix and Charlotte; the first one consists of around 60 000 inhabitants and was divided into 6 districts.

In the second manual published by SEDESOL in the year 2009 related to the thematic of the subregion districting, the collector truck aspects are considered as the number of trips, total distance traveled by route, number of users to be attended, total time and total waste collected. The above allows to calculate the optimal number of routes and vehicles needed to attend a particular village [11].

By the year 2012 in Venezuela [12], it shows model revisions for macro and micro routing of the MSW, however, the content in general is very similar to the model proposed by SEDESOL [11].

In Colombia by the year 2015, despite of the absence of a pre-established methodology, a model to optimize the routing of the MSW collector vehicles in Zarzal, Valle del Cauca was designed. During the first phase of this model [13], the authors divide the collection area into homogeneous zones by applying the sweep method, taking into account the characteristics of residues generation, topography, waste type as well as the limits were delimited by geographical accidents or urban installations, using the municipality of Zarzal map, obtaining reductions of about 18% in displacement distances.

By the year 2016, in Mexico [14], García et al., proposed a new model of Integer Linear Programming (IP) based on the Eulerian path [15] for the subregion districting problem, later they combined it with a micro-routing algorithm by arcs within each district. As a result, the model provides relatively homogeneous districts with a certain tolerance, ensuring the balance of workload and compactness between them. This can be applied in postal delivery operations, electrical consumption readings, winter deliveries, road maintenance, and MSW collection. The model successfully worked in networks which contains until 401 nodes and 764 arcs.

2 Materials and Methods

By means of the study of the collection routes of MSW that are carried out in the residential and commercial zones of the Ibarra city, it was possible to demonstrate the lack of definition of the districts for the MSW collection system, and even more that these are currently overlapped, as shown in Fig. 1.

For the definition of an optimal MSW collection network, it is imperative to know the features of the collection points in which information has been compiled together with the companies related to the waste management in the Ibarra city and they raised the internal logistics processes of the MSW collection trucks.

Then, data, characteristics and parameters were collected in conjunction with related institutions and the Territorial Ordering Plan of the Ibarra city 2015-2023 (PDOT) [16]. Afterwards, the modelling calculations were carried out which are used to elaborate the improvement proposal, as shown in Fig. 2.

Fig. 1. Urban routes of MSW collection in the Ibarra City, current situation.

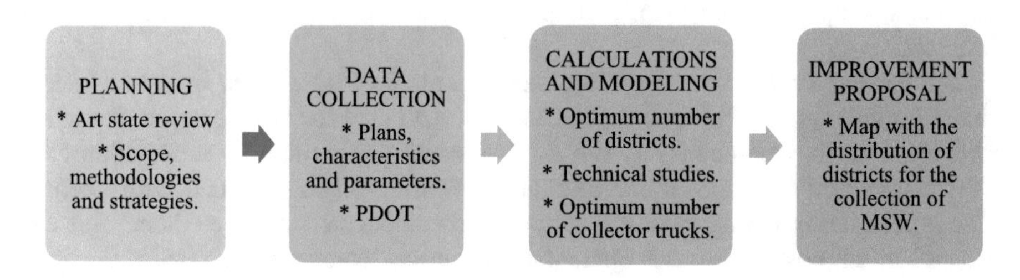

Fig. 2. Procedure proposed to define the MSW collection routes.

After the bibliographical review, the proposal of a subregión districting is made by calculating the optimal number of routes by means of the WHO model [9], including also the suggested elements for the collection network of a city; it means, the implementation of an transfer station and a final landfill or disposal [3, 4], as shown in Fig. 3.

2.1 Input Data

In order to use QGIS 2.18.2 it is required to introduce an initial database, which it will be the starting point for the tool. It will carry out several analyses that will serve as aid in the decision making for the optimization of the MSW collection system.

Within the field research and data collection, we worked with the Public Company of Housing and Industrialization of Solid, Arid and Stony Waste (VIRSAP-EP), warehouse public workers of the municipality, the Environmental Management Department of the municipality and the PODT, which provided the official data.

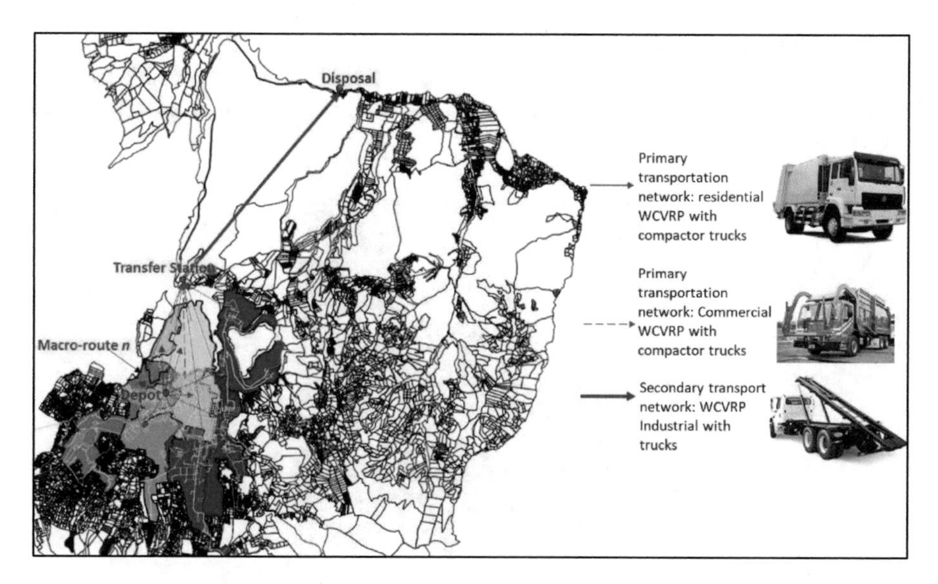

Fig. 3. Concept of optimal configuration of the MSW Collection Network.

2.2 Information Processing

First, the road structure, the property plane, and the truck parking lot of MSW were stablished. All of them were located within the same plane with overlapping and georeferenced layers. Therefore, it was possible to establish a geographic database to continue adding information.

It is necessary to know the locations of houses and buildings in the city with nodes, that is way it was necessary to install the complement of Google Earth [17], in order to obtain satellite photos that allow going placing a node in each one of the houses and buildings. The urban parishes of the Ibarra city are: Alpachaca, San Francisco, El Sagrario, Priorato and Caranqui [18].

Definition of the Commercial Zones

According to the PODT [18], a series of calculations were carried out based on the proportionality of the area of the zones whose number of inhabitants available was known. It could be established that the commercial zone comprises 3.64 km^2 and it contains about 29,587 inhabitants, which gives around 8123.94 inhabitants/km^2.

Grow Population forecast of Inhabitants in City Residential Parishes

From the data obtained in the population and housing census executed in the year 2010 by the INEC [19] and considering the growth rate of 2.02% and 2.5% for the urban sectors and San Antonio respectively according to the current PODT [16], in that way, the population could be projected in the residential sectors.

Delimitation of Zones for the Subregión Districting

After the lifting of the collection processes and meetings with institutions related to the MSW management, the commercial and residential areas of the city were limited to carry out the subregión districting.

Surface Characteristics of the Ibarra City

By means of the Global Mapper software [20] and QGIS [21], it was possible to obtain a layer with information about the area relief included, what is represented as level curves, as shown in Fig. 4.

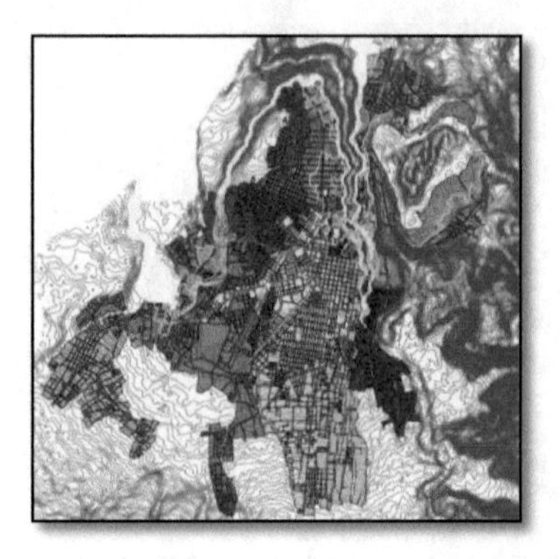

Fig. 4. Residential, commercial zone and level curves in the Ibarra city.

2.3 Determination of the Optimal Number of Districts

From the bibliography analysis, it was decided to apply the model suggested by the WHO [9], since the requirements observed in the local environment were satisfied. The specific data to determine the optimal number of subzones are shown in Table 1.

Table 1. Characteristics of the subzones of the city of Ibarra.

Factors	Value	Units
Residential zone area	31,97	km^2
Commertial zone area	3,64	km^2
Residential and commercial areas	**35,62**	$\mathbf{km^2}$
PPC (waste production per person)	0,685	kg/inhabitant-day
Collector Truck Capacity (Kc)	10	Ton
Collection frequency (Fr)	2–7	Days/week
Residential Area Density	4282,80	Inhabitant/km^2
Commercial Zone Density	8123,94	Inhabitants/km^2
Number of trips by district (Vss)	2–3	Trips/district-First collection
Working days	3–7	Days/week

Calculation of the Optimal Amount of Commercial Districts

$$\text{Kss} = area * density * PPC * \frac{Work\ day\ at\ week}{Week\ frequency} \tag{1}$$

Where;

Kss	: Load per district
Area	: Commercial area
Density	: Commercial Zone Density
PPC	: Waste production per person

$$\text{Kss} = 3,64 * 8123,94 * 0,685 * \frac{7}{7} = \mathbf{20,26}\ \boldsymbol{ton/day}$$

$$\text{Nsz} = (kss * Kc)/Vss \tag{2}$$

Where;

Nsz	: Number of Districts
Kc	: Truck Capacity
Vss	: Number of trips by district

$$\text{Nsz} = \frac{20,26 * 10}{2} = \mathbf{1,01\ districts}$$

Calculation of the Optimal Amount of Residential Districts

$$\text{Kss} = \text{ar}ea * density * PPC * Fr \tag{3}$$

Where;

Fr	: Frecuence Recollection
Area	: Residential zone area
Density	: Residential zone Density
PPC	: Waste production per person

$$\begin{aligned}\text{Kss} &= 31,97 * 4282,80 * 0,685 * 3 \\ &= \mathbf{281,37}\frac{\boldsymbol{ton}}{\boldsymbol{subzone - first\ recollection}}\end{aligned}$$

$$\text{Nsz} = (kss/Kc)/Vss \tag{4}$$

Where;

Kc	: Truck Capacity
Vss	: Trips by district -First collection

$$\text{Nsz} = \frac{281,37/10}{3[5]} = \mathbf{9,38\ districts}$$

2.4 Design of the Districts of Collection

It could be deduced that, although there are currently several heuristic models for the determination of the optimal form of collection districts, there is no consensus in a pre-established methodology for the subregion districting still. That is the reason that the manual method remains a solid approach that has shown good results [10, 13].

In order to develop the collection districts, the guidelines of SEDESOL [5] and WHO [9] were followed; in addition, some factors were considered, such as:

1. Production of MSW contrasting commercial and residential areas.
2. Existing network road.
3. Areas of the districts.
4. Existence of houses or buildings that generate MSW.
5. Area relief through the level curves.

In the Fig. 5, the subregion districting proposal for the residential and commercial zones of the Ibarra city is shown. It implies the implementation of nine residential districts and a single commercial district.

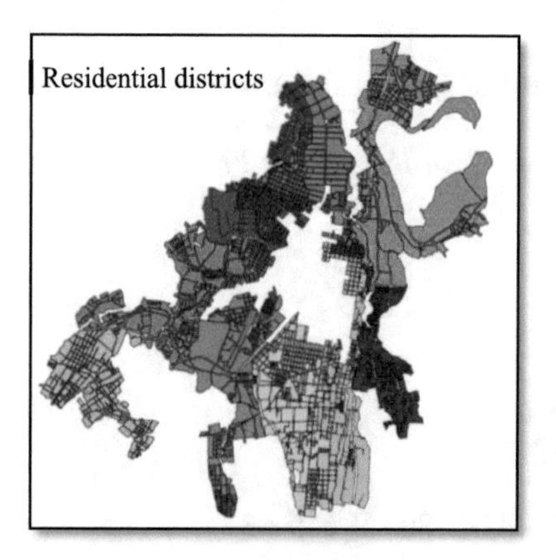

Fig. 5. Proposal of nine residential districts and one commercial district.

2.5 Model for the Calculation of Optimal Number of Trucks Needed

Decision Variables

Xi = Number of trucks type i required
Y_i = Number of vehicles type i to acquire

Parameters

Ci = Cost per hour truck type i
n = Total number of trucks type i employed
P = Total population of the city
PPC = Daily MSW generation per capita

B = Generation of MSW from other sources
S = Daily MSW generation in the city; S = P * PPC + B
ki = Daily cost per employee for each type of truck employed
W = Daily maximum operating cost of the municipality budgeted
P_i = Purchase price of a new vehicle type *i*
l = Annual budget for acquisition of new equipment
a_i = Current number of trucks type *i*

Objective function

$$Min\ Z = \sum_{i=1}^{n} K_i * X_i \leq W$$

Restriction 1

The first restriction will vary depending on the desired collection frequency, as shown in Table 2.

Table 2. Quantity of waste to be collected based on the collection frequency [22].

Waste inventory			Waste left in the city (Sunday)	Maximum waste to be collected	Extraordinary days of collection	Restriction
Collection Frequency	Waste left	Waste that is collected				
Diary	0	S	S	2S	Monday	$\sum_{i=1}^{n} w_i x_i N_i \eta_i \geq 2S$
Every third day	$\frac{S}{2}$	S	$\frac{3S}{2}$	$\frac{3S}{2}$	Monday, Tuesday	$\sum_{i=1}^{n} w_i x_i N_i \eta_i \geq \frac{3S}{2}$
Twice a week	S	S	2S	$\frac{4S}{3}$	Monday, Tuesday and Wednesday	$\sum_{i=1}^{n} w_i x_i N_i \eta_i \geq \frac{4S}{3}$

Restriction 2

$$\sum_{i=1}^{n} K_i * X_i \leq W$$

Restriction 3

$$X_i * Y_i = a_i; \forall i = 1, 2, \ldots, n$$

Restriction 4

$$P_i * Y_i = l$$

Positivity restrictions

$$X_i, Y_i \geq 0$$

Model Applied

The calculation of the number of trucks needed for the waste collection considering as x_1 to the T370 Kenworth truck of 16yd^3, and as x_2 to the van truck of 10yd^3 is shown below. For this, three scenarios were taken into account: in the first, a collection frequency of twice a week, so that the maximum inventory in the first collection would be 4/3*S. In the second, a collection frequency every third day, then the maximum inventory in the first collection would be 3/2*S. In addition, in the third scenario, daily, causing the maximum inventory in the first collection to be 2*S. It is important to mention that for this calculation; only the residential parishes of Ibarra (Alpachaca, Priorato, Sagrario, San Francisco, Caranqui, and San Antonio) were considered; as shown in Table 3.

Table 3. Parameters of the model [22].

Parameter	Description	Value	Unit
P	Total population	166528	inhabitants
PPC	Daily production per capita	0,685	kg/inhabitant-day
B	Other sources generation	21588,78	kg/day
S = P*PPC + B	Daily waste generation	135660,46	kg/day
4/3*S	Collection frequency twice a week	180880,61	kg/day
3/2*S	Collection frequency every third day	203490,69	kg/day
2*S	Daily collection frequency	271320,92	kg/day
W_1	Type 1 truck capacity in kg	10000	kg
N_1	Number of trips per day of the type 1 truck	2	daily trips
n_1	Type 1 truck filling efficiency	0,95	%
$x_1*(W_1*N_1*n_1)$	Type 1 truck values in restriction 1	19000	kg/day
W_2	Type 2 truck capacity in kg	6000	kg
N_2	Number of trips per day of the type 2 truck	2	daily trips
n_2	Type 2 truck filling efficiency	0,7	%
$x_2*(W_2*N_2*n_2)$	Type 2 truck values in restriction 1	8400	kg/day

3 Results and Discussion

This paper proposes the division of the residential areas homogeneously and indicates the measurements made on the proposed subzones which demonstrates the homogenization between the as well as implying this a balancing in the workload for the MSW collection crews, The evaluation of the model is shown in Tables 4 and 5.

Table 4. Characteristics of the proposed districts.

Zone	Nodes	People for each zone	Estimated garbage by district (Ton/day)
Residential district 1	3172	15296,70	10,48
Residential district 2	3160	15238,83	10,44
Residential district 3	3167	15272,59	10,46
Residential district 4	3166	15267,76	10,46
Residential district 5	3178	15325,63	10,50
Residential district 6	3162	15248,48	10,45
Residential district 7	3176	15315,99	10,49
Residential district 8	3161	15243,65	10,44
Residential district 9	3123	15060,40	10,32
Total	28465	137.270,03	94,03

Table 5. Specifications of residential and commercial zones.

Specifications	Residential Zone	Commercial Zone
Inhabitants	137.270,03	29.247,94
PPC	0,685	0,685
kg/day	94.029,97	20.034,84
ton/day	94,03	20,03
ton/truck	10	10
trucks/day	9,40	2,00
Districts number	9,38	1,01
Inhabitants/district	14.633,70	–
kg/district-day	10.024,08	–

During the analyses of the MSW collection network with the use of the QGIS software, it was possible to realize the existence of a particular route of lesser distance in the voyages connecting the transfer station (which will also function as the MSW classification plant) and the final landfill or disposal. However, this route does not guarantee the conditions for running the operations safely.

To calculate the optimal number of trucks needed in waste collection operations in the Canton Ibarra, the model suggested by the SEDESOL was used, and it was programmed in the software Lingo V.17.0. It was performed in three different scenarios: considering a collection frequency twice a week, every third day and daily. As shown in Fig. 6.

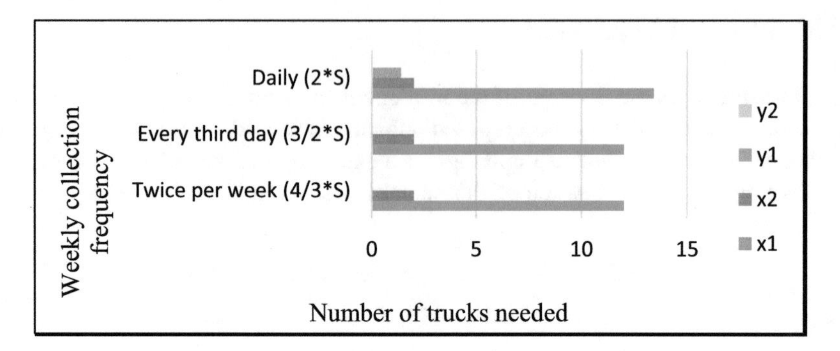

Fig. 6. Optimum number of trucks according to collection frequency.

4 Conclusions

This work proposes a model of territorial distribution for the optimization of the distribution of the routes of the collector trucks of MSW in the commercial and residential sectors of the Canton Ibarra. It is based on population data and waste generation as well.

Once the proposal of the georouting is approved, it is imperative a socialization campaign to the Canton Ibarra's inhabitants before the implementation of the proposal; therefore, it is essential the commitment and work of the institutions related to the Management of the MSW in the canton and the higher educational institutions.

After the analysis developed, it is suggested to carry out other related studies such as: districting considering other type of constrains like allocation of frequencies and schedules of operation, lifting of the collection, schedule, and movements process as well as the measurement of the environmental impact caused. In addition, to an adequate cost analysis.

Once this research has been executed, it is clear the lack of adequate technical definition in the rates of generation-production per capita (PPC) in the Canton Ibarra and the absence of an optimal system for waste collection routes planning. In addition, a high level of acceptance of the proposal developed in the institutions related to waste management was observed.

By means of the execution of the mathematical model for calculating the optimal number of truck. It could be concluded that, if it is desired a collection frequency of two times per week or every third day, it should be enough with the current fleet of trucks; however, if the service offer want to be reached to a daily collection, a new vehicle acquisition has to be executed around 1.395 16yd compactor trucks.

Within the future work it is recommended to carry out a study of cost reduction that would imply the implementation of the proposal for the city of Ibarra, considering all the factors previously presented in the model.

References

1. Herrera I, Collaguazo G, Lorente L, Montero Y, Valencia R (2016) Revisión: Optimización de rutas de recolección de residuos sólidos municipales en países en vías de desarrollo. In: Tecnologías aplicadas a la ingeniería, 1st edn. Editorial Universidad Técnica del Norte, Ed. Ibarra, p 301
2. Wilson DC (2015) Global Waste Management Outlook 2015, 1st edn. United Nations Environment Programme, Austria
3. Empresa Pública Metropolitana de Gestión Integral de Residuos Solidos (2014) Estación de Transferencia Sur
4. Diario El Mercurio (2013) Estación de transferencia de residuos
5. Secretaría de Desarrollo Social de México (2000) Manual para el diseño de rutas de recoleccion de residuos sólidos municipales, p 50
6. Male JW, Liebman JC (1978) Districting and routing for solid waste collection. J Environ Eng Div 104(1):1–14
7. Lorente-Leyva LL, Herrera-Granda ID, Rosero-Montalvo PD, Ponce-Guevara KL, Castro-Ospina AE, Becerra M, Peluffo-Ordóñez DH, Rodríguez-Sotelo JL (2018) Developments on solutions of the normalized-cut-clustering problem without eigenvectors. In: Huang T, Lv J, Sun C, Tuzikov A (eds) Advances in Neural Networks – ISNN 2018. LNCS, vol 10878. Springer, Cham, pp 318–328 https://doi.org/10.1007/978-3-319-92537-0_37
8. Herrera-Granda ID, Lorente-Leyva LL, Peluffo-Ordóñez DH, Valencia-Chapi RM, Montero-Santos Y, Chicaiza-Vaca JL, Castro-Ospina AE (2018) Optimization of the university transportation by contraction hierarchies method and clustering algorithms. In: de Cos Juez F, et al (eds) Hybrid Artificial Intelligent Systems. HAIS 2018. LNCS, vol. 10870. Springer, Cham, pp 95–107. https://doi.org/10.1007/978-3-319-92639-1_9
9. Sakurai K (1980) Diseño de las Rutas de Recolección de Residuos Sólidos. Div. Protección la Salud Ambient. Cent. Panam. Ing. Sanit. y Ciencias del Ambient. pp 32–34
10. Lin HY, Kao JJ (2008) Subregion districting analysis for municipal solid waste collection privatization. J Air Waste Manage Assoc 58(1):104–111
11. SEDESOL (2009) Manual Técnico sobre generación, recolección y transferencia de Residuso Sólidos Municipales, pp 1–139
12. Márquez J (2011) Macro y microruteo de residuos sólidos residenciales, pp 1–91
13. Henao B, Piedrahíta J (2015) Diseño de un modelo de ruteo de vehículos para la recolección de residuos sólidos en el municipio de Zarzal Valle del Cauca
14. García-Ayala G, González-Velarde JL, Ríos-Mercado RZ, Fernández E (2016) A novel model for arc territory design: promoting Eulerian districts. Int Trans Oper Res 23 (3):433–458
15. Chartrand G, Zhang P (2011) Discrete mathematics 1st edn. Illinois
16. Castillo A (2015) Plan de desarrollo y ordenamiento territorial del cantón Ibarra 2015–2023, Ibarra
17. Wuthrich D (2006) Google earth pro. Geospatial Solut 16:30–32
18. Martínez J (2012) Plan de Desarrollo y Ordenamiento Territorial del cantón Ibarra, p 300
19. Base de Datos-Censo de Población y Vivienda 2010 – a nivel de manzana (2010) Instituto Nacional de Estadística y Censos
20. Blue Marble Geographics (2017) Global Mapper - All-in-one GIS Software
21. QGIS Development Team (2014) QGIS Geographic Information System. Open Source Geospatial Foundation Project. Qgisorg. http://qgis.osgeo.org
22. SEDESOL (2000) Manual para el diseño de rutas de recoleccion de residuos sólidos municipales, p 50

IT Management

Importance of ICT's Use in Business Management and Its Contribution to the Improvement of University Processes

Johanna Rosalí Reyes Reinoso(✉) and Deisy Carolina Castillo Castillo(✉)

Universidad Católica de Cuenca, Cuenca, Ecuador
{jreyesr,dccastilloc}@ucacue.edu.ec

Abstract. The ICT's use in public and private sectors is an increasingly important need, especially in education, training and professionalization. For this reason, this research consisted in an analysis of results in ICTs' use in business management and its contribution to improving university processes, proposing these three categorical elements, in response to university management under the systemic approach get contributions from the business world and be accompanied by ICT's to optimize their responses. It was based on several methodological procedures: documentary research and descriptive research, using a SIPOC procedure and a cause-effect diagram to determine the various relationships that occur in the management process under assessment. Among the results, a characterization of the processes was obtained that are carried out and a list of the causes that affect the weaknesses of the management on university processes. One of the most important was the staff training element, weaknesses in leadership and lack of strategic planning, as well as inadequacies in the applications implemented for the automation in processes and a low sense of belonging by the staff, whose performance does not show commitment to the organization. This article is outlined for the scientific Track in the thematic axis of ICT Management.

Keywords: ICT's · Business management · University process management

1 Introduction

The present research consists of a study in which three theoretical and empirical elements of investigative interest are addressed; namely: Information and communication technologies (hereinafter, ICTs), business management, and the implementation of both aspects in university processes, as an option for their improvement and optimization. This study is based on a concern regarding the needs evidenced in university sector, from which effective and efficient responses are still required in the context of a constantly changing and updated social context.

M. Botto-Tobar et al. (Eds.): TICEC 2018, AISC 884, pp. 241–252, 2019.
https://doi.org/10.1007/978-3-030-02828-2_18

1.1 Use of Information and Communication Technologies

In the accelerated and globalized society we live in, information and communication technologies (ICTs) have played a leading role in promoting substantive changes in cultural modes in today's world. These technologies have had a fast evolution, denoting a society whose rhythms and interests move in the midst of new rationalities and codes.

Individually and collectively, we must adapt to this new world, whose languages and modes translate other ways of organizing dynamics. Therefore, incorporating ICTs becomes a mandatory practice for those who must remain articulated with the surrounding reality. It should be noted that ICTs "are developed from the scientific advances produced in the fields of information technology and telecommunications (…) are the set of technologies that allow access, production, processing and communication of information presented in different codes" [1].

From this perspective, this new reality starts from a new communicational logic, which also makes use of visual elements and links that extend the conventional vision of the world and reality. So:

> In general terms we could say that new information and communication technologies are those that revolve around three basic media: information technology, microelectronics and telecommunications; but they rotate, not only in an isolated way, but the most significant thing is, it can be done in an interactive and interconnected way, which allows to achieve new communicative realities [2].

In this case, organizations, whether public or private, make sense in a social context, and are due to it. For this reason, it is vitally important that they develop an adequate capacity to adapt to their environment, and be prepared to give the answers their public requires, within the guidelines of the area in which they work.

ICT use has been making the difference between successful and unsuccessful organizations; those that offer services and quality products, and those that are far from generating satisfaction in their users and customers. This creates new scenarios of competitiveness, renewed demands that require evaluating the processes themselves and updating them with a view to providing better responses to these new needs.

Starting from the General Theory of Systems, and going through all the technical aspects towards which this new technology evolves, ICTs have opened a broad scenario of alternatives, concepts, techniques and procedures that can be applied in the organizational field, and that offer utility tools to optimize processes within each organization. This has been demonstrated in numerous studies, whose contributions rebound in describing the advantages, which include the incorporation and good use of ICTs in businesses and institutions, and that positively impact the various processes that are carried out exponentially [3–7]. It is worth saying that ICTs consist of:

> Technological devices (hardware and software) that allow editing, producing, storing, exchanging and transmitting data between different information systems that have common protocols (…) enable both interpersonal communication (person to person) and multidirectional communication (one to many or many to many). [8]

Seen in this way, ICTs offer very diverse alternatives that are irreversible nowadays, especially when society, the public, users and customers, live an increasingly accelerated reality, in almost instantaneous communications and scarce time for paperwork.

Everything evolves towards an increasing use of automated and interactive applications and tools. Every organization and every institution is obliged to adapt to this reality, or become obsolete.

1.2 ITCs in Business Management

In the field of administration and management, nowadays the notion of business management has been incorporated and understood as a line that addresses organizations, their structure and functioning, as integrated wholes that operate in a systemic way, with interdependence between their components.

A change in business thinking has occurred from classical perspective, wherein the company was understood as "an organization (…) whose objective is the achievement of a benefit through the satisfaction of a market need. It specifies the offer of products (…) with the consideration of a price or economic value" [9]. Subsequently, and thanks to updates in the dynamics of organizations, the company is seen now as "the interrelation of various elements and variable attributes that interact in a complex way with each other and the environment in which they develop" [10].

From this latter perspective, business management is closely linked to the way the organization is conceived, in such a way there must be a correlation between one and the other for its advancement in time. So, we can characterize the types of companies and types of management as proposed by Machado [10] in the following Table 1:

Table 1. Types of business and types of management

Types of business	Types of management
Closed organization: it is characterized by being strongly centralized and hierarchical. Principle of causality	Mechanistic and reactive management, experience and common sense are privileged as forms of knowledge
Anticipatory organization through the provision of information that enables the determination and control of relationships. Random paradigm (statistic)	Proactive management that emphasizes the function and the fact to build an image of reality. The data must be verified and the information corroborated
Organization as an open system characterized by imbalance, non-linear relationships and emerging properties. Paradigm of complexity	Systemic management that must be creative, innovative and strategic through the language that allows to account for relationships. Daily activity and management are objects of control (surveillance)

Source: Machado [10]

From this classification, it is common to give priority to the systemic approach to management, whereby "all the interrelation and interdependence of the different elements that make up the organization are observed, including the relationship of the organization with its environment" [11]. We speak thus about an approach also denominated administration by processes in which each aspect as a whole is in permanent interrelation with the rest and the whole at the same time, so that each element must be analyzed and valued according to its relations with all the system. Special emphasis is placed on all the processes that are carried out, and the multiple factors that affect their implementation.

This approach is summarized in the following figure, which includes all the elements that are part of a business system (Fig. 1):

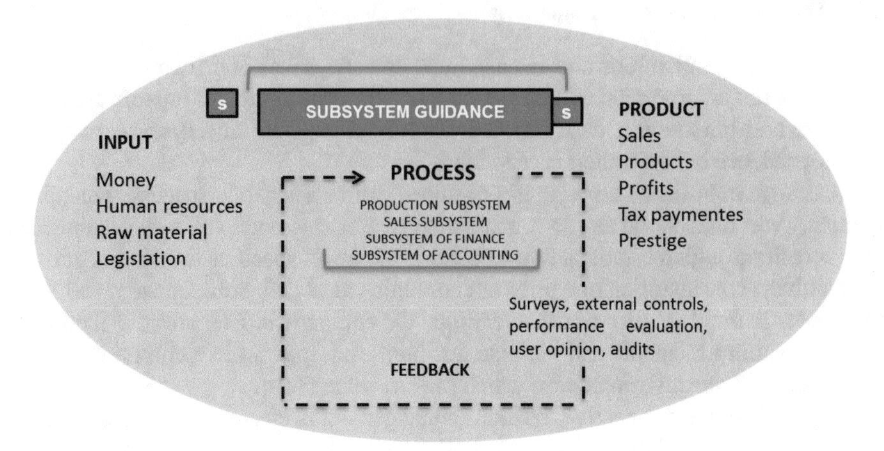

Fig. 1. Elements that make up a business system, Hernández [12]

This same author talks about the advantages of the automation of processes under the systemic approach, which can order and process information generated during the processes, and guide the system towards satisfying needs that must be attended. It places special emphasis on the need for control units, whose purpose is to "maintain the system process variables within the desired and pre-established terms" [12].

This approach has been widely implemented today, to the extent that ISO 9000-200 standards establish where companies must be designed and organized under the systemic approach and therefore have control units, and the documentation and staff must be aware of this approach and know clearly the purposes and dynamics of each of the processes that are carried out.

For this norm, every process is understood as a set of interrelated activities transforms the inputs into products. Likewise, every product is the result of a process and, in addition, every service is also a product, since the latter is also the result of a process [12].

This last point is important because of the interest in this work, which revolves around university processes that are considered more services than products. In this sense, the systemic approach in this case will be applied to the service, understood as the result of an interrelated process.

Similarly, automating processes within the organization incorporates in this sense the use of ICTs as a tool for the optimization of processes and decision making in the university context. Addressing these processes through the use of ICTs from a business management approach offers contributions that can result in the qualitative and quantitative improvement of higher education in the country.

Of course, the incorporating ICTs in business environments by itself does not guarantee the optimization of processes or better results. Its use must be accompanied by other elements of utmost importance, among which are having a detailed and exhaustive knowledge of each and every one of the processes and procedures that are carried out in the company or organization, knowing the general and specific needs, from the human point of view and from the operational point of view, strategic management plans and technical development actions that will be implemented, among others.

1.3 ICTs in Higher Education, a Vision from Business Management

Higher education is one of the most complex areas within social systems today, since there are elements of a cultural, political, economic, institutional, governmental and intersubjective nature, among others. The importance of this sector in each country lies in the responsibility to provide professional training, academic and scientific production required to develop nations.

It is for this reason that higher education institutions continuously adjust, aiming to implement methods to achieve the objectives that are attributed to them. One of the greatest challenges at present in higher education is to encourage a management that, in a business sense but service-oriented, significantly improves the processes such as those that contribute to the professionalization, research, scientific and academic development of all those who make life in the university campus in order to offer significant contributions to the social environment, inasmuch as university demands the greatest of contributions for the development of a country.

Effectively incorporating ICTs in universities implies a thorough review of how the processes are structured within this system in general and within each university in particular. This means that it is essential to orient the structure, forms of organization and procedures towards optimizing the processes, educational quality, personnel and users satisfaction, administrative and teaching staff update, work improvement and educational environment. In this vein, Benjamin and Blunt, cited by Cano-Pita [3] affirm,

> The emergence of a new organizational configurations has coincided and has been aided by the development of ICTs, which are usually attributed a fundamentally flexible function and a dynamic character of the organization, and the need of contrast action and direction do ICTs affect the features of organizational design and how do both elements fit together to respond to the crucial issue of organizational change?

In this effort to guide university processes under the systemic approach, today we talk about the fundamental processes that must be developed in this sector, around which their organization and activities revolve. These processes are aimed to obtain "results from developing training processes, through the permanent updating of its teaching staff. The development of a research culture and the promotion links with society" [13].

These processes must be oriented from a new management approach, one that is developed

> through an extensive system of relationships and interactions of a social nature that are established among the subjects involved in it, aimed at creating, developing and preserving, in an adequate working environment, human talent, competent and motivated to perform with relevance, impact and optimization of their processes and thus achieve the organization goals [13]

This implies the obvious need to incorporate the ICTs use in management processes in higher education to optimize the services that must be fulfilled, thereby benefitting administrative staff, teaching staff, researchers, students and society in general.

2 Methodologies

In response to the objective of this work to analyze the effect of ICTs use in business management and its contribution to improving university processes, the study used several methodological procedures. First, documentary research was used, which consisted in a theoretical revision of the three categories involved in the study; namely: Information and communication technologies, low business management in a process management approach, university management processes, all limited to the systemic approach to management.

In this sense, documentary research is understood as "an analysis of written information about a certain topic, with the purpose of establishing relationships, differences, stages, positions or current state of knowledge regarding the subject matter of study" [14].

Additionally, it was a descriptive investigation; it is about "selecting the fundamental characteristics of study topic and its detailed description parts, categories or classes" [14]. This typology was used to analyze the management approach implemented in management of university processes in Ecuador, and to determine if ICTs are used or not in university processes.

For such purposes, a characterization of management process in higher education system was carried out, using the procedure known as SIPOC (for its acronym in English that refers to: Supplier-provider, Inputs-resources-entry, Process-processes, Output-output and Customer-client). Finally, a cause-effect diagram was made to determine the various relationships that occur in the management process subject to valuation.

3 Results

Below, results on application of SIPOC to the management of higher education processes in Ecuador will be observed, characterizing which general management process was sought, taking into account the general aspects that must be developed in the university context and its concrete dynamics (Table 2).

Identifying these general elements allowed detailing the weaknesses through a cause-effect diagram, in which a series of factors that directly affect the management of university processes were obtained, with the understanding that the focus has been on the work in terms of advantages of using ICTs in these processes. The result was the following (Fig. 2):

Table 2. General process of management in the university

Suppliers	Entries/Means	Processes	Exits	Customers
– Secretary of Education – Principal ship – Vicerectorade – Decanato – Faculties and graduate/ postgraduate programs – Organizations and companies – National and international organizations that govern higher education	– Projects and financing – Organic law of higher education – Documents and decrees on development of sciences – Students – Research professors – Agreements and contracts	– Schedules for predetermined periods (annual and per academic period) – Formulation of specific projects and work plans – Academic calendar execution – Execution of administrative activities – Execution of general service activities and maintenance – Execution of research activities – Execution of social bonding activities – Execution of academic, scientific and community events	– Administrative closing of the academic calendar (reports, grades, defenses minutes, among others) – Closure of research activities – Completion of research products: research papers, publications, papers – Closure of financial and non-financial administrative procedures – Reports and project renditions and executions of assigned resources	– Students – Teachers – Investigators – National and international organizations and institutions – Government – Secretariat of Higher Education, Science, Technology and Innovation – General society

Source: own elaboration (2018)

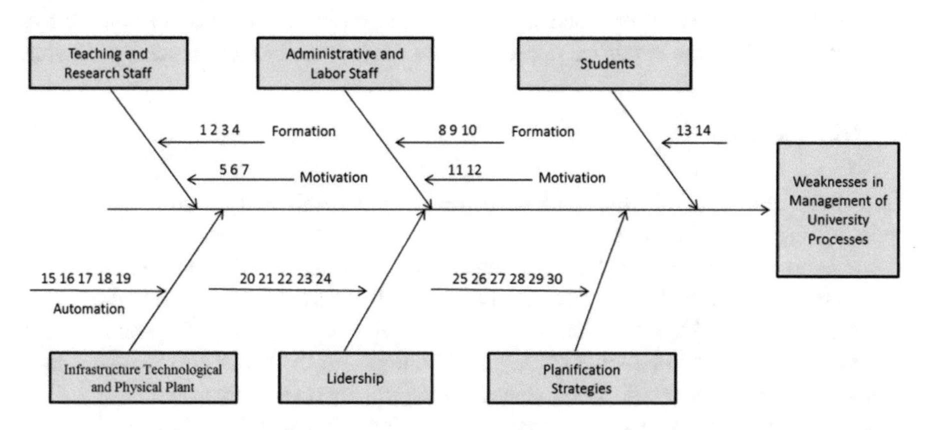

Fig. 2. Cause-Effect diagram on management of university processes

To better understand these results, causes are identified according to the caption described below (Table 3):

Table 3. Causes identified in the Cause-Effect diagram

1. There is no training in Process Management 2. There is no training in Strategic Planning 3. ICTs use in management, teaching and research 4. Software management (basic packages, interactive tools, networks, internet) 5. Lack of institutional stimulus 6. Low motivation to scientific-academic achievement 7. Little sense of belonging to the institution 8. Sensitization towards optimization of processes 9. Role of each department in system operation 10. ICTs and computer tools use 11. Little sense of belonging 12. Lack of clarity regarding the processes	13. Web tools are not used for academic and scientific purposes, only for social purposes 14. Updating is required in applications to manage their academic-administrative requirements 15. Lack of budget to improve the infrastructure of networks and physical plant 16. Insufficient computer equipment in each unit or department 17. Insufficient design of databases and automated systems that allow the attention of the different requirements and the realization of the necessary processes 18. Failures in data connectivity 19. The systems implemented do not generate data required to optimize decision making	20. Lack of training in process management 21. Low vision of general processes of the university 22. Little clarity regarding the information required for decision making 23. Limited use of ICT tools for process optimization 24. Little interest in staff training in ICT tools 25. Poorly formulated strategies 26. Work plans with few objectives 27. Insufficient control of processes 28. Lack of clear management indicators 29. Badly formulated purposes 30. Insufficient recognition of results

Source: Own elaboration (2018)

These causes were determined from the bibliographic review carried out for this research, as well as the empirical observation of the university management processes.

4 Discussions

This study initiated from the evidenced needs in the university sector, in terms of updating its processes in order to adjust itself to requirements that society has nowadays. The relevance of the three aspects studied has been demonstrated in a large number of studies that deepen the importance of an adequate ICTs use in organizational environments (public and private). This leads to a growing optimization of processes and responses that result in benefits of the organization and the social group which their goods and/or services are directed, generating an impact on society as well.

In the university context, it is considered that the effective incorporation of ICTs in their management processes will have a positive impact on training of up-to-date and suitable professionals for the real work they have to face, as well as in development of the country. All this thanks to the possibility of having an effective tool for decision making, management and planning, from a systemic perspective of processes.

Based on these ideas, a bibliographic and empirical study was carried out in which the procedure known as SIPOC (for its acronym in English referring to: Supplier, Input, Process, Output and Customer) was applied, and a cause-effect diagram to determine the relationships that occur in university management process. From these procedures the following elements can be obtained to contribute to the discussion:

Personnel (teacher, research, administrative and workers): they do not receive adequate training or the adequate motivation in managing of process management, optimization and automation, handling of computer packages, use of networks, which weakens any initiative for improvement of processes on educational system and every single university through the incorporation of ICT's as a tool for university management.

Students: although they have the knowledge to handle web tools and applications, this is largely oriented to social networks and entertainment applications, and not towards academic and research use. There is no guidance on this.

Technological Infrastructure and Physical Plant: lack of adequate budgets for the implementation of automated systems, absence of equipment and technological infrastructure to support automation, presence of some systems that do not generate enough information to influence adequate decision making from the different levels of management.

Strategic Planning: no clear plans are drawn up since there is no precise information about the real requirements of the process, the purposes are not well defined, no management indicators are proposed to guide the execution of plans, there is no proper follow-up of processes and activities, insufficient training in ICT tools aimed at optimizing processes for academic and research purposes.

5 Conclusions

ICTs use is increasing, since the automation of processes in any organization results in an immeasurable series of benefits, such as improvement of procedures and internal operations, covering a greater number of customers, users and/or beneficiaries, opening doors to new audiences, offering fluid communication about the internal and external aspects of the organization, and increasing efficiency by reducing response times. The information needed to stimulate decision making is guaranteed, as well as a broader vision of processes in their complexity.

Applying the postulates of the private sector in public sector could help the focus on their priorities and increase their responsibility, effectively providing the services demanded by people, with efficient costs [11]. In the same way, the idea that a new education system, appropriate to these times, should be based on absolute clarity with respect to the purposes, mission, vision and goals to be achieved, is firmly held so each action contributes to coherent progress towards the paths that have been traced, and the participation of each and every one of people who make life within the system. This new approach does not mean biased or partial changes, but requires a deeper and more conscious structural transformation, committed to a new society.

This approach has found strong resistance, despite its advantages and adaptability, especially in the public sector and very specifically in the public university environment. This is a situation that draws great attention, since universities are called to take the lead in innovation, technology, scientific and academic advances. These resistances are translated into important weaknesses, evidenced in results obtained on this work. Among them, the following can be mentioned:

- Lack of clarity with respect to the objectives and purposes of the institution, beyond its mere function of giving classes and multiplying content
- Absence of a strategic vision by the leaders of the organization
- Little motivation to participate in training and updating activities by the staff in general
- Little attention to internal communication processes
- The different levels of leadership do not pay much attention to the need to train and update staff
- A significant resistance to change
- Very little dedicated to monitoring processes
- Many applications implemented, do not give the necessary answers and are not replaced or updated
- The procedures are too bureaucratic in general, and this prevents further progress towards a real automation of processes, leaving many processes still manually run or to be resolved at discretion of the official in charge.

In this scenario, the challenge is really to combine the management of university processes with systemic approach of business management and the assertive and appropriate ICT's usage, not only as a teaching tool, but as a resource to optimize the entire university, with its diverse and complex processes, with a view to raising the quality of the academic and scientific response, its ultimate purpose is the professionalization of the country.

It is important to use the appropriate technologies, which are up-to-date and that contemplate applications and automated systems that really cover the complexity of the processes and procedures that are developed in each unit of the organizational structure, and that also throw the information so in each hierarchical level the correct decisions are taken according to the plan within a permanent monitoring. For this, the digital platform must be solid, democratic, secure and stable.

6 Recommendations

In response to what is contemplated in this work and its results, several interesting recommendations emerge. In the first place, a more detailed study can be carried out within several university campuses, if you want a comparative type in which analytical tools such as benchmarking are applied, whose results yield more specific data about weaknesses in human resources and in technological matters, which are evident in this sector and which hinder their progress towards management models under systemic approaches, as well as the implementation of efficient ICT tools to achieve this. The need to initiate a process of training and sensitization that allows the different levels of leadership to understand the urgent need to undertake solid and constant actions in this direction, will allow us to move towards a new quality educational paradigm.

It is also recommended that services offered through ICTs be as democratic as possible, accompanied by information and training campaigns on how to use them and their multiple benefits. Since every automated system must have different levels of security, it is important that the information reaches each level users.

All teaching and administrative staff must have clear the general objectives of the institution as well as the partial objectives of each operating plan for specific periods. Likewise, each of them must have clear commitments to these objectives and to the specific planning; therefore, each leader, according to his/her hierarchy, must keep his/her staff informed and motivated so he/she fulfills his/her responsibilities and process in a committed manner.

There must be continuous monitoring of processes. It is advisable to evaluate the possibility of incorporating the Integral Control panel as a tool that allows continuous monitoring of the whole processes, in order to take actions in prudent times to prevent failures or weaknesses.

It is important to handle conflicts in an appropriate way and turn them into irreversible contradictions. Complex organizations, addressed under systemic approach, will always present levels of conflict and will require changes in strategies and permanent decision-making in emergent situations. All leaders, regardless of their hierarchical level, must be trained to handle conflicts in complex scenarios, so they can give the most effective responses in each case.

The ICTs implementation in the university environment should be oriented towards equity, accessibility and the efficient use of time and resources. Because of this it is convenient to remember that it is educational field and its main function is to form, in an equitable and democratic way, in respect to the differences and attention to a multicultural scenario. Therefore the application that is used, should be in favor of everyone equally, students, teachers, researchers, administrative workers, academic units, administrative units, management units.

For future research it would be important to talk about "Business Intelligence", which consists in structuring and processing on large volumes of data and its analysis for decision making.

References

1. Belloch C. https://www.uv.es/. https://www.uv.es/~bellochc/pdf/pwtic1.pdf. Accessed 14 Feb 2018
2. Cabero J (1998) Impacto de las nuevas tecnologías de la información y la comunicación en las organizaciones educativas. Enfoques en la organización y dirección de instituciones educativas formales y no formales, Granada, Grupo Editorial Universitario, pp 197–206
3. Cano-Pita G (2018) Las TICS en las empresas: Evolución de la tecnología y cambio estructural en las organizaciones. Revista Científica Dominio de las Ciencias 4(1):499–510

4. FEMP (2012) Experiencias de éxito en e-Administración de las entidades locales. Ministerio de Industria, Turismo y Comercio, Madrid
5. Feijoó A (2009) La e-administración: El paso de la gestión tradicional a la gestión electrónica. Revista General de Información y Documentación 19:161–171
6. Hernández B, Jiménez J, Martín M (s/a) Influencia de las TIC en la gestión de la información empresarial. Conocimiento, innovación y emprendedores, 1972–1986
7. Nores C (2001) La administración electrónica: Teoría y práctica. Econ Ind 11(338):97–103
8. Cobo J (2009) El concepto de tecnologías de la información. Benchmarking sobre las definiciones de las TIC en la sociedad del conocimiento. Zer 14 (27):295–318
9. Hernández H (2011) La gestión empresarial, un enfoque del siglo XX, desde las teorías administrativas científica, funcional, burocrática y de relaciones humanas. Escenarios 9(1): 38–51
10. Machado M (2009) Contabilidad y realidad: Una relación crítica bajo el enfoque de la representación. Actual Contab 19:38–55
11. Rubio P (2008) Introducción a la gestión empresarial. Instituto Europeo de Gestión Empresarial, Madrid
12. Hernández S (2006) Introducción a la administración: Teoría general admisnitrativa. Origen, evolución y vanguardia. Mc Graw Hill, México
13. Albán M, Vizcaíno G, Tinajero, F (2014) La gestión por procesos en las instituciones de educación superior. UTCiencia. Ciencia y Tecnología al Servicio del Pueblo 1(3):140–149
14. Bernal C (2010) Metodología de la investigación: Administración, economía, humanidades y ciencias sociales. Pearson Educación, Colombia

ICT and Business Inclusion in the Southern Communities of the City of Bogotá – Colombia

Camilo José Peña Lapeira(✉) and Cliden Amanda Pereira Bolaños

Corporación Universitaria Minuto de Dios - UNIMINUTO,
Cra. 24 # 45 A – 71 Sur, Bogotá, Colombia
{cjpena,cliden.pereira}@uniminuto.edu

Abstract. Competition is today's business world is the result of the dynamic markets working to satisfy the rapidly changing habits of a consumer-based society. It has become evident that business practices also follow these trends, yet preserving the important role that knowledge and technological management has in the process of forming a value-chain-management to increase their competitive advantage. Information, Communication, and Technologies (ICTs) play an important role in the business-management process, since these are essential components in the merger of intangible assets as a result of the dynamic business capabilities. Then, based on administrative principles such as those by Adams, Terry, Fayol, Koontz & O'Donnell, an effective response must be given to business problems related to the use of technology and related to available resources, so that both can become strategic assets.

Keywords: ICT · Business · Technological management
Dynamic capabilities · Intangible assets

1 Introduction

The contemporary theory of organizations has based its research agenda on the development of theories that seek to overcome the shortcomings of the neoclassical theory of the firm and of the economics' field known as Industrial Organizations Economics. The main contributors of this theory are Marshall, Coase and Shumpeter. The current proposals for contemporary theory are recognized in multiple works from Walras, Jevons, Menger, Edgeworth, Pareto, among others; they are also corroborated in organizational-management research. The weaknesses of the dominant economic theories are as follows [1]: simplistic treatment of the firm and its technological and organizational changes; the apparent understanding of human rationality; the emphasis on static analysis; the absence of an active role for the entrepreneur and manager along with the lack of knowledge on the institution's fundamental market structure.

In the same way, [1] Michael Porter's proposal (the Five Forces model) is not able to identify and develop organizational skills resulting in sustainable, competitive advantages according to the neoclassic theory; in addition, empirical studies show that differences between firms in the same industry have an impact on the performance of earnings within the industry.

M. Botto-Tobar et al. (Eds.): TICEC 2018, AISC 884, pp. 253–265, 2019.
https://doi.org/10.1007/978-3-030-02828-2_19

The contemporary theory of the firm considers organizations essentially as heterogeneous entities and capable of learning [1]. From this perspective, the variety of proposals for organizational analysis has evolved in a prolific fashion; the following two major perspectives can be clearly identified: the theory of governance and the theory of competencies. The theory of governance is based on economic reasoning, as expressed by O'Connor, Nohlen, Huntington, Crozier, Watanuki and the theory of competencies focuses on organizational analysis. Subsequent paragraphs, however, are indented.

1.1 The Dynamic Capabilities

Dynamic capabilities typically evaluated to create intangible assets within companies include:

1. Organizations and their employees need the ability to learn quickly to be able to develop strategic assets that will result in a competitive advantage. Therefore, continuous training and adequate management of knowledge contribute to the strengthening of this learning ability.
2. New strategic assets such as the ability to receive feedback, the use of ICTs, and customers have to be integrated with the core competencies of the organization. Therefore, management of these assets as a whole and not independently is critical, and should be part of the company's administrative practices and strengths [2, 3].
3. Existing strategic assets must be found in a dynamic and transformative process of reconfiguration. This process results in constant change and evolution within the organization, mainly due to market dynamics, competition, and changing preferences of customers [4].

Thus, the use and absence of technological resources, ICTs, and effective means of communication can make a difference in the organization's competitiveness. In the same way, these factors can make a difference in the ability or inability to access new markets for Micro, Small and Medium Enterprises (MSMEs) such as those that make up the business network within the southern communities of the City of Bogotá, Colombia. According to data developed by the City's Chamber of Commerce, these businesses are characterized for originally emerging from family nucleus in search of auto employment opportunities. This source of income provides them with the ability to establish a capital that eventually allows them grow and expand into other markets, different from the local markets. Unfortunately, the City's MSMEs are left behind due to lack of training, investment, strategy, opportunities, and other similar factors; destined to vanish due to the fierceness of foreign markets and the competition that goes along with rapid growth resulting from their access to cutting edge technology.

Internal management is important for the families that are dedicated to a particular productive activity according to [5]; over time, this families become a new productive force characterized by a relatively low working capital, use of low-skill yet highly productive workforce, and fragmentation of raw materials in favor of long term yield of profits. However, [6], [24], it must be pointed out that there is a lot of work to be done so that these small companies can become competitive while maintaining sustainable schemes in order to keep or increase their current market share. This competitive model should be based on new business strategies, sustained growth in productivity, ability to participate in negotiations with other companies within their industry, technological improvement, ICT implementation plans and policies introduced by national governments and regional economic alliances.

An objective from the research paper entitled “Business Analysis of MSMEs in the south of Bogotá” included the study of market orientation strategies implemented by MSMEs in that part of the City. The paper evaluated the usage of ICTs and made an attempt to analyze their influence in business management. The analysis was based on the hypothesis that by understanding the future behavior of companies, an environment for corporate and organizational improvement can be created as a starting point for the process of development and modernization of the companies within a specified sector. For this purpose, it must be taken into account that this business network is located within low income communities with few opportunities for formal employment, and few resources to create and develop a company. This is also an area with limited government support and commitment.

2 Related Work

This paper is based on the review of works from authors that have contributed to the firm’s contemporary theory; and who also have contributed in providing ways for business analysis research with a high degree of uncertainty and informal scenarios, such as the Colombian case. [1, 2, 7, 8], are the starting point that allow the establishment of lines of thought that complement the practical experiences of some educational entities in Colombia. The conception of all organized activity implies the existence of actions with a repetitive nature; characteristic that is expressed in the essence of the capabilities [1]. This must be made consciously in the process analysis, since the overlap between routine and capacity appears continuously.

The conception of dynamic capacities implies considering each of them as a dynamic and complex process; hence, in this third order there is room for ambiguity and the overlapping of concepts. A more apprehensive description: the respective dynamic capacity corresponds to a process, a learning curve conditioned by a context and consciously defined in the search for a purpose. This author proposes to study the capacities through the study of the organizational and managerial processes, the procedures, systems and structures that underlie each type of capacity and capacity in itself [7].

For [9], the success of SMEs lies in the investment they make in strengthening financial capabilities, investment in updated technological resources, training to have a good competitive marketing team, management and quality management in the area of human resources together with investment in innovation and ICT.

3 Methodology

The investigative approach for this project can be considered as mixed, since it presents some phases of qualitative analysis and others of a quantitative nature, which are marked by a path defined by the characteristics of the problem formulated, the empirical object, and the paradigm from the one that is studied, theoretical object [10–12].

Based on the previously proposed methodology, the study is structured in three phases.

Phase I. Documentary research to be able to establish the state of the art of the literature produced in Colombia around the Dynamic Capacities and their relationship with ICT.
Phase II. Characterization of the MSMEs located in the study locations applying non-probabilistic sampling by defined quotas according to the distribution of the total number of companies.
Phase III. Design of the protocol as application of the theory of dynamic capabilities that contains the methodological guidelines and research instruments for the study of MSMEs.

Mixed methods data were used: government statistics, institutional documents; archive sources; observations; personal interviews; surveys and focus groups. Information will be gathered from different stakeholders, e.g., students, faculty, administrators, government officials, opinions of entrepreneurs and business owners.

The diversification of the productive structure of Bogotá and its surroundings is complex. In 2017, a total of 380,265 companies (29% national) exist in the City. 87% of this companies are micro-enterprises, 9% are small companies, and 3% are medians [25]. Part of this rich business network is located in southern communities of the City as a result of family effort or of community associations due to geographical proximity as shown in Fig. 1.

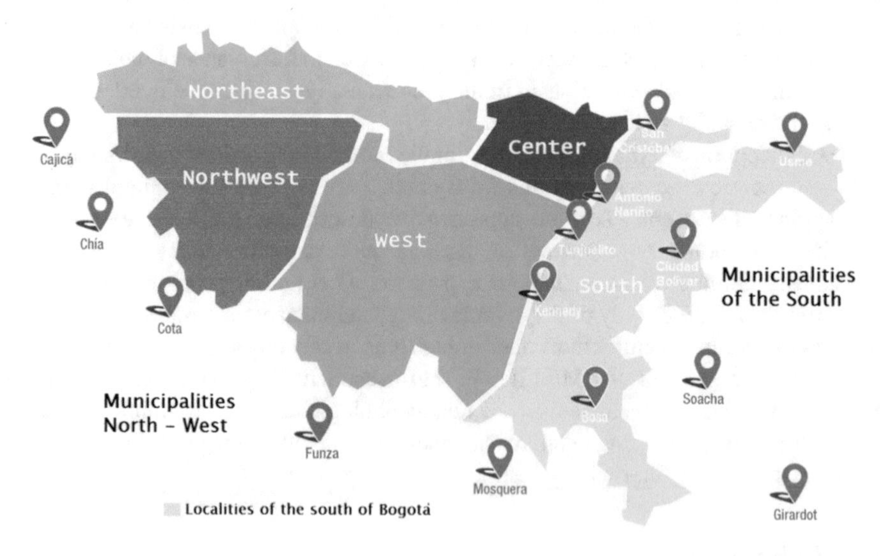

Fig. 1. Southern communities of the city of Bogotá

The population consists of the MSMEs of the localities in the south of Bogotá, representing approximately 18.8% of the total of SMEs in the city, which are approximately 35,600 companies that would constitute the total population [25]. For the calculation of sample size when the universe is finite, that is to say accounting and the variable of categorical type, the sample size was 115 companies given the scope of the project, the categories of stratification were determined by the different sectors in which these companies are located that are 9 in total (commerce, industry, real estate and rental services, transportation, storage and communications, restaurants and hotels, trade and repair of vehicles, construction, community and social services), which are represented by percentage in Fig. 2.

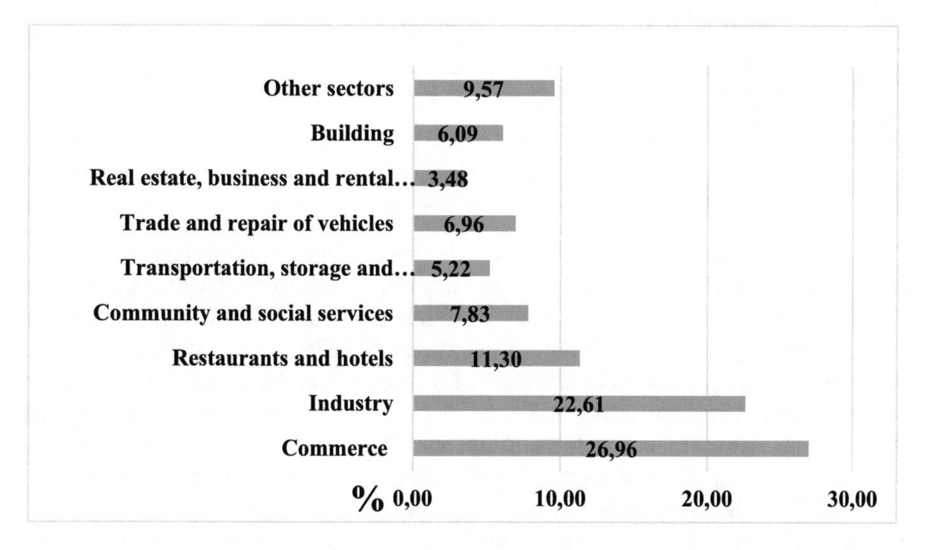

Fig. 2. Sectors of the economy where the SMEs of southern Bogotá are located

Resources used for data collection include: documentary review of government statistics and institutional documents as well as the structured interviews, field observations, surveys and focus groups. Information was collected from different stakeholders such as students, professors, government officials, entrepreneurs and business owners. Triangulation and documentary matrixes were used for data analysis.

4 Results

The business analysis of the weaknesses, strengths, threats and opportunities contribute to the constant restructuring and dynamism that is required by companies today. When requesting data from companies regarding the main factors that can affect a business, it was found that new technologies and ICTs are the third concern on threats for companies with 12.5% showing that it is an important factor to take into account (Table 1).

Table 1. Main threats of the business or company

Aspects considered	%
Informality	8,93
Large stores and large companies	8,33
Thieves and insecurity	15,48
Unfair competition and low prices of others	25,60
Increase in prices of inputs and labor materials (Dollar)	11,90
New technologies and ICT	**12,50**
They have not identified	2,38
Trust in the business	2,98
The vehicular restriction	1,19
The contraband	7,14
Regulations and taxes	3,57

As part of the process of building strategic assets, it was inquired about the progress in the constant training of employees, because if it is wanted to reinforce the consolidation process of these then investing in training is a must. However, as shown in Table 2, approximately 23% of companies have not included training within their media plans, it is also appreciated that the issue of technology, systems and ICT, despite not being among the first concerns, remains among the most important, taking into account that the first ones exists due to the fulfillment of the norms and legal dispositions that force these training to be done, limiting the resources and training spaces reason why only 12.5% of the companies considered in the last 12 months to address this topic, which usually represents a limitation for companies that want to excel in the market given the complexity of the social environment where they are located.

Table 2. Training in the last 12 months

Have your employees received any type of training in the last 12 months? On what topic?	%
They have not received	22,91
Electricity	5,56
Customer service and marketing	13,89
Handling of food and substances	16,67
Technology, systems and ICT	**12,50**
Safety and health at work, first aid	18,06
Pedagogy and free time	2,08
Motivation, trade and sales	5,56
Work under pressure and assertive communication	0,69
Heights and confined spaces	2,08

The management of any type of technology in companies is also important since it leads to great changes in the organizational structure, in the way information is managed and conditions the competition work [13–15]. However, it was found that close to 37% of these companies did not handle any technology, which may be unthinkable for these times where everything moves by technological means; but as most of these companies come from the manufacturing sector. May be the case that they have many technological limitations due to the costs that this implies and the few gains that are obtained in the face of competition from emerging markets.

Within the remaining 63% of companies, 26% use computers with internet connection and some office applications, in other cases the business management is done through social networks or simply as an advertising strategy to continue exploring markets. 21% use some type of specialized and updated equipment, which allows them to have an improvement in innovation and production management, making them much more competitive; and 16% makes use of some specialized software in their administrative management, which allows to advance in the streamlining of processes and feedback of them as evidenced in Fig. 3.

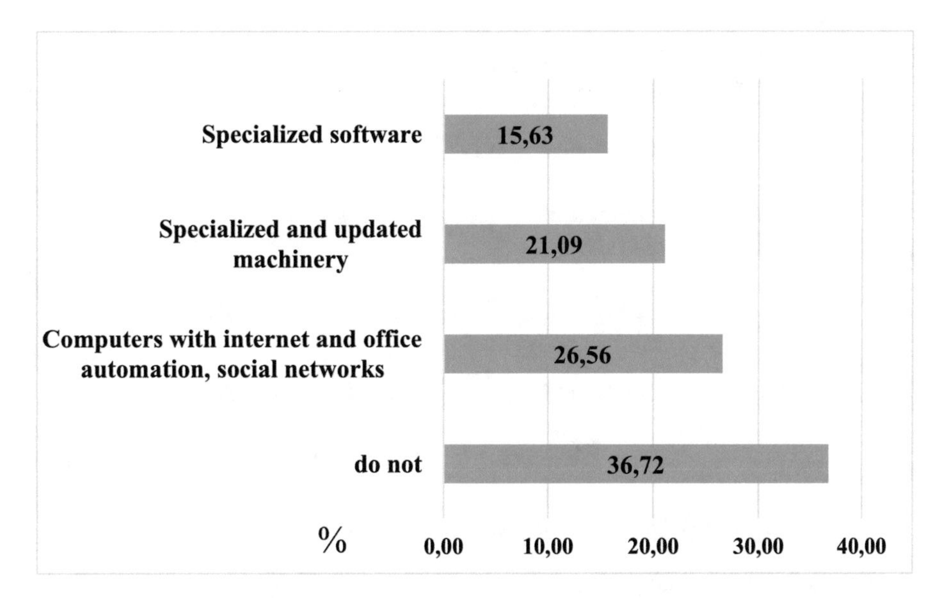

Fig. 3. Use or implementation of any type of technology in the company

According to [16] with the help of informational and computational resources for SMEs, they have access to a number of tools that allow the development of simple applications such as templates, macros and applications that serve the development of specific cases that an entrepreneur has to face along with their work teams, this will allow the company to reduce costs, increase productivity, be more competitive in the market, increase the quality and efficiency of different business processes. At least 72% of the companies consulted handle some type of information system, which can range from the use of simple databases made in MS Office software to the use of accounting, financial, inventory and commercial applications, which is a clear index of the use of business computing.

The analysis of the organization identifies the resources in relation to its contribution to make sustainable the competitive advantage; that in this aspect allows the organization to stay in the market and enjoy a preeminent position. What characteristics do we look for in the resources to give them this position? They must generate value, be rare, inimitable and non-substitutable [7]. Today, information is a valuable resource. Those who have it look for the best and most efficient way to manage it, so the treatment with the clients and the storage of their contact data in databases deserves a special care and treatment such as its protection, its effective management and its constant enrichment. It was found that 28% of the companies did not have any type of system that would allow them to effectively manage the information, the others had some kind of implementation in databases or committed to information management through the use of accounting software, financial and administrative, inventory control, commercial, sales and communications, as evidenced in Fig. 4.

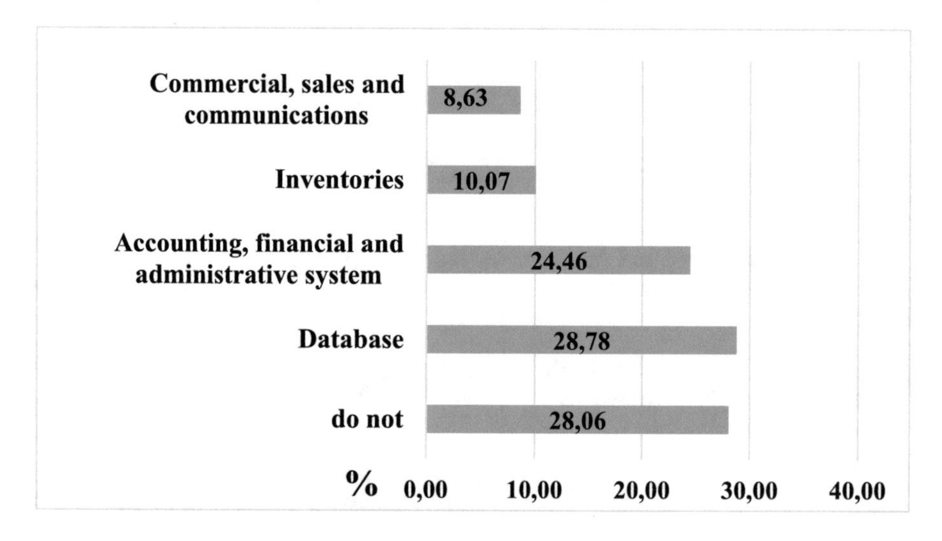

Fig. 4. Use in the company of some system for information management

When consulting whether the company had any customer database or communicated with them using some kind of technological means, 63% of them said that they did, although in most of these cases the management of these was not performed in an appropriate manner since there are no specialized elements or software available for its care and protection due to the costs involved in implementing and maintaining such systems.

5 Discussion

New technologies and new forms of communication are a source of information with great potential in the creation and development of the business model, so it is very important for the company to incorporate technologies and information systems, especially those businesses that are developed online taking advantage of the web; business options such as teleworking, e-commerce, Apps development, video mapping, prototypes of robotics, home automation and urbotics [A set of public services and facilities that are automated to improve the city and public services, safety, well-being, communications and energy management.], have opened a door to new markets, but technological limitations also limit the future of the company.

The use of technologies and information systems in companies favor communication and marketing processes, serve as support for the improvement of processes within organizations, improve efficiency and quality, contribute to the increase of market globalization, facilitate the Diagnosis, analysis and business projection, increase visibility in social networks, facilitate the management of physical and financial resources, improve relationships with customers, facilitate decision making in favor of increased productivity. All of which in summary lead to generation of new business opportunities, coinciding with the conclusions and affirmations of other published authors [17–19].

For [20] "the philosophy of the companies with the best practices requires that the organizations learn and respond quickly to the changes in the market and that rationalize the operations, with the purpose of facilitating modifications of workflows at the moment" (p. 71), puts special pressure on companies to automate all their processes and be a bit more efficient with respect to the use of technology, the development of automated applications that can be configured and regulated, as well as adaptable to the needs; this then demands the design of a Resistant Architecture for Computer Technology (ARTI).

An important aspect that has prevented companies from facing new markets, doing business in other countries, and being more competitive internationally is the certification of their processes, services or products in a technical standard, whether national or international. Although it is true that this does not prevent it from growing and competing in the local market, when it is necessary to expand the business to the international level it becomes a limitation and a market entry barrier, which in turn results in a competitive disadvantage [21]. When these management systems are implemented in SMEs, they seek to be a little more efficient with the management of physical resources and raw material, improvement of planning, administration, organization, as well as having an efficient control over the use of technological and human resources.

The main obstacles for SMEs to develop in Colombia, according to [13], are: credit restrictions, difficulties in identifying and accessing appropriate technology, formalization and absorption of new technologies, technical and competitive limitations imposed to the production scales, the deficient physical infrastructure, the lack of business association networks, the lack of managers with managerial capacity and strategic thinking, and the difficulty of articulation within the sector and large companies and state purchase systems. In competitiveness at company level, many factors influence, among others the following: a stable macroeconomic environment; a solid financial system; the ability to use, adapt and create new technologies; the ability to attract, train and retain human capital [22, 23].

6 Conclusions

In the business world, the evaluation of dynamic capacities is gaining strength and value because it allows the in-depth study of organizations, allows measuring the true potential of the company to compete in the market, as well as evaluating its capacity technique and its ability to learn from the different challenges imposed by the modern business world and globalization. You must first establish a plan for operational and managerial development, based on the disposition of the same organization, resources such as personnel and the main asset that it is information. The organization must also have the disposition to learn to do, to learn to manage its information, and to assign it the it the value it really has.

The reality found in the SMEs that are located in the southern area of Bogotá indicates that we must work to improve the following points:

- Take advantage of business opportunities and market growth.
- Effective incorporation of ICTs, giving it the importance that current business trends demand.
- Updated marketing plans that involve the use of new technologies and social networks to make the business more visible. This should be implemented by inclusion in the strategic planning stages and should be accompanied by specific goals in the short, medium and long term.
- Improvement of technologic support, technical infrastructure, management tools that facilitate the improvement of processes from the administrative and financial side.
- Trained personnel, investing in training plans is contributing in the build-up of strategic assets for the company.
- Study of the market and competition, especially in those aspects that are making a difference, which allows the future to venture into different markets and innovative technological means.
- Formalization of those who are still in the informal sector, as an opportunity to access state support plans for technological improvement and innovation.

On behalf of state and governmental entities supporting MSMEs, the work plan must address:

- Improvement in financing planning, training and permanent advisory services for those who want to start their business legally.
- Promote competitiveness by regulating markets efficiently
- Promote legalization
- Plans to overcome the financing barrier by improving these support source, especially by promoting those that refer to investment in technological upgrades, incorporation of ICTs and new technological developments as part of innovation processes.
- Provide opportunities to entrepreneurs who are located in areas such as the study that is quite neglected because they are people with limited resources that move more in the world of informality

- Training programs for managers and business owners around administrative, financial, technological, innovation, leadership, occupational health and safety issues, with a view to improving competitiveness
- Facilitate the possibility for the state and local government entities to implement a QMS that will open doors to other markets
- Regulate the import market of low quality and low price products that compete with the local market in an unfair manner, generating a destabilization in the market.

The adoption of a strategic model defined by the organization helps to meet the goals and objectives set by the person who conceived or directs it. However, it is necessary to complement the process with periodic monitoring and evaluations along with a plan of business development, which should include technological updating processes, personnel training, and competitive analysis.

In the business world, the guideline is always marked not only by those who are at the forefront, but also by the technologies that they use and develop, to the extent that they try it out and put it in the knowledge of others, because society will advance as well of knowledge and the sector.

Part of the study also concludes that the lack of adoption of appropriate technological and communication tools in the companies of this area of the city continues to hold them back, so that their introduction to new markets such as international markets where the highest incomes are, the greatest sources of financing, innovation and cutting-edge technology, it is in these conditions very unlikely, thus showing in official figures in the decline of the competitiveness of the region and the country, low quality productivity and economic decline, that in the future it becomes less money for investment in different sectors of state responsibility such as social, support for research, innovation, education and health.

To give continuity to the investigative process, the formulation of a new project is staged where the impact it has had in different areas such as the General Management, Administration and Human Resources, Production, Finance and Accounting, Advertising and Marketing, is evaluated. Low appropriation of ICTs, as well as the impact that the application of public policy has had on the subject in the study area in particular due to the vulnerability condition of the majority of its inhabitants.

References

1. Dávila L (2009) Capacidades dinámicas un acercamiento a las teorías contemporáneas de la firma. (Vol. Monografías de Administración). Universidad de los Andes, Facltad de Administración, Bogotá D.C.
2. Teece DJ (2007) Explicating dynamic capabilities: the nature and microfoundations of (sustainable) enterprise performance. Strateg Manag J 28(13):1319–1350. https://doi.org/10.1002/smj.640
3. Nelson RW (1982) An evolutionary theory of economic change. The Belknap of Harward University Press, Cambridge
4. Barreto I (2010) Dynamic capabilities: a review of past research and an agenda for the future. J Manag 36(1):256–280. https://doi.org/10.1177/0149206309350776

5. Hirsch J (2007) Familleet enterprise en histoire. En: Dictionnairehistorique de l´economie-droitXVIIIe – XXesiecles. Droit et Societe 17. L.G.D.J. Paris
6. Solleiro J (2005) Competitiveness and innovationsystems: ThechallengesforMéxico'sinsertion in the global contex, Articulo electrónico Planificación estratégica y niveles de competitividad de las Mipymes del sector comercio en Bogotá. Recuperado de: http://www.sciencedirect.com/science/article/pii/S0123592314001600
7. Wang CL (2007) Dynamic capabilities: a review and research agenda. Int J Manag Rev 9. https://doi.org/10.1111/j.1468-2370.2007.00201.x
8. Helfat C (2009) Understanding dynamic capabilities: progress along a developmental path. Strategic Organization. Los Angeles: Sage Publications
9. Sánchez AA, Bañón AR (2015) Factores explicativos del éxito competitivo: el caso de las pymes del estado de Veracruz. Contaduría y administración, (216). http://www.cya.unam.mx/index.php/cya/article/view/568
10. Creswell JW, Klassen AC, Plano Clark VL, Smith KC (2011) Best practices for mixed methods research in the health sciences. Bethesda (Maryland): National Institutes of Health, 2013, pp 541–545. http://www2.jabsom.hawaii.edu/native/docs/tsudocs/Best_Practices_for_Mixed_Methods_Research_Aug2011.pdf
11. Arias FG (2012) El Proyecto de Investigación. Introducción a la metodología científica. 5ta. Fidias G. Arias Odón
12. Brannen J (ed) Mixing methods: qualitative and quantitative research. Routledge (2017)
13. Saavedra García ML, Tapia Sánchez B (2013) El uso de las tecnologías de información y comunicación TIC en las micro, pequeñas y medianas empresas (MIPyME) industriales mexicanas. Enl@ ce: Revista Venezolana de Información, tecnología y conocimiento, 10(1). http://www.redalyc.org/html/823/82326270007/
14. Johnston MW, Marshall GW (2016) Sales force management: Leadership, innovation, technology. Routledge
15. Porter ME, Millar VE (1985) How information gives you competitive advantage (1985)
16. Paredes P (2012) Informática para Pymes. Aplicaciones para aumentar su productividad. Editorial Macro 351(1):2–11
17. Stadtler H (2015) Supply chain management: an overview. In: Stadtler H, Kilger C, Meyr H (eds) Supply chain management and advanced planning. Springer Texts in Business and Economics. Springer, Heidelberg. From: https://doi.org/10.1007/978-3-642-55309-7_1
18. de Pablos Heredero C, Agius JJLH, Romero SMR, Salgado SM (2012) Organización y transformación de los sistemas de información en la empresa. Esic Editorial
19. Riascos Erazo SC, Aguilera Castro A (2011) Herramientas TIC como apoyo a la gestión del talento humano. Cuadernos de administración 27(46). http://www.redalyc.org/html/2250/225022711011/
20. Kerr JM (2009) Las mejores prácticas para empresas exitosas. Editorial Panamericana, pp 71–78
21. Acosta Prado JC, Longo-Somoza M, Fischer AL (2013) Capacidades dinámicas y gestión del conocimiento en nuevas empresas de base tecnológica. Cuadernos de Administración 26(47): 35–62. http://www.scielo.org.co/scielo.php?script=sci_arttext&pid=S0120-35922013000200003&lng=en&tlng=. Accessed 3 June 2013
22. Nabi I, Luthria M (eds) (2002) Building competitive firms: incentives and capabilities. World Bank Publications
23. Favaro Villegas D (2013) Enfoques de la teoría de la firma y su vinculación con el cambio tecnológico y la innovación. Revista Cultura Económica 31(85). https://dialnet.unirioja.es/descarga/articulo/5089778.pdf

24. Lozano AXC (2015) El impacto de la tecnología en el ámbito social y en la desigualdad (The Impact of Technology in the Social Sphere and Inequality). Inclusión & Desarrollo 2(2):16–20. http://dx.doi.org/10.26620/uniminuto.inclusion.2.2.2015.16-20
25. C.D.C. de Bogotá (2017). Balance de la Economía de la región Bogotá-Cundinamarca. http://bibliotecadigital.ccb.org.co/handle/11520/18795

Edition, Publication and Visualization of Geoservices Using Open-Source Tools

Pablo Landeta(✉), Jorge Vásquez, Xavier Rea, and Iván García-Santillán

Software Engineering, Faculty of Applied Sciences, Universidad Técnica del Norte, Ibarra, Ecuador
{palandeta,jwvasquez,mrea,idgarcia}@utn.edu.ec

Abstract. Currently, there is a need to share and reuse geographic information with the academic and business sectors through the use in different application fields, such as: agriculture, tourism, geology, cadaster, so on. Thus, this paper introduces the development of a software for the edition, publication and visualization of geographic layers as geoservices, using open-source tools. This module is an add-on to the Geoportal software previously developed at Universidad Técnica del Norte (Ecuador). For the editing phase, Atlas Styler was used to establish connection to the WMS server (Web Map Service) and then modify the published geoservices using SLD code (Styled Layer Descriptor). For the publication phase, a Geoserver software was installed to allow WMS service to publish the geoservices. To visualize the published information, the Spatial Data Infrastructure (SDI) software from Universidad Técnica del Norte - GEOPORTAL UTN was used. In this interface, the Leaflet library was used to display geoservices on the map. Finally, to evaluate the WMS server performance the following elements were analyzed: concurrent users, response times, response errors obtained and their percentage. The results showed a high correlation between number of users and response error and a very high correlation between number of users and processing time of a request. Although, there are some response errors when the number of concurrent users exceeds 5000, the server still works well.

Keywords: GIS · Geoportal · Geoservices · Geoserver · Mapserver
Atlas styler

1 Introduction

1.1 Problem Statement

A Geographic Information System (GIS) is a set of tools to collect, store, retrieve, transform, and display real-world spatial data for some purposes [1].

The information contained in an GIS repository has no academic or scientific validity if the stored data remains undisclosed [2]. For this reason, it is necessary to create a software which publishes geographic layers and its associated alpha-numeric information so that researchers and the community can easily access and benefit from geo-spatial information for the visualization and the analysis of the generated maps.

M. Botto-Tobar et al. (Eds.): TICEC 2018, AISC 884, pp. 266–280, 2019.
https://doi.org/10.1007/978-3-030-02828-2_20

In Ecuador, access to this kind of software is still limited [3], and universities and other institutions must address this problem by encouraging research related to the creation of new tools and/or the improvement of existing ones in order to facilitate the broad use of this technology. The performance of these software is something that should be improved in terms of response times, accuracy, ease of use, updated information, etc. [4]

The present work aims to complement the work developed in [5], detailed in the next section. It is proposed to deploy a WMS (Web Map Service) server that allows users to edit, publish and visualize geoservices thanks to several open source software technologies. To verify the smooth operation of the WMS, a WMS client software is included to make sure that the layers can be called as geoservices from other geoportals. Finally, a server performance analysis is performed based on response times with multiple concurrent users. This is the main contribution of the present work which is further elaborated in the following sections.

1.2 Review of Literature

The task of exposing and publishing geo-spatial data has been carried out by the academic and research communities over the past few years. The work proposed by Landeta et al. [5] consisted in providing the academic community with a geoportal containing both local and national geo-statistical information for different fields, for example: precision agriculture [6], environmental change [7], renewable energy [8], geology [9], population studies [10], urban cadaster [11], transit [12], etc. Wei et al. [13] provided an overview of some important works in the area of GIS that can benefit from the work proposed in [5]. In addition, in [5], it was possible to visualize geographic layers sorted by type and subtype of projects, as well as a thematic map viewer which allows the geographic information to be combined and showed on a map in an integrated way. Additionally, a management software was created to automate some processes such as geographic layers loading to the database and their management in a way that permitted greater independence between the system manager and the software.

An important research was the one developed by Evangelidis et al. [14], who propose a model that besides generating a data acquisition engine, adds an engine for the analysis of these geo-spatial data, taking advantage of the potential offered by cloud computing using satellite images.

In Hu et al. [15], the problem caused by the large amount of data generated from geographic-topographic information is considered and addressed by a supercomputing-based architecture, focusing on data usability and evaluating the performance of the servers to use geo-spatial data.

In Gong et al. [16], a sensors-based platform was developed. It collects in real time information from the environment (weather, temperature, air quality, etc.) and displays it as web services. The result is a geographical portal where it is possible to expose as services the environment information showing very specific statistics of each analyzed parameter.

In Juan and Lee [17], the research focused on the efficient management of geospatial data, in which there is no collaborative work with very scattered data. They proposed an architecture working with the OCG standards by storing data in GeoJSON format in a non-relational MongoDB database.

These existing works were relevant for the development of the current proposal because they show the importance of GIS and some possible fields of application. Also, they determine what has been done in this area until now. Thus, this knowledge allows that current research to focus on new findings. Below is a brief theoretical rationale of the main concepts used in the development of the study.

1.3 Theoretical Foundations

Spatial Data Infrastructure. The Spatial Data Infrastructure (SDI) is a collection of relevant technologies based on political and institutional structures that facilitate the availability and access to spatial information. The purpose is to achieve the democratization of information [18].

WMS Geo-services. The services of an SDI are the functionalities accessible through Internet resources. They include WMS map query service, geographic data download service with all its attributes (WFS - Web Feature Service), access service to raster coverage data or photographic images (WCS - Web Coverage Service) [19].

Map Server for Publication of Geoservices. Map servers are software components whose function is to deliver raster and vector data on the Web in the form of maps. They are available through the OWS protocols (WMS, WFS and WCS). Map servers enable to centralize spatial information to avoid duplication of information and to allow different users to reuse it. They are a major component of the SDI and they offer data as web services through the implementation of OWS protocols [20].

REST API. To facilitate the use of OWS and create better accessibility, the concept of REST (Representational State Transfer) can provide a service layer where part of the complexity is resolved. REST can also be used to create links between data sets that are normally isolated from each other [21].

REST provides a representation of the current state of something through URI (Uniform Resource Identifier), which is, if required, editable and saved in the URI where the state of the object is modified [22]. In order to get the geoservices information published in Geoserver, this API uses the GET operation to collect the data and present it in JSON format.

SLD (Styled Layer Descriptor). Styled Layer Descriptor is an XML schema specified by the OGC (Open Geospatial Consortium) as a standard language to describe several layers defining a map. It defines a coding that extends to WMS to allow user-defined symbolization and colors. It allows users to have control over the visual part of the geospatial data. It can define style conventions that both the server and the client can understand [23].

2 Materials and Methods

2.1 Tools

The project described in Landeta et al. [5] works with the operating system Centos 7 [24], the PostgreSQL 9.5 database [25] and the PostGIS 9.2 module [26] for the support of spatial objects in the object-relational database. The map server used was Mapserver 7.0.1 [27]. Pmapper 5, a framework based on PHP and MapScript, was used to visualize the maps [28]. For the development of the management module, the Yii framework [29] was used, which allows to apply the MVC architecture (Model View Controller). For the View layer, the Bootstrap framework [30] was used to integrate HTML5 technology. Figure 1 shows the architecture of the Geoportal UTN.

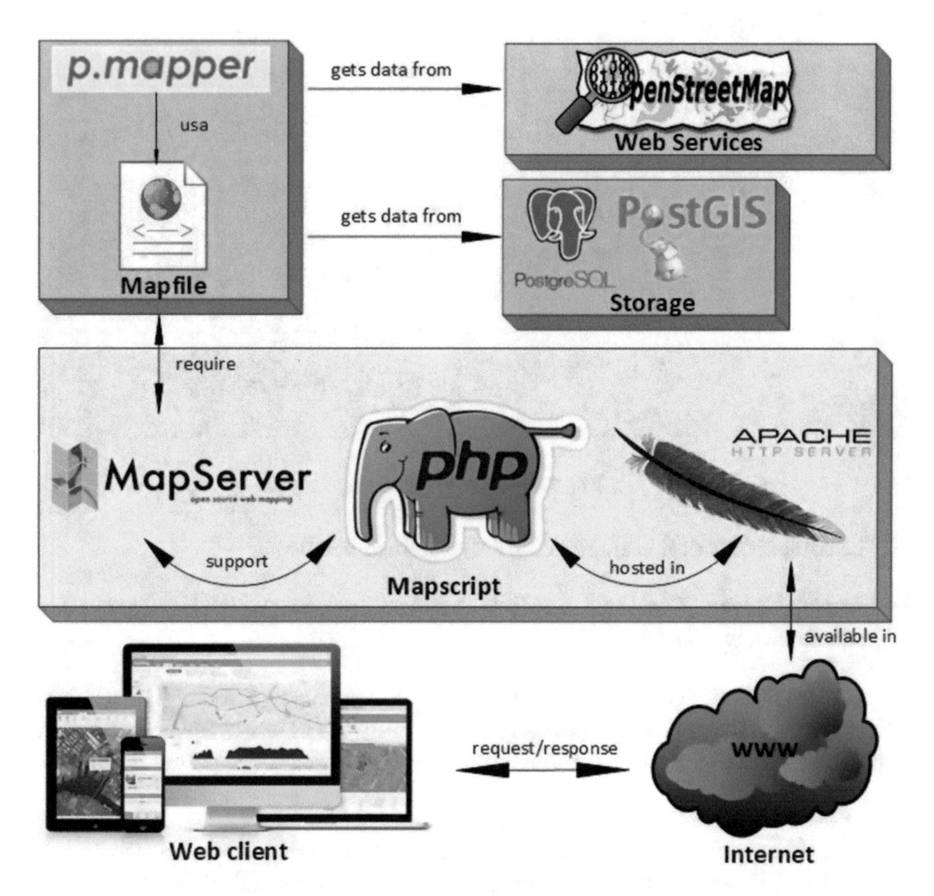

Fig. 1. Architecture of the GEOPORTAL UTN [5].

In the implementation of this project, open-source tools were used because they comply with international standards and have good performance in execution [31]. The use of free software tools is a policy of the Ecuadorian government described in

Executive Decree No. 1014 issued in 2008 [32]. This regulation disposes to use open source software within computer systems and equipment in the public administration of the Ecuadorian state.

Figure 2 shows the architecture used to edit layers using SLD and to publish Geoservices.

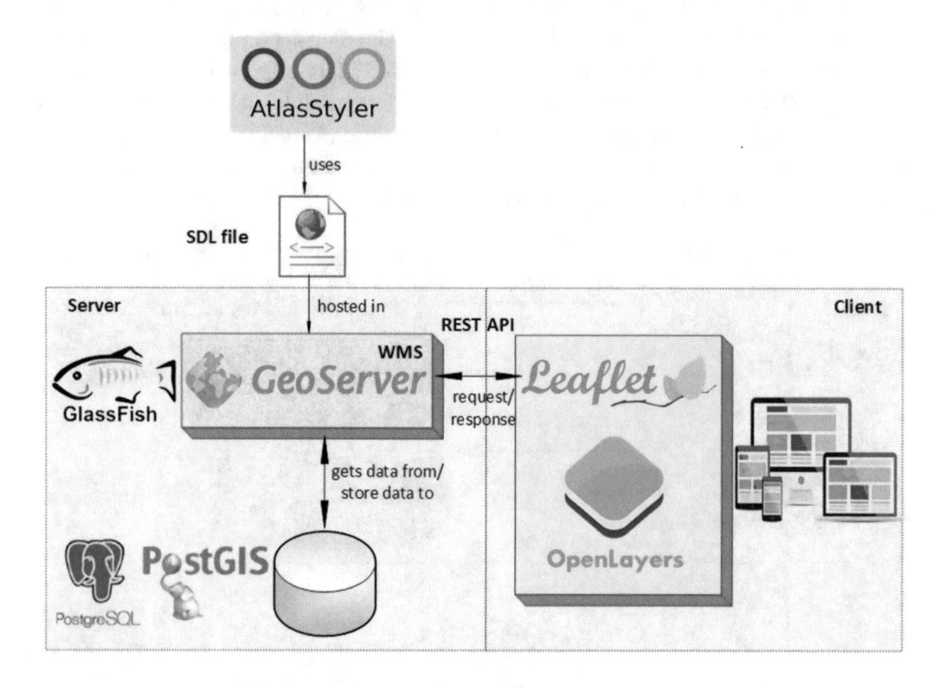

Fig. 2. Project architecture.

The components of this architecture are explained below:

Application Server. Oracle GlassFish 4.1 application server which implements Java EE technologies was installed [33].

Geoservices Server. Geoserver 2.12.1 was used to implement several OGC standards such as WFS, WMS, WCS, etc. [34]. In this case, only WMS was used because it is the service that allowed the editing, publishing and visualization of the geoservices.

Geoserver REST API. This API requires the client to authenticate itself for security purposes. The type of content used was application/json; this method consists in validating a requirement and if it is satisfactory, it returns a status code that acknowledges it as valid [35].

Figure 3 shows the flowchart used to access from the web application the published layers provided by the Geoserver server as WMS services.

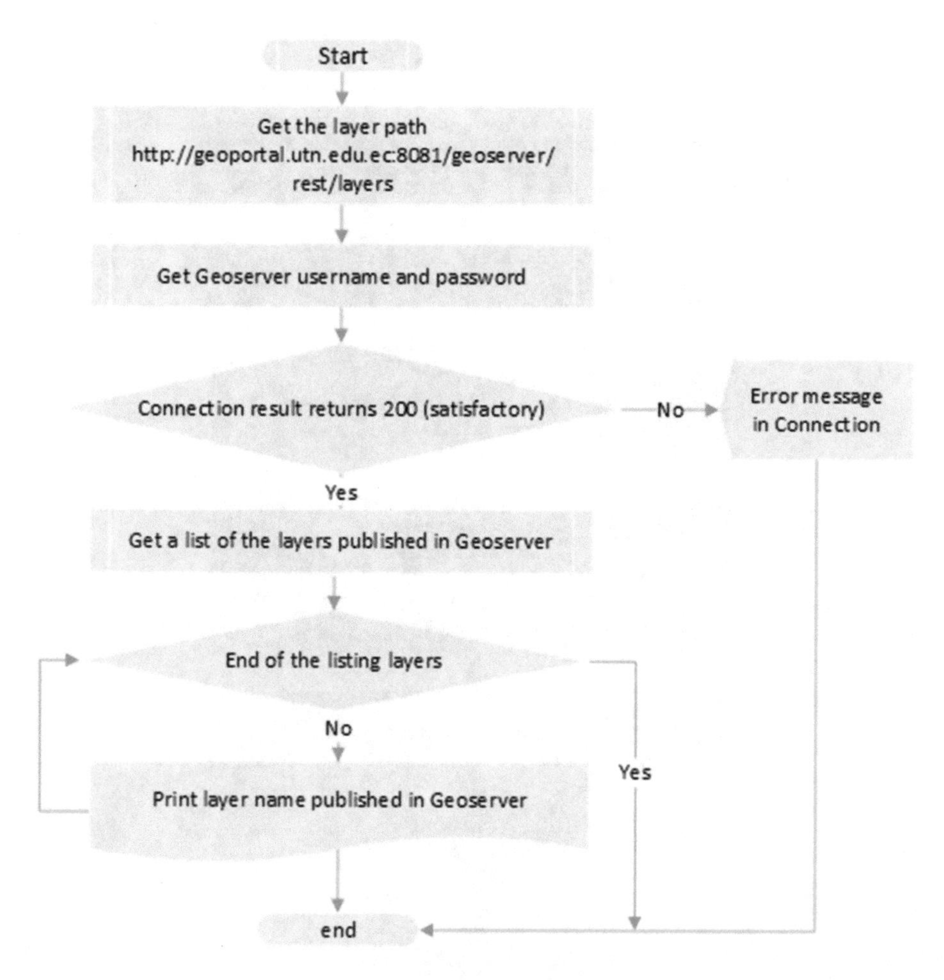

Fig. 3. Flowchart to access the Geoserver REST API.

Leaflet 1.3.1 JavaScript Library. Leaflet is a JavaScript library used to create interactive web-based maps [36].

Figure 4 shows the flowchart of the OpenStreetMap base layer as well as the layer that is selected from the set of layers available on the server as a WMS service.

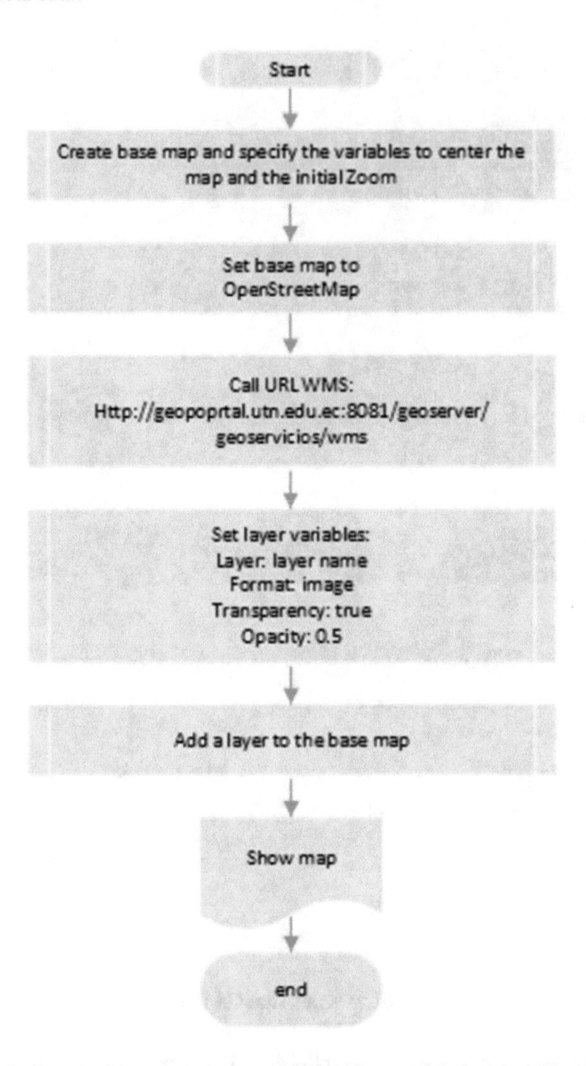

Fig. 4. Flowchart used to preview the published layer with the use of Leaflet library.

Software for Editing Layers in SLD Format. For the edition of the layers and the creation of the file in SLD format, Atlas Styler 1.9 [37] software was used. It allows users to connect to the WMS server and visually modify the style of any published layers. Code is generated in SLD format which must then be copied into the Geoserver interface to complete the editing of the style of a published layer.

Figure 5 shows the Geoserver WMS server architecture, where data is uploaded from the database, style rules are applied, and image is rendered and transformed into a selected map image to be displayed in a WMS client software.

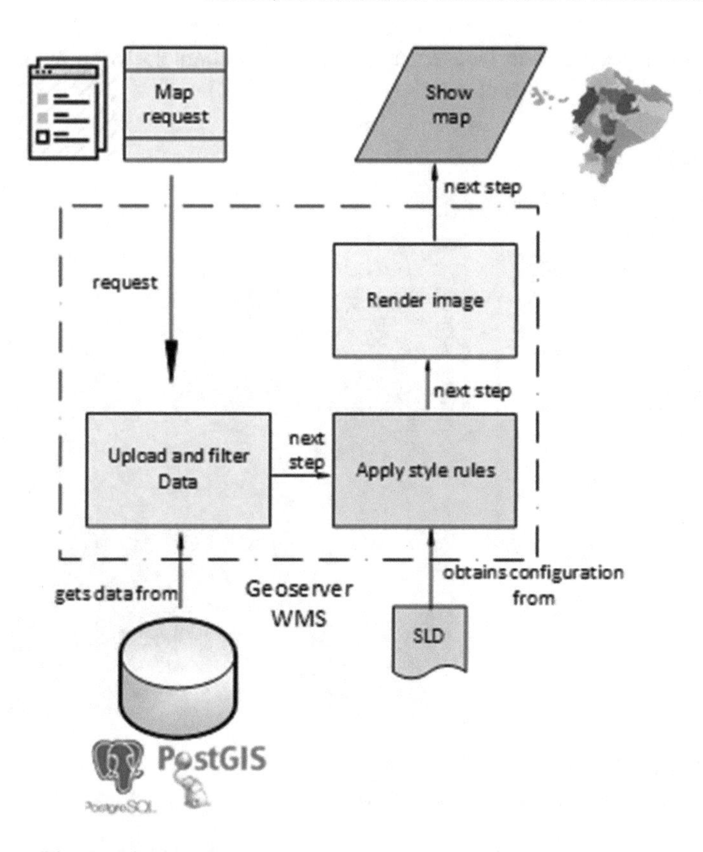

Fig. 5. The WMS server accepts the request and shows the map

3 Experimental Results

3.1 Verification of the Correct Operation of the Geoservices

Once the platform had been configured correctly, the following results were obtained:

- The Geoserver as a WMS server correctly publishes the geographic layers.
- The PHP-based language software successfully provides visualization of the published layers through the Geoserver REST API.
- The preview of the geoservices using the Leaflet library is successfully performed, showing a base layer as a background and superimposing the layer selected from a list of published layers.
- Style changes that are made in the Atlas Styler software are displayed successfully when a layer is selecting as geoservices; this shows that the software in charge of generating SLD code works correctly.

Figure 6 shows the display obtained from the published layers as WMS services and the map with the use of the Leaflet library. The example shows the WMS server URL, a list of layers published in Geoserver and the WMS client software with the preview of the selected map. This application is available in [38].

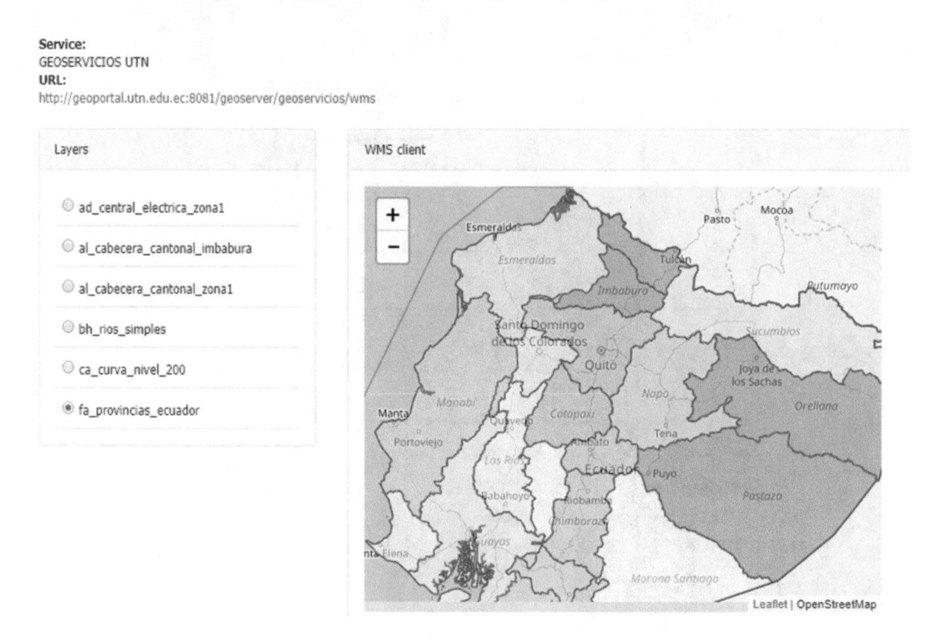

Fig. 6. Map resulting from the edition, publication and visualization of geographic layers for the Geoservice related to the provinces of Ecuador.

3.2 Performance Analysis of Geoservices

The objective for the experiment was to prove the WMS server's performance. The research question is established as follows: what is the required number of concurrent users for the WMS server to start giving failed responses? In order to check the successful operation of the geoservices, the methodology proposed by Naranjo [39] was applied: The WMS Geoserver server was submitted to performance tests in order to measure the response times of the web services while simulating several concurrent users. The Apache JMeter tool [40], which allows load testing of Web applications, was used for this purpose. To confirm the obtained data, it was done three repetitions with each configuration and then the average was calculated. To apply the experiment, the scenario included the following test parameters:

- Server: geoportal.utn.edu.ec
- Port: 8081
- Route:/geoserver/geoservicios/wms?
- Service: WMS
- Version: 1.1.1
- Request: GetMap
- Layer: geoservicios:fa_provincias_ecuador

The variables analyzed in the test were:

- Concurrent users
- Response time to requests received (mean and standard deviation)
- Size, in bytes, of the expected file i.e. 789 bytes. This value is the number of bytes of the geoservice with which the tests were performed. It is detailed in the test parameters.
- Errors, i.e. the server's failed responses or server's responses showing byte size discrepancy.
- Percentage of errors, i.e. ratio between the number of errors and the number of concurrent users.

Table 1 shows the results obtained in the performance measurement:

Table 1. Results of the WMS server performance measurement

# Users	Response time (ms)				
	Mean	Standard deviation	Bytes	Error	% Error
500	946	191.2	789	0	0.00
1100	923	294.19	789	0	0.00
2100	4114	1475.32	789	0	0.00
2900	4409	1375.14	789	0	0.00
4000	8953	3356.39	789	0	0.00
5000	10153	4261.76	789	0	0.00
6000	11456	4914.37	789	642	10.70
7000	12003	5426.24	789	1695	24.21
8000	12053	5667.06	789	2587	32.34
9000	14433	6419.22	789	3087	34.30
10000	12784	5762.48	789	4925	49.25

Because the number of bytes is a constant, it has been disregarded. Based on the variables analyzed in Table 2 and from the analysis of the Pearson's correlation coefficient [41], it is demonstrated that there is a high positive correlation of 0.881 between the number of users and the error, which means that the more users consuming geoservice, the greater the number of error. It is also shown that there is a high positive correlation of 0.740 between the standard deviation and the error, which indicates that the higher standard deviation, the greater the number of errors.

Figure 7 shows the graph of results obtained from the response times compared to the number of concurrent users:

Table 2. Correlations between variables measured in the performance test

		Users	Time	Deviation	Error
Users	Pearson correlation	1	0.949[a]	0.964[a]	0.881[a]
	Sig. (bilateral)		0.000	0.000	0.000
	N	11	11	11	11
Time	Pearson correlation	0.949[a]	1	0.994[a]	0.699[b]
	Sig. (bilateral)	0.000		0.000	0.017
	N	11	11	11	11
Deviation	Pearson correlation	0.964[a]	0.994[a]	1	0.740[a]
	Sig. (bilateral)	0.000	0.000		0.009
	N	11	11	11	11
Error	Pearson correlation	0.881[a]	0.699[b]	0.740[a]	1
	Sig. (bilateral)	0.000	0.017	0.009	
	N	11	11	11	11

[a]Correlation is significant at 0.01 (bilateral).
[b]Correlation is significant at 0.05 (bilateral).

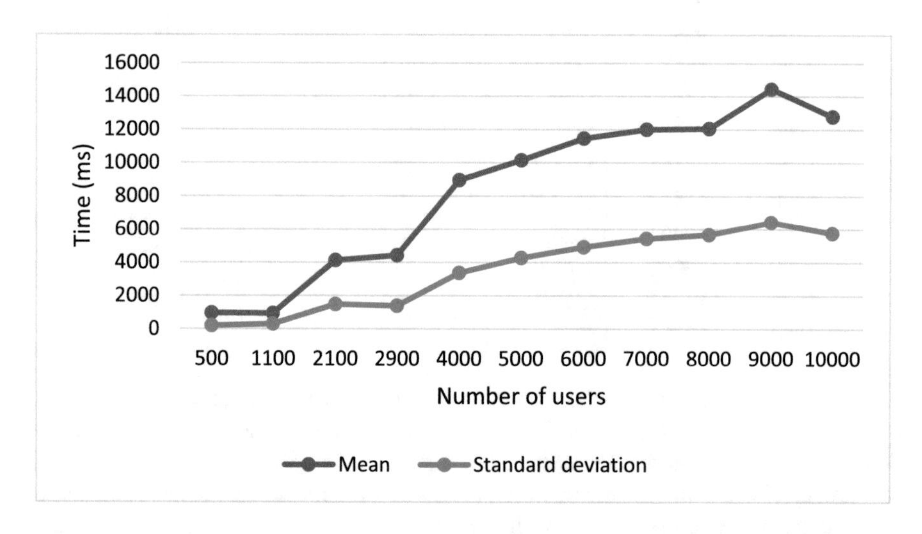

Fig. 7. WMS server performance analysis results graph.

There is a certain linear tendency in the relationship, which also confirms the analysis and calculation of Pearson's correlation coefficient [41]. There is a very high positive correlation between the number of users and the processing time of the request. From 9000 users an up, the service starts showing errors, the mean and the deviation begin to decrease and the error to increase.

It is also important to analyze the number of occurring errors and from what point this phenomenon starts. Thanks to this information, it is possible to predict the behavior of the server under specific circumstances.

Figure 8 shows a double-axis graph between response times and response errors with respect to the number of users.

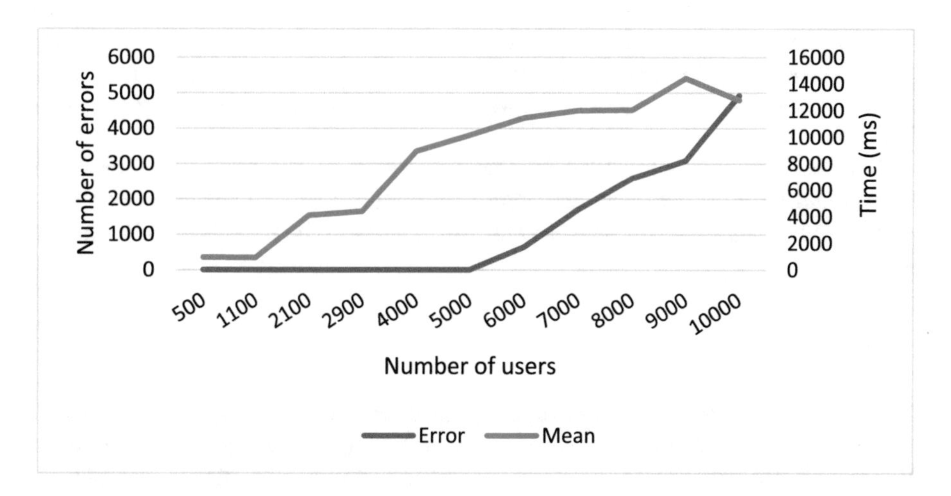

Fig. 8. WMS server performance analysis results graph.

It shows in a better way the behavior of the number of errors and the response times against the growth of the number of users. Although, there are response errors, the application's performance continues to work well.

4 Discussion

The results obtained by analyzing the response times and the number of response errors compared to the number of concurrent users show us that the WMS server has a good performance. This means that the server will have no problems when displaying geoservices with a number of users that exceeds 5000.

However, this number of errors represents a threat to the performance of the server One of the possible solutions could be to implement a load balancer to face this threat, which is proposed as future work.

Gui et al. [42] have an investigation of interest because it conducts a performance analysis of several WMS public services. The advantage of that proposal is that it makes use of a distributed monitoring framework to track WMS services and capture information required for the analysis. Nevertheless, its disadvantage is that performance analysis focuses only on response times without considering concurrent users. In the present study, however, this variable is considered as a key element when making performance analysis.

On the other hand, one of the flaw of this study is that it does not work with a non-relational database that is capable of handling large amounts of information from different sources such as Big Data, as proposed in [43]. The latter for example, could process raster-type information necessary for the analysis of satellite photography. This future work is left for further investigation as it would enhance the capacity of the current platform.

The research presented by Moncrief et al. [44] shows the advantage of using styles for the layers, which affects the way data is displayed. This feature can make a difference when interpreting information exposed in a geoportal. This proposal embraces this advantage and integrates styles using SLD, with which the published layers can be formatted.

Finally, it is estimated that the developed application will have a strong impact in the medium term. In the meantime, alternatives related to load balancing and high-performance computing will be analyzed, which will also be the subject of a separate study.

5 Conclusions

In this research it has been possible to bring together several open-source technologies to create a software that allows users to edit, publish and visualize geospatial layers. This module can display these layers on the web as services, increasing the initial functionalities of the one proposed in [5].

The Atlas Styer software correctly generates the SLD file to edit the published geographic layers as WMS services. This is verified by the layer display in a WMS client. It also shows that the Geoserver REST API can be an intermediary between the layers published as WMS services and the presentation of the information in the web interface. The Leaflet library enables to show the layers published in Geoserver in a WMS client.

Regarding the results of the performance analysis, it is considered that they are in the range of expected acceptable values (Fig. 8). There is a high positive correlation between the number of concurrent users and the number of response errors (Table 2). This correlation is given for a number of users greater than 5000, since below to this number no errors appear. As for the results of the response time analysis, they are also considered to be within the acceptable values (Fig. 7). The analysis also shows a very high positive correlation between the number of users and the response times. (Table 2)

The experimentation (Table 1) shows efficiency in the response to requests as over 5000 users could be handled without response errors. When tested with 10000 users, the level of errors only reaches 50%.

References

1. Burrough P, McDonnell R, Lloyd C (2015) Principles of geographical information systems, 3rd edn. Oxford University Press, Oxford
2. Bishop W, Grubesic T (2016) Geographic information, maps, and GIS. In: Geographic information, pp 11–25. Springer, Cham
3. Pacheco D (2013) Infraestructuras de datos espaciales (IDE) Universidad del Azuay. In: II Congreso Binacional de Investigación, Ciencia y Tecnología de las Universidades, Loja
4. Longley P, Goodchild M, Maguire D, Rhind D (2015) Geographic information science and systems, 4th edn. Wiley, Hoboken

5. Landeta P, Vásquez J, Saltos T, Rea X (2018) Geoportal con visores temáticos para visualización de información Geoespacial del Ecuador. IDEAS Innov Dev Eng Appl Sci 1(1): 31–40
6. Oshunsanya S, Oluwasemire K, Taiwo O (2017) Use of GIS to delineate site-specific management zone for precision agriculture. Commun Soil Sci Plant Anal 48(5):565–575
7. Wei Y, Ye X (2014) Urbanization, urban land expansion and environmental change in China. Stoch Env Res Risk Assess 28(4):757–765
8. Resch B, Sagl G, Törnros T, Bachmaier A, Eggers J, Herkel S, Narmsara S, Gündra H (2014) GIS-based planning and modeling for renewable energy: challenges and future research avenues. ISPRS Int J Geo-Inf 3(2):662–692
9. Porwal A, Carranza E (2015) Introduction to the Special Issue: GIS-based mineral potential modelling and geological data analyses for mineral exploration. Ore Geol Rev **71**:477–483
10. Greger K (2015) Spatio-temporal building population estimation for highly urbanized areas using GIS. Trans GIS 19(1):129–150
11. Lateş I, Luca M, Chirica S, Iurist D (2017) The use of the GIS model on the implementation of urban Cadastre. Present Environ. Sustain. Dev. 11(2):163–172
12. Czioska P, Mattfeld D, Sester M (2017) GIS-based identification and assessment of suitable meeting point locations for ride-sharing. Transp. Res. Procedia 22:314–324
13. Wei F, Grubesic T, Bishop B (2015) Exploring the GIS knowledge domain using citespace. Prof Geogr 67(3):374–384
14. Evangelidis K, Ntouros K, Makridis S, Papatheodorou C (2014) Geospatial services in the Cloud. Comput Geosci 63:116–122
15. Hu H, Hong X, Terstriep J, Liu Y, Finn M, Rush J, Wendel J, Wang S (2016) TopoLens: building a CyberGIS community data service for enhancing the usability of high-resolution national topographic datasets. In: Proceedings of the XSEDE16 conference on diversity, big data, and science at scale, pp 39:1–39:8. USGS Publications, New York
16. Gong J, Geng J, Chen Z (2015) Real-time GIS data model and sensor web service platform for environmental data management. Int J Health Geogr 14(2) (2015)
17. Jun S, Lee S (2017) Prototype system for geospatial data building-sharing developed by utilizing open source web technology. Spat Inf Res 25(5):725–733
18. Veintimilla-Reyes J, Avila F (2015) Análisis e implementación de una Infraestructura de Datos Espaciales (IDE). Caso de estudio: Gobierno autónomo descentralizado municipal del cantón Guachapala. Rev. Tecnológica - ESPOL 28(2):79–99
19. Víquez-Acuña O, Víquez-Acuña L, Treviño-Villalobos M, Chaves-Álvarez M (2017) Developing applications using spatial data infrastructure geoservices. Use Cases: Directorio Comercial SC, AgroMAG, IDEHN Mobile. Revista Tecnología en Marcha 30(3):85–96
20. Zader P (2016) Implementación de geoprocesos como servicios web. M.S. thesis, Universidad Nacional de Córdova, Argentina
21. Lundkvist A (2016) Development of a web GI system for disaster management. M.S. thesis, Lund University, Sweden
22. How to add layers to GeoServer using the REST API, https://boundlessgeo.com/2012/10/adding-layers-to-geoserver-using-the-rest-api/. Accessed 3 Mar 2018
23. Mishra S, Sharma N (2017) WebGIS based decision support system for disseminating NOWCAST based alerts: OpenGIS approach. Global J Comput Sci Technol 16(7) (2017)
24. CentOS Project. https://www.centos.org/. Accessed 26 Oct 2016
25. PostgreSQL: The world's most advanced open source database. https://www.postgresql.org/. Accessed 26 Oct 2016
26. PostGIS — Spatial and Geographic Objects for PostgreSQL. http://www.postgis.net/. Accessed 26 Oct 2016

27. MapServer - MapServer 7.0.7 documentation. http://mapserver.org/. Accessed 2 May 2018
28. p.mapper - a MapServer PHP/MapScript Framework. http://www.pmapper.net/. Accessed 26 Oct 2016
29. Yii framework. https://www.yiiframework.com/. Accessed 2 May 2018
30. Bootstrap. https://getbootstrap.com/. Accessed 2 May 2018
31. Swain N, Latu K, Christensen S, Jones N, Nelson E, Ames D, Williams G (2015) A review of open source software solutions for developing water resources web applications. Environ Model Softw 67:108–117
32. Subsecretaría de Informática: Decreto Ejecutivo No. 1014 (Executive Decree No. 1014). http://cti.gobiernoelectronico.gob.ec/ayuda/manual/decreto_1014.pdf. Accessed 22 Apr 2018
33. GlassFish: the open source Java EE reference implementation. https://javaee.github.io/glassfish/. Accessed 7 Mar 2018
34. GeoServer. http://geoserver.org/. Accessed 5 Mar 2018
35. REST configuration API reference — GeoServer 2.14.x User Manual. http://docs.geoserver.org/latest/en/user/rest/api/index.html. Accessed 6 Mar 2018
36. Crickard P (2014) Leaflet.js essentials, 1st edn. Packt Publishing Ltd, Birmingham
37. AtlasStyler SLD editor. http://www.geopublishing.org/. Accessed 7 Mar 2018
38. IDE UTN – Geoservicios. http://geoportal.utn.edu.ec/ide_utn/index.php?a=Geoservicios. Accessed 18 June 2018
39. Naranjo A (2013) Evaluación del rendimiento de los servicios WMS de Mapserver y Geoserver para la implementación IDE, M.S. thesis, Escuela Politécnica del Ejercito, Sangolquí
40. Apache JMeter. https://jmeter.apache.org/. Accessed 19 Apr 2018
41. Meyers L, Gamst G, Guarino A (2016) Applied multivariate research: design and interpretation, 3rd edn. SAGE Publications, California
42. Gui Z, Cao J, Liu X, Cheng X, Wu H (2016) Global-Scale resource survey and performance monitoring of public OGC web map services. ISPRS Int J Geo-Inf 5(6):88
43. Lee J, Kang M (2015) Geospatial big data: challenges and opportunities. Big Data Res 2(2): 74–81
44. Moncrieff S, Gulland E (2018) Dynamic styling for thematic mapping. In: Free open source software for geospatial (FOSS4G) conference proceedings 15(47)

Web Technologies

LOD-GF: An Integral Linked Open Data Generation Framework

Víctor Saquicela[1(✉)], José Segarra[1(✉)], José Ortiz[1(✉)], Andrés Tello[1(✉)], Mauricio Espinoza[1(✉)], Lucía Lupercio[1(✉)], and Boris Villazón-Terrazas[2(✉)]

[1] Department of Computer Science, University of Cuenca, Ave. 12 de Abril and Agustín Cueva, Cuenca EC010201, Ecuador
{victor.saquicela,jose.segarra,jose.ortizv,andres.tello, mauricio.espinoza,lucia.lupercio}@ucuenca.edu.ec

[2] Fujitsu Laboratories of Europe, Madrid, Spain
boris.villazon.terrazas@gmail.com

Abstract. Linked Open Data (LOD) generation is a common activity within organizations due to its advantages for sharing and reusing information. Since these technologies require specialized knowledge, the development of technological and methodological tools that allows its implementation is limited. Most of the current solutions are built on top of specific tools which require considerable effort to consolidate into an integral solution. Moreover, those tools work on specific domains, or they do not support some of the phases required for LOD life cycle (e.g., data cleaning, data exploitation). In this paper, we present a framework for LOD management which follows methodological principles presented in the state of the art in scientific literature and provides an unified software tool for publishing LOD for multiple domains and technologies. Our platform leverages a modular ETL processor, allowing a transparent and flexible integration, providing an integral environment for LOD. This framework was tested, successfully, using data sources from different domains, e.g., digital repositories, libraries.

Keywords: Linked Data · Framework · Data integration Methodological guidelines · LOD life cycle

1 Introduction

Currently, a large number of enterprises and organizations are joining the Linked Open Data initiative, with the goal of take advantage of the latest technological advances and making their data to the next level (web). Joining LOD comes along with many benefits such an improvement in the interoperability, reutilization and standardization of data from both inside and outside of the institutions. Moreover, LOD offers the capability of gaining a higher level of information from the data through linkage processes with other LOD repositories. All these improvements are intended to be exploited by both humans and machines according to Semantic Web point of view. However, a weight of the growing

M. Botto-Tobar et al. (Eds.): TICEC 2018, AISC 884, pp. 283–300, 2019.
https://doi.org/10.1007/978-3-030-02828-2_21

volume of data available in RDF (Standard to share information in LOD), the problems that limit its widespread adoption are still evident. The main problems are the lack of formal methodologies that help the LOD application process, as well as the lack of platforms to support all phases of these methodologies. Deficiencies that are complemented with the high level of conceptual and technical knowledge required for the correct generation and publication of LOD.

Proposals such as [13] and [11] present methodologies to support the LOD life cycle through different phases, which try to overcome the lack of a general guide that leads the generation and publication of Linked Data. Those works, besides providing the baseline for LOD maintenance and application, propose several tools that support, totally or partially, the needs presented within its different phases. Most of those tools help to minimize, considerably, the implementation effort of the activities proposed within the methodologies. However, most of those tools arose from individual efforts and were designed to solve specific problems. Hence, its integration, with the aim of providing an integral solution for LOD generation and publication, imposes new barriers when applying them in real use case scenarios. To overcome the integrability problem of the current tools, we propose a framework to give technological support for the implementation of the different phases for LOD generation and publication proposed in [13]. Our framework is built on top of ETL (Extract, Transform, and Load) Pentaho Data Integration[1] (Kettle), which flexibility allows delivering an extensible framework that can be used in multiple domains as shown Sect. 4.

In this document, we present a detailed description of the components of the framework which support the different phases of a LOD publication methodology, as well as real use cases over different application domains. The remainder of this paper is as follows. In Sect. 2, we present the related work about different platforms for supporting LOD management and application. In Sect. 3, we describe the components of our proposed framework and its use in the support of the different phases of LOD publication. In Sect. 4, we present several successful use cases of our framework within different domains, showing its validity and applicability. Finally, in Sect. 5, we present the conclusions and the future research in this field.

2 Related Work

There are several tools which support the different phases of the LOD generation and publication process. For example, CSV Import, R2R, D2R, RML and Karma for data extraction and RDF generation, LOD Refine and Onto Refine for data cleaning, Virtuoso, Fuseki for LOD publication, etc. In addition, Linked Data communities have developed more specific tools for certain domains such MARiMbA for the bibliographic domain [12] and Bio2RDF for the Life Sciences domain [1], which attempt to provide LOD integration solutions bounded within these fields. These tools and others, together, may support the whole LOD

[1] http://community.pentaho.com/projects/data-integration/.

generation process. However, we still require a significant manual effort and specialized knowledge of these tools to deliver integral Linked Data solutions.

In addition, the heterogeneity of the data sources brings additional problems to deliver a platform that supports different domains. Nonetheless, there are outstanding tools such as RML Editor [6] and Karma [8] which may become generic solutions for most of the users. These tools provide graphical web interfaces, mapping languages (e.g., RML [5] and R2RML [3]), and support to heterogeneous data sources. However, even these tools provide a rich functionality, they focus on a specific phase of the LOD publication methodology, RDF generation. Therefore, they are only a partial solution to the LOD publication process.

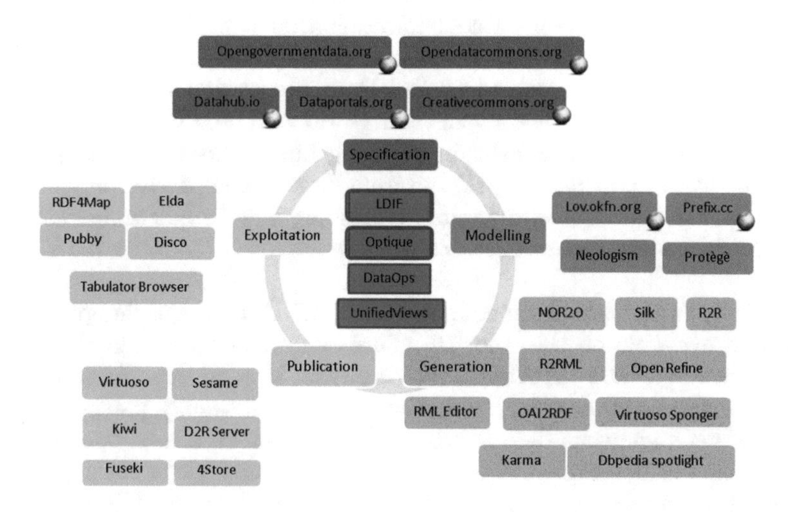

Fig. 1. Overview of current LOD tools

Figure 1 shows an overview of the current tools to support the phases of the methodology for LOD application presented in [13]. The large number of available tools corresponds to the diversity of data formats to be translated to RDF, and other tools to support other processes (e.g., publication, linking, etc.) [11]. The edges of the figure show the standalone tools, each of them aligned to the phase that they support. On the other hand, the software projects that aim to provide an integral solution to the LOD application process are at the center of the figure. In the next paragraphs we describe with more detail the last ones.

The solutions focused on supporting the whole process of LOD application fall mainly into two categories: ETL managers with Linked Data support, and *Frameworks* for Linked Data Life Cycle management. The software tools that have been developed under these two approaches are described below. This review intends to describe the main characteristics and limitations of these tools from the perspective of the Linked Data Life Cycle methodology.

The ETL approaches for Linked Data have had an special relevance due the popularity and massive adoption of ETL tools within the IT community.

Therefore, there have been many attempts for bringing ETL solutions to the Linked Data realm. For instance, tools such OntoRefine and Allegro Integration for Talend were created with the only aim of dealing with legacy data problems. However, the most prominent works that have recently arose within this category are UnifiedViews and its successor LinkedPipes [4]. These provide a graphical interface to define data flows, which allow extraction, RDF transformation and Linked Data publication. In addition, these tools may be extended by means of new Data Processing Units (plugins). However, since they are a green-field software, the number of available DPUs is limited. Other limitations of these tool are the restricted support to different data sources (databases, CSV, RDF, and SPARQL), limited cleaning data functionality (lack of string operations, filters, etc.), and functionalities for data exploitation are not available.

There are several contributions regarding the Frameworks for Linked Data Life Cycle management such as Linked Data Integration Framework (LDIF), Information Workbench (DataOps), and Optique. These solutions attempt to provide a all-in-one environment for handling Linked Data. Each tools address the Linked Data management problem from its own perspective providing components for extracting, posprocessing, publishing and visualizating data. These tools and their main features are described below.

Linked Data Integration Framework (LDIF). LDIF is a tool designed to integrate data available on the web [10]. This tool supports a limited number of data sources, e.g., RDF, SPARQL, Web crawlers. Most of these data sources are based on semantic standards; hence, this tools supports only partially the integration of heterogeneous data sources. This framework also supports merging and matching of ontologies using R2R and Silk[2], but it does not support data cleaning. It allows data publication directly to RDF files or SPARQL Endpoints. The framework does not support data exploitation. Moreover, this tool does not provide a graphical interface and its configuration is complex.

Information Workbench (DataOps). DataOps is a framework that supports enterprise data integrations through ontologies [9]. This framework extracts and transforms different data sources to RDF; however, it focus mainly on semantic sources (i.e., RDF, OWL, SPARQL) and a limited number of non semantic formats (i.e., XML, CSV, Data Bases). It provides data cleaning through open refine and data linking through Silk Workbench. However, this framework just imports the data from those other tools which has to be manipulated in a standalone way. The framework has an embedded triplestore for data publication which reduce its flexibility. It relies on Widgets for data exploitation which requires additional configuration. This framework provides a web interface and its configuration is simple, but limited for advanced functionality.

Optique. It is a framework for Big Data built on top of Information Workbench. Optique virtually integrates relational data sources using ontologies OWL2 [7].

[2] http://silkframework.org/.

This tool supports data extraction exclusively from databases. Optique does not support data cleaning because it considered that data is cleaned at the data source side. Data linking is at data model level though a mapping between relational schemas and ontologies. Optique allows users to query the data using a SPARQL Endpoint and a graphical wizard for creation sparql queries. However, this Endpoint executes virtual queries which are translated to the native query language of the data source through the technology ONTOP [2]. Also, it provides a friendly user web interface, which requires knowledge of OWL2 and R2RML.

3 Framework to Support the LOD Life Cycle

In this work, we present a framework to support the LOD life cycle according to the methodology proposed in [13]. This methodology proposes well defined phases (specification, modeling, generation, publication and exploitation) with the aim of becoming and ordered guide for continuous Linked Data publication.

The proposed framework relies on Pentaho, a proved data integration platform, which allows delivering an integral solution to support the whole LOD life cycle. We leverage the Pentaho ETL process manager, Kettle, to allow the generation of complex data flows of data transformation. We implemented different plugins to extend Kettle main functionality.

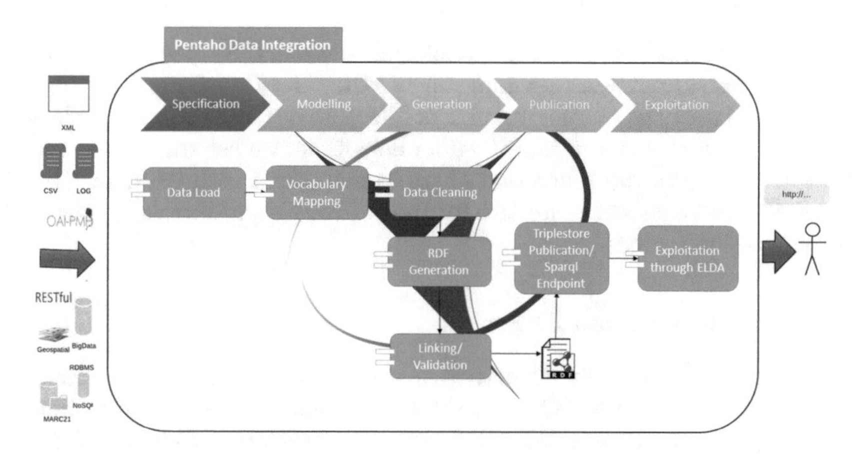

Fig. 2. Architecture of the proposed framework

Figure 2 shows a general overview of the proposed framework. In the figure we observe that each component of the framework is grouped under the corresponding phase that it supports. In some cases we implemented additional plugins to meet the requirements of a particular component. However, for other phases we used the native plugins provided by Kettle. For example, we can use native or ad-hoc plugins during the specification phase, depending on the data format at the source. In the next sections we present a detailed description of the implemented plugins that support each phase of the methodology.

Support for Specification Phase

Plugins for the specification phase must be able to support different input data formats, which depend on the available data sources, as shown in Fig. 2. Usually, we may read the input data using the native plugins provided by Kettle, which supports different data formats (e.g., csv, excel, relational data bases, web services, etc.). However, the flexibility and scalability provided by Kettle allow extending the tool to support other input data formats by implementing ad-hoc plugins. Generally, the plugins of this phase support data loading from the different sources and deliver them as data tables, which is the data structure used within Kettle.

Figure 4 of Sect. 4 shows a example of configuration interface of one of the implemented plugins. It allows data extraction from the sources using the OAI-PMH protocol which is an input format not supported by Kettle. The OAI-PMH protocol is widely used to interchange information among digital repositories. Hence, it may be used to extract bibliographic resources that use such protocol. The configurable options for the "OAI-PMH Loader" are path to service, data format (i.e., OAI-DC or XOAI), output path. We developed ad-hoc plugins for library data sources which we describe in Sect. 4.

Support for Modeling Phase

In this phase we select and load the ontologies to use in the semantic data model. For our proposed framework we implemented the "Get Properties OWL" ad-hoc plugin. It requires the ontology prefix or the ontology file to load its elements (i.e., classes and properties) into the data flow, then, we use the elements of the ontological model in the generation phase to map them with the data from the sources. Figure 5 shows the interface for the ontology elements loader plugin and its configuration options.

Support for Generation Phase

This phase comprises different activities which go beyond data conversion to RDF. One of the main activities is the data cleaning, which aims to improve the data quality and standardization. Pentaho provides a rich set of plugins for data cleaning designed to support data transformation and manipulation, making this functionality one of its most remarkable features. We may use the predefined plugins for data filtering and transformation depending on the errors founded in the data. Next describe the common plugins used for data cleaning:

- **String Operations:** It allows operations with strings, e.g., remove numbers, delete special characters, remove spaces, etc.
- **Replace String:** It replaces strings withing other strings which allows deleting unknown or corrupted characters.

- **Value Mapper:** This plugin allows mapping field names with other names, which is very useful for data standardization.
- **Split Fields:** It allows separating fields with multiple data types.

Once the data does not contain errors, we transform the data to a standard format for data interchange within Linked data, i.e., RDF. In this phase, we map the extracted and pre-processed data from the sources with the selected ontology vocabulary. Figure 6 depicts the ad-hoc "Ontology & Data Mapping" plugin, which allows specifying the relation between the extracted data and the selected vocabulary. We define three types of mappings depending on the relation between the data and the vocabulary.

- Classification or Type Mapping: In this mapping we associate a resource to a specific data type.
- Properties or Annotation Mapping: It allows relating the properties of a resource to a defined vocabulary.
- Relation Mapping: It allows defining a semantic relation among the resources.

The relations defined by the plugin are processed and mapped in an R2RML configuration file. This file is the input for the next plugin "R2RMLtoRDF", which automatically transforms the data stored temporary in a data base into the expected RDF format, e.g., XML or Turtle. The temporary data base it is also configurable through an ad-hoc plugin which is located in the last step of the data cleaning and transformation process.

The next step in the methodology is *Linking* the generated link with external data sources. We develop the ad-hoc "Silk plugin" which allows integrating and exploit the Silk Workbench capabilities to find similar resources among two data sources. Finding similar resources among different sources increases the amount of available information and brings the Linked Data capabilities required by LOD. We provide two configuration settings for the data Linking process.

- Provide a configuration file in Silk Link Specification Language (Silk-LSL).
- Configure the main parameters of the sources to be linked allowing the plugin to generate a linking file automatically.

The data to be linked must be accessible through a SPARQL Endpoint independently of how we configure the plugin. Hence, it is recommended the linking is executed after the publication phase. However, to keep the same order as proposed in the reference methodology, we describe the linking plugin in the generation phase. Another important consideration is the configuration parameter to define the threshold of the similarity metric between resources (e.g., *owl:SameAs*, *skos:closeMatch*, *skos:nearMatch*, etc.), as well as the validation process. Figure 8 shows a configuration interface of this plugin to Link two data sources.

One of the main features of this plugin is that besides finding similar links through syntactic metrics, it also allows validating those links by a establishing a semantic relation between related attributes.

Support for Publication Phase

At this stage the data is generated and we provide a public access by publishing them in a triplestore. In this phase we developed the plugin "Fuseki Loader". It allows deploying the Apache Fuseki Triplestore which serves as a storage, query, and data interaction through its SPARQL Endpoint. The advantages of Apache Fuseki against other similar technologies such as Virtuoso or Apache Marmotta are its small size and its support to special features as federated queries and text indexing.

The configurable options for this plugin (Fig. 7) are the data graph definition, the port to access the SPARQL Endpoint Service, access permissions to the Endpoint (i.e., read, update, write), and endpoint field indexing.

Support for Exploitation Phase

The proposed framework allows consuming and navigating through the generated data leveraging the LDA open API by means of its java implementation ELDA[3]. The data required to deploy ELDA's web interface are self-contained and it requires some basic configuration how entity and property selection to be displayed to user and label definition to rename the selected elements.

Although the visualization features provided by ELDA would be enough for most use cases, some implementations could require more advanced exploitation strategies (e.g. georeferencing and fulltext search). For those particular cases additional software will be needed on top of the components already provided by the framework. For instance, the semantic search tool FEDQuest[4] has been used in bibliographic data applications in order to ease users performing federated queries, fulltext search and data visualization through graphs. These type of specific setups implies manual configuration and customization according to the requirements of each case. Nevertheless, the incorporation of such software within the exploitation components of the framework could be performed on demand of the community with relative ease due the modular architecture of the framework.

Overview of aforementioned plugins are presented in the Fig. 3 and illustrates how they, together, gives the support for the whole Linked Data Generation process. The plugins that make up this platform can be found in the following web address https://ucuenca.github.io/lodplatform/, as well as more details of its operation and configuration.

---

[3] http://www.epimorphics.com/web/tools/elda.html.

[4] https://github.com/ucuenca/sparql-fedquest.

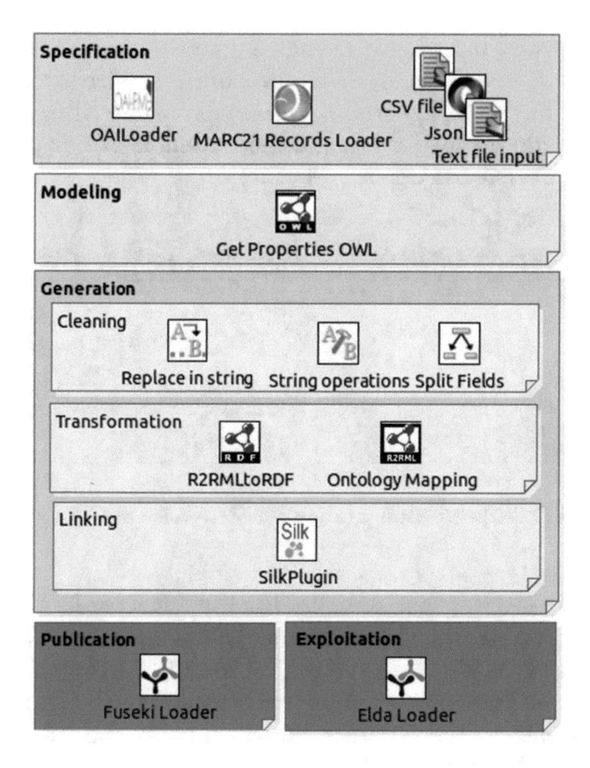

Fig. 3. Framework plugins to support the LOD generation

4 Use Cases

In this section we present different applications scenarios of our framework on which we assess its validity. We apply our framework to generate Linked Data within three different scenarios: digital repositories, library resources, enterprise data base. We present a complete detailed description only of the first use case to avoid redundant information. For the next uses cases we show a table summary pointing out the main characteristics supported by our framework for each use case.

4.1 Use Case A - Digital Repositories

This use case describes the transformation and publication of Linked Data from Dspace digital repositories. Particularly, the input data sources are the Dspace repositories of different Ecuadorian universities.

Specification. We use the standard OAI-PMH to read the data from the Dspace repositories. This is a well known mechanism for metadata extraction in the domain of digital repositories. It allows accessing the data through

HTTP requests. Although Kettle include a set of plugins for reading different input data formats, due to the particularities of the service invocation we developed an ad-hoc plugin. This guarantee the reliability of loading and reading this type of input. Figure 4 shows a configuration example of this plugin to extract data from the University of Cuenca repository.

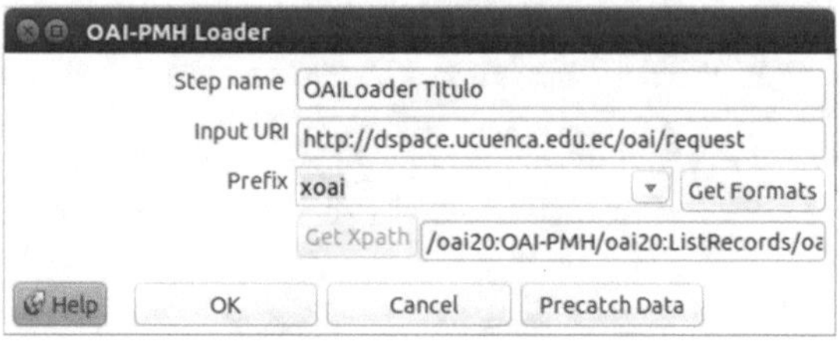

Fig. 4. OAI Data Load configuration for University of Cuenca repository.

The URI's template of this examples has the following format.

http://gtrdata.cedia.edu.ec/{DataSource}/{ResourceType}/{ResourceName}

The first part of the URI is the address of the API ELDA service. This allows addressing to a web page with the resource description when it is accessed. The rest of the elements in the URI are the resource id and its provenance. For instance, the following example shows an URI of a person who is the author or contributor of some resource.

http://gtrdata.cedia.edu.ec/ucuenca/contribuyente/ESPINOZA_MEJIA__JORGE_MAURICIO

Modeling. In this phase, we select ontologies which cover the bibliographic domain such as como **BIBO**, **DCTERMS**. In addition, for data that does not belong to this domain (e.g., persons), we select different ontologies. For example **FOAF** which allows to describe semantically the authors or contributors of a bibliographic resource, and **RDAA** for special relations between resources (e.g., *has contributed in, has created, is part of*). We use the "GetPropertiesOWL" plugin to load those vocabularies within the data flow by declaring the prefixes of the selected ontologies. Regarding other tools for LOD generation, only Optique, RML Editor and Karma provide support for data sources ontology modeling. These tools allow to pre-load vocabularies and suggest concepts during the mapping process. Other tools just ignore this phase and they require intense changes in its design to incorporate this functionality. Figure 5 shows the configuration for this plugin.

Generation. The first step of this phase is to review and analyze the state of the data sources. We do such analysis with the aim to ensure the quality of the data and correct them, if necessary, before the transformation process. After we

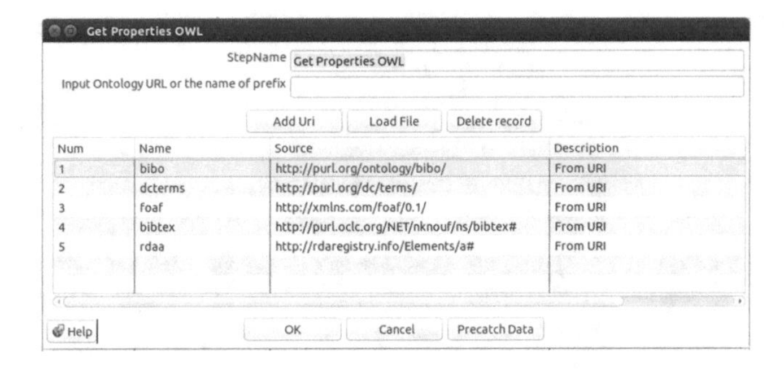

Fig. 5. Plugin configuration for ontology loading.

identify problems within the data we define data cleaning workflows using the native plugins of Pentaho's Kettle. The most common problems we found within the data are show in the Table 1:

Table 1. Commons problems - Use Case A

Description	Example
Additional information which does not belong to the author's name	Recalde Moreno, Dr. Celso
Two names belonging to different authors written in only one field	Castro Muñoz, Celia María;Chang Gómez, José Vicente
Full name without a separation character between the first and last name	Poma Salazar Mónica Eulalia
Codification problems of some names	Ca??izares Aguilar Aurelio Ernesto

The remarkable features of our framework for this phase are the ability to define complete cleaning and normalization workflows before the RDF generation. Other tools (e.g., Karma) for LOD generation do not provide data cleaning functionality or at most they allow a basic normalization of the data. Although platforms such as UnifiedViews may be extended by developing ad-hoc plugins, such implementations requires a great effort.

After the cleaning stage, we define a set of mapping rules to transform the data sources at Dspace to RDF using the "Ontology Mapping" plugin. We define the mapping rules based on the available sources and the ontologies loaded in the previous phase. Figure 6 shows the mappings for all the resources found at University of Cuenca Dspace. We identify different entities or resources within this phase. People, which may be author's or co-authors of a document. Documents which represent bibliographic resources. Collections which contain a set of documents.

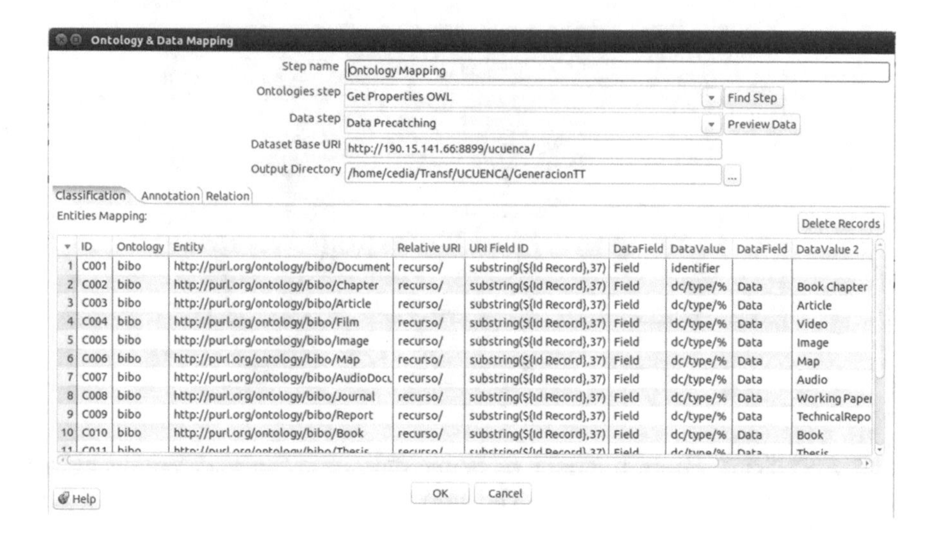

Fig. 6. Semantic Mapping for the University of Cuenca Dspace.

Table 2 shows an example of the mappings properties for bibliographic resources. The left column of the table shows the name of the field at the source which is associated with a semantic vocabulary shown in the right column.

Table 2. Mapping of Bibliographic Documents properties

Fields	Semantic vocabulary	Fields	Semantic vocabulary
Date/accessioned	dcterms:dateSubmitted	Subject	dcterms:subject
Date/available	dcterms:available	Title	dcterms:title
Date/issued	dcterms:issued	Language	dcterms:language
Abstract	bibo:abstract	Handle	bibo:handle
Provenance	dcterms:provenance	License	dcterms:license

Publication. We configure "Fuseki Loader" plugin to publish the data transformed to RDF from each source repository. The settings for this process are the graph definition for each endpoint, read only permissions to avoid unauthorized modification to the data, and indexing of the resource properties to be used in the exploitation phase such as Title, Abstract, Author's Names, Collection Names, etc. Specifically, the data published for the repository of the University of Cuenca is available publicly in a SPARQL Endpoint[5].

Our platform allows to deploy new SPARQL services (pre-loaded together with the data) automatically, without any additional software installation or any manual configuration. Current platforms are limited to generate RDF

[5] http://fedquest.cedia.org.ec:8891/.

(e.g., Karma, LDIF, RML), load the generated data to pre-existing Endpoints (e.g., UnifiedViews), or just publish the data to an embedded Triplestore (e.g., DataOps, Optique). The ability to create services on demand is useful when it is required to automate the data publication process or when we work with independent data sources. Figure 7 shows the configurations of the "Fuseki Loader" plugin to publish the data of the University of Cuenca.

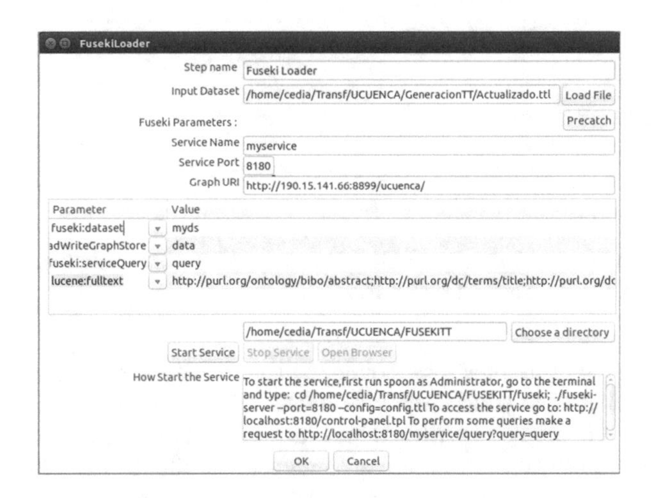

Fig. 7. Settings for publishing the data of the University of Cuenca

Linking. We do the linking process, as mentioned in Subsect. 3, after the publication phase. This way we can use the data available at the Endpoint which is needed as input to the Linking plugin. We use the default configuration parameters for linking, i.e., url address of the SPARQL Endpoints to link, its graphs, and the comparison and validation thresholds, as shown in Fig. 8. In this test we perform the linking between the data from the University of Cuenca and CEDIA repositories. Table 3 shows the authors founded and validated after linking the data from the University of Cuenca and CEDIA using the "Silk Plugin".

Table 3. Results of the Linking process between the repositories of University of Cuenca and CEDIA

University of Cuenca	CEDIA
Delgado Suconota, María Fernanda	Delgado, María Fernanda
Andrade, Gabriela	Andrade, Gabriela
Saquicela, V	Saquicela, Víctor

Currently, all the LOD generation platforms, except Optique, provide support for the linking process. However, the configuration tasks are not flexible for

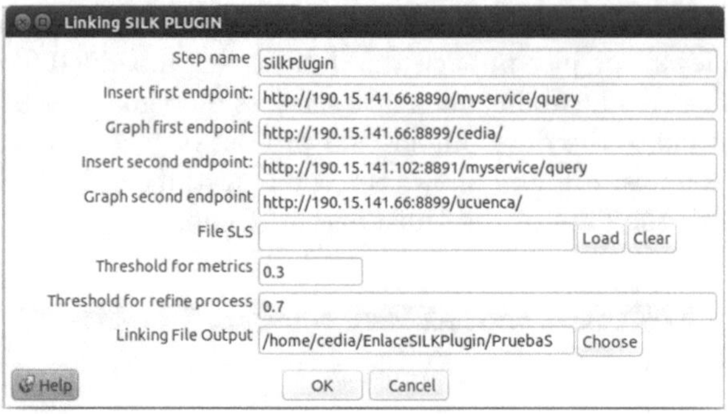

Fig. 8. Silk Plugin Configuration Example

some platforms (e.g., DataOps, LDIF) because they use their own fixed repository as one of the sources to compare.

Exploitation. In this phase we configure the "Elda Loader" plugin to generate an ELDA instance which provides a user friendly interface to visualize the resources (See a example of author's information in http://gtrdata.cedia.edu.ec/ucuenca/contribuyente). We create a configuration file for each Dspace source. However, at the end we deploy all the configuration files in just one ELDA service for resource optimization.

Regarding the other platform for LOD generation only DataOps and Optique provide predetermined exploitation data functionality. Those tools provide web pages for resource visualization and text based searches. In addition, such functionality can be extended by widgets implementation and configuration (e.g., graphic query generator). However, one of the drawbacks of those platforms is that they do not allow to consume external sources (Endpoints outside the platform) for exploitation tasks. This reduce flexibility when we require to exploit external data.

In addition to the data visualization interface, we use a search engine of RDF bibliographics resources to facilitate the users to find information. With this functionality we exploit the data indexing performed in the publication phase. We can access to the searching tool trough http://fedquest.cedia.org.ec.

4.2 Use Cases B and C - Library Resources and Enterprise Data

In the same way that we described the previous use cases, we used our proposed LOD generation framework for library data sources and a enterprise data base. In the case of the library resources, we use the data from the "Juan Bautista Vásquez" library using its bibliographic resource catalog. On the other hand, in the case of the enterprise data base, we used data from employs of different enterprises which represent a common scenario in the business domain. Table 5 shows a summary each use case. In the such table we point out how our framework supports the particularities of both cases.

The first row of the table describes the specification phase. This phase is the one which presents the most variability because it depends on the format of the data to be extracted from the sources. We implemented a Marc21 reading plugin for the library data use case because the input data has a special format. It has a number of descriptors which indicate the position of each field. On the other hand, we use a native Kettle's plugin to read data from a relational database for the enterprise data use case.

During the modeling phase we use the "GetPropertiesOWL" plugin to load the ontologies in both use cases. The variation here depends on the domain, we use different ontologies for each use case. We use the same ontologies that we used for the Dspace digital repositories in the library data use case. While in the case of the enterprise data we use an ontology which allows to describe enterprise-related resources (e.g., employee, branch office relations, enterprise, etc.). In the generation phase we perform several activities such as error detection and data cleaning. In the library resources use case we found the following errors: Table 4.

Table 4. Commons problems - Use Case B

Description	Example
Additional author's information which does not belong to author's name	Beryl [et al.]
Many authors in the same author's name field	Curtius, Ernst Robert. Frenk Alatorre, Margit, traductor
Wrong information at author's name field	1572-1631

We do not perform data cleaning for the enterprise data use case because the data we used test data specifically for this experiment and we know the data status. Then we perform the data transformation using the corresponding ontology for each case to obtain the RDF representation of the data.

The publication phase is similar in both use cases. We use the "Fuseki Loader" plugin to load the RDF generated data, store it in a triplestore, and support querying it through a SPARQL Endpoint later on.

We perform the linking process only for the library resources use case. We perform the linking between the "Juan Bautista Vásquez" library and the University of Cuenca. That library belongs to that university, increasing the chance to find common data between both sources. The results as shown in the table is 41 links found. On the other hand, the enterprise use case data was a created only for the test and thereby there were no similar sources in this case.

The exploitation phase is the same in both use cases. We use the "Elda Loader" plugin to create an ELDA instance and provide a user-friendly visualization interface of the RDF data.

Table 5. Summary of the application of use cases for library resources and business database.

	Bibliographic source	Business database
Specification	Development of Marc 21 plugin	Application of "Table Input" plugin for database reading
Modelling	Loading of vocabulary using developed plugin: – bibo – dcterms – foaf	Loading of vocabulary using developed plugin: FOAF and SCHEMA – foaf – schema.org
Generation	Transformation tasks for standardization and error correction. For example: – Deleted strange characters into authors field – Split two authors names in one field	Cleaning was not applied since the correct state of the data was known
	Resources found and described using the semantic vocabulary selected through the developed plugin: – Persons – Documents – Collections – Cataloging center	Resources found and described using the semantic vocabulary selected through the developed plugin: – Persons – Organizations
Publication	Storage and publishing of RDF using the developed plugin, accessible through SPARQL Endpoint (http://gtrdata.cedia.edu.ec/fuseki/)	
Linking	Links were sought between the librarian repository and the University of Cuenca. As a result 41 links were discovered and validated	Linking resources are not found
Exploitation	Elda plugin was applied to facilitate the visualization and navigation of the data by the users	

5 Conclusions and Future Work

In this work, we present LOD-GF a framework for LOD Generation, which follows the state of the art methodological guidelines for LOD publication. This platform is implemented on top of the ETL Pentaho Data Integration process manager, Kettle. The components (plugins) we develop within our framework can be used either independently or jointly in order to support the LOD life cycle. Moreover, its modular architecture facilitates its extensibility for heterogeneous data sources from multiple domains and technologies. This is a feature that not many platforms include because they were designed to solve specific problems, e.g., enterprise data integration, linked data management, etc.

To assess our framework capabilities we presented three uses cases over three different domains which prove its applicability on solving real problems. In each use case we provided an overview of main requirements and implementation issues we found. As a result of we summarize the LOD generation process using our framework and pinpoint how the features of the framework facilitated the overall work.

In the future work, we plan to extend our platform to support other domains through the implementation of new plugins. In addition, we aim to develop an ontology and concepts recommendation module within the semantic mapping phase. Furthermore, we plan to include components which allow a native support of semantic technologies within the ETL manager, i.e., reading, manipulation, and storage of RDF-SPARQL.

Acknowledgments. This is a project of the Computer Science Department of the Universidad of Cuenca (https://www.ucuenca.edu.ec/la-oferta-academica/oferta-de-grado/facultad-de-ingenieria/dptos/dcc) with the support of the "Grupo de Repositorios" (GTR) and "Repositorio Ecuatoriano de Investigadores" (REDI) of the "Corporación Ecuatoriana para el Desarrollo de la Investigación y la Academia" (https://www.cedia.edu.ec/) (CEDIA, Spanish Acronym).

References

1. Belleau F, Nolin M-A, Tourigny N, Rigault P, Morissette J (2008) Bio2RDF: towards a mashup to build bioinformatics knowledge systems. J Biomed Inform 41(5):706–716
2. Calvanese D, Cogrel B, Komla-Ebri S, Lanti D, Rezk M, Xiao G (2015) How to stay ontop of your data: Databases, ontologies and more. In: The semantic web: ESWC 2015 satellite events - ESWC 2015 satellite events, revised selected papers, Portorož, Slovenia, 31 May–4 June 2015, pp 20–25
3. Das S, Sundara S, Cyganiak R (2011) R2RML: RDB to RDF mapping language (W3C working draft)
4. Dimou A, Heyvaert P, Maroy W, De Graeve L, Verborgh R, Mannens E, Van de Walle R (2016) Towards an interface for user-friendly linked data generation administration. In: Proceedings of the 15th international semantic web conference: posters and demos. CEUR-WS, pp 1–4
5. Dimou A, Vander Sande M, Colpaert P, Verborgh R, Mannens E, Van de Walle R (2014) RML: a generic language for integrated RDF mappings of heterogeneous data. In: Proceedings of the 7th workshop on linked data on the web, April 2014
6. Heyvaert P, Dimou A, Herregodts A-L, Verborgh R, Schuurman D, Mannens E, Van de Walle R (2016) RMLEditor: a graph-based mapping editor for Linked Data mappings. In: Sack H, Blomqvist E, d'Aquin M, Ghidini C, Ponzetto PS, Lange C (eds) The Semantic web - latest advances and new domains (ESWC 2016), vol 9678. Lecture notes in computer science. Springer, Cham, pp 709–723

7. Kharlamov E, Jiménez-Ruiz E, Zheleznyakov D, Bilidas D, Giese M, Haase P, Horrocks I, Kllapi H, Koubarakis M, Özçep Ö, Rodríguez-Muro M, Rosati R, Schmidt M, Schlatte R, Soylu A, Waaler A (2013) Optique: towards OBDA systems for industry. In: Cimiano P, Fernández M, Lopez V, Schlobach S, Völker J (eds) The semantic web: ESWC 2013 satellite events, vol 7955. Lecture notes in computer science. Springer, Heidelberg, pp 125–140
8. Knoblock CA, Szekely P, Ambite JL, Goel A, Gupta S, Lerman K, Muslea M, Taheriyan M, Mallick P (2012) Semi-automatically mapping structured sources into the semantic web. In: Proceedings of the extended semantic web conference, Crete, Greece
9. Pinkel C, Schwarte A, Trame J, Nikolov A, Bastinos A, Zeuch T (2015) DataOps: seamless end-to-end anything-to-RDF data integration. In: Gandon F, Guéret C, Villata S, Breslin J, Faron-Zucker C, Zimmermann A (eds) The semantic web: ESWC 2015 satellite events, vol 9341. Lecture notes in computer science. Springer, Cham, pp 123–127
10. Schultz A, Matteini A, Isele R, Bizer C, Becker C (2011) LDIF: linked data integration framework
11. Van Nuffelen B, Janev V, Martin M, Mijovic V, Tramp S (2014) Supporting the linked data life cycle using an integrated tool stack. In: Auer S, Bryl V, Tramp S (eds) Linked open data – creating knowledge out of interlinked data, vol 8661. Lecture notes in computer science. Springer, Cham, pp 108–129
12. Vila-Suero D, Gómez-Pérez A (2013) datos.bne.es and marimba: an insight into library linked data. Libr Hi Tech 31(4):575–601
13. Villazón-Terrazas B, Vilches-Blázquez LM, Corcho O, Gómez-Pérez A (2011) Methodological guidelines for publishing government linked data. In: Wood D (ed) Linking government data. Springer, New York, pp 27–49

Semantic Architecture for the Extraction, Storage, Processing and Visualization of Internet Sources Through the Use of Scrapy and Crawler Techniques

Ramiro Leonardo Ramírez-Coronel(✉), Ana Cristina Cárdenas, María-Belén Mora-Arciniegas, and Gladys-Alicia Tenesaca-Luna

Departamento de Ciencias de la Computación y Electrónica, Universidad Técnica Particular de Loja, San Cayetano Alto y Marcelino Champagnat S/N, Loja, Ecuador
{rlramirez, accardenas, mbmora, gtenesaca}@utpl.edu.ec

Abstract. The collection of structured data on the web involves a significant problem at the time of its abstraction in HTML pages, subsequently the processing of information for the reuse of any user and finally send it to a semantic process involves a difficult task to find an architecture that fulfill all these objectives. The present researching work has two main objectives that give solution to two of the major problems of the web of today. (a) Information overloaded: To provide a solution to the data collection hosted on the WEB in HTML format by merging data collection tools (Scrapy, Selenium) involving the user to perform a monitoring of the data to be collected. In addition, the existing limitations within tools that provide a similar service are taken into consideration. (b) Conceptualization of the data: To afford the user with a work space where the transformation of structured data to semantic data is allowed, taking into account the principles of Linked Data, moreover, the process of giving semantics to the data where aspects are taken into consideration Important such as: reuse of vocabularies, for covering this aspect it is made use of online catalogs that help to search existing vocabularies.

Keywords: Web data · Semantic web · Ontologies · RDF · Crawler Scrapy

1 Introduction

The Web since its inception has become one of the main bearers of information, since the arrival of Web 2.0 the user receives the most important role so it becomes the main information manager and the interpretation process is only carried out by users accessing a website, one of the main problems of this website is the overload of information that is generated by the user, most websites provide information in an "unstructured" format which limits the use of the data, in addition the current Web has heterogeneous structures and the terminology used in each website leads to a problem of synonymy.

M. Botto-Tobar et al. (Eds.): TICEC 2018, AISC 884, pp. 301–313, 2019.
https://doi.org/10.1007/978-3-030-02828-2_22

One of the biggest challenges is to represent the information that can be understood by machines and people, that is why Web 3.0 is created it seeks to add semantics to data through ontologies, an optimal solution is that most Web pages use syntactic standards within of its HTML, one of the recommendations is to use the <meta> tag that helps in the communication process between those who publish data and those who consume them. The present researching focuses on two main objectives such as:

(i) Information overload: To give a solution to the collection of data that are hosted on the WEB in HTML format by merging data collection tools (Scrapy, Selenium) involving the user to perform a monitoring of the data to be collected In addition, the existing limitations within tools that provide a similar service are taken into account and
(ii) Conceptualization of the data: To provide the user with a space work, where the transformation of structured data to semantic data is allowed, considering the principles of Linked Data, as well as the process of giving semantics to the data, where important aspects are taken into consideration as: reuse of vocabularies, for covering this aspect it is made use of online catalogs that help to search existing vocabularies.

This work continues, in Sect. 2, presenting the theoretical foundation related to Web Architecture. The main data extraction processes of internet pages are identified.

In Sect. 3, it describes the methodology that this research is based on to perform data recovery, to model vocabularies and the generation of RDF. Section 4 presents the Web architecture implemented to solve a case of application in a specific scenario of the teachers of a Latin American University. Finally, in Sect. 5 the conclusions are described.

2 Theoretical Fundamentals

2.1 Web Architecture

The architecture of the Semantic Web is composed of a set of layers in which the different representation standards are described, such as XML, XML Schema, RDF, RDF Schema, OWL among others [1]. Table 1 presents the description of each of the layers that are part of architecture.

2.2 Vocabularies and Ontologies

The architecture of the Semantic Web is composed of a set of layers in which the representation of knowledge is one of the most important tasks which this Web is responsible, ontologies are the way to represent this knowledge, since it is considered as an explicit specification of a domain or area of the real world that wants to be described in a structured way [2].

Table 1. Layers of the architecture of the Semantic Web.

Layer name	Description
Unicode + URI	UNICODE is a standard that provides a unique number for each character, no matter the platform or the program, that is, it encodes the information in any language It is the most technical layer of the Semantic Web in which are grouped the different technologies that make communication possible XML offers a common format for document exchange
XML + NS	**NS** provides a method for qualifying names elements and attributes.
XML Schema	Defines the universal language with which we can express different ideas of the semantic Web, which defines a data model to describe resources using triplets (subject-predicate-object)
Ontologies	They are formal specifications of a knowledge domain, which is identified with a taxonomy
Logic	These are the formal rules that allow determining if a reasoning is followed from its premises

2.3 Data Extraction Process

The extraction of data is the discipline that intends by means of various algorithms, processes and tools, to identify pieces of information or subchains of characters that represent relevant facts in such a way that the semantics of the content of the documents can be represented.

(1) **Framework Scrapy:** It is an Open Source Python framework. This framework makes “web scraping”, that is to extract information or data from some websites, where web scraping is allowed, it is a fast-high-level web crawl framework, which is used to browse websites and extract data from their pages. It can be used as a wide range of purposes, from data extraction to data mining for automated monitoring and testing [3].
Scrapy’s architecture is based on “spiders,” which are autonomous trackers that are given a set of instructions, also it makes easy to build and scale large tracking projects by allowing developers to reuse their code. Scrapy also provides a web tour shell that developers can use to test their assumptions about the behavior of a site [4].

(2) **Selenium** WebDriver: is a set of tools to automate web browsers on many platforms. It is popularly known as Selenium 2.0 [5]. WebDriver was developed to better support dynamic web pages where elements of a page can change without reloading the page [6]. The objective of WebDriver is to provide a well-designed object-oriented API that provides enhanced support for modern advanced web application testing [7].

(3) Selenium WebDriver is used in the following context: (i) Testing of multiple browsers, including improved functionality for browsers that is not well supported by Selenium RC Selenium1.0; (ii) Management of multiple frames, multiple browser windows, pop-ups and alerts; (iii) Complex page navigation; iv) Advanced user navigation, such as drag and drop [8].

Webdriver comes with a better API set that fulfill the expectations of most developers by being closer to object-oriented programming in terms of its implementation [9].

2.4 Related Work

Currently there are no tools that integrate the data collection to achieve the conversion in RDF format, this research shows related cases of tools that have similarity in data collection and tools that generate RDF files through CSV.

One of the jobs developed is the Web Scraping Online project, import.io is an online application that automatically analyzes the structure of the web page and displays the data in a table format, it is possible to extract pagination data. Import.io extracts exactly what you need, when you need it [10]. Another of the initiatives is DataScraping, which is considered as an application that allows the tracking of web dates as the extraction process and combination of contents of interest of the Web in a systematic way. It is a software agent-type process, also known as a web robot, mimics the navigation interaction between web servers and humans in a conventional web tour [11]. Finally, we mentioned Extractly, a project in which an online application has been developed in which the total control to the user, where you can select the elements you need to extract, this tool offers to extract the data from any website.

3 Methodology

It is proposed to divide the problem into two major functionalities such as: data collection and transformation to RDF, Fig. 1 illustrates three stages that have been considered convenient to use, each stage contains the necessary tools to perform the semi-automation of the collection of data until the conversion to RDF, which are: Data extraction, Modeling Vocabulary and RDF Generation.

3.1 Data Extraction

The heterogeneity of web pages and the inappropriate use of HTML tags is one of the major problems of the current Web, this problem is mitigated with the participation of the user since this collection process must be coupled to any website, the user It assumes the most important role where it can establish parameters or areas that it requires to collect in an interactive way.

In this stage the scope of data collection is established, the Web has heterogeneous structures in the representation of information where it is necessary to define criteria where the application can perform the collection in a successful manner:

- Web pages that have filtering or information searching through a criterion.
- To perform a successful crawler and show the items that can be collected it is necessary that inside of each link the information is in a semi-structured format has been found.

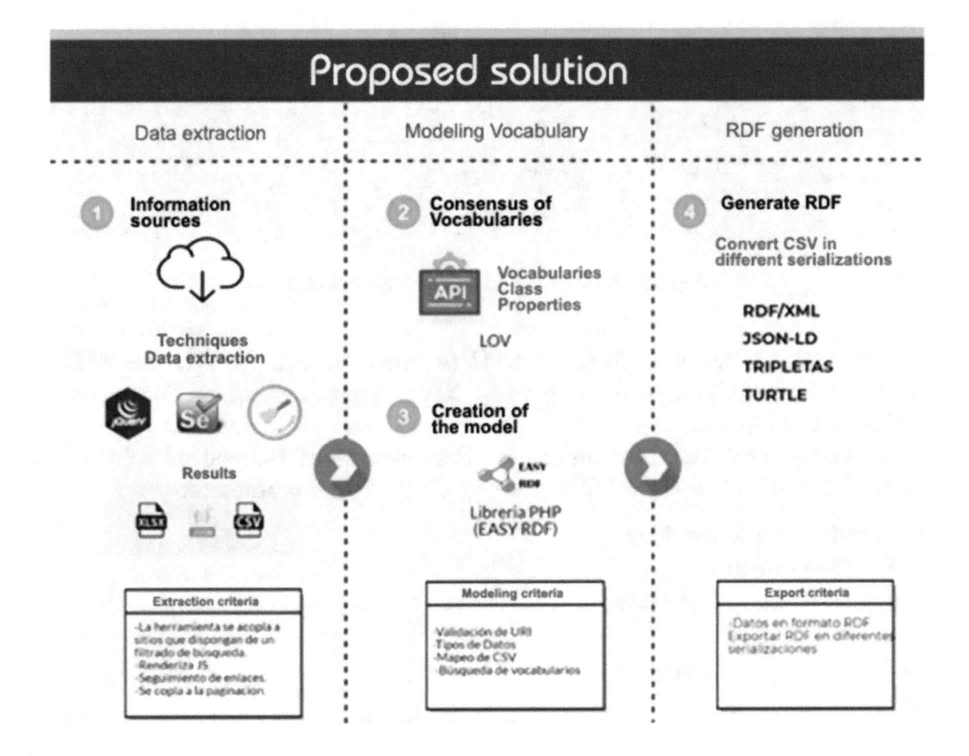

Fig. 1. Semi-automation stages.- solution

For the collection of information, multiple techniques are used, such as: Scraping, Selenium and Jquery, where it is convenient to merge these techniques for the fulfillment of the proposed objectives, which facilitate the extraction of information that is hosted on the Web, from this point a web application is created that allows the user to extract information taking into account selection criteria such as: relevant data, open data, use of information.

Solution flow: One of the main features covered in this web solution is the user's manipulation with the application, that's why it is necessary to use user-friendly technologies such as HTML Jquery, JavaScript, which allow the user to perform the development focused on user iterations, in Fig. 2 the flow of the proposed solution for data collection is shown, taking into account different aspects such as:

- The inappropriate use of HTML tags in the representation of the content is not standard
- Most web pages load their content on a dynamical way, which makes data extraction difficult.
- The pagination within each website is used in different ways, making it difficult to locate automatically to track links.

Obtain HTML: To develop an optimal way to get the HTML code with their respective style sheets and images.

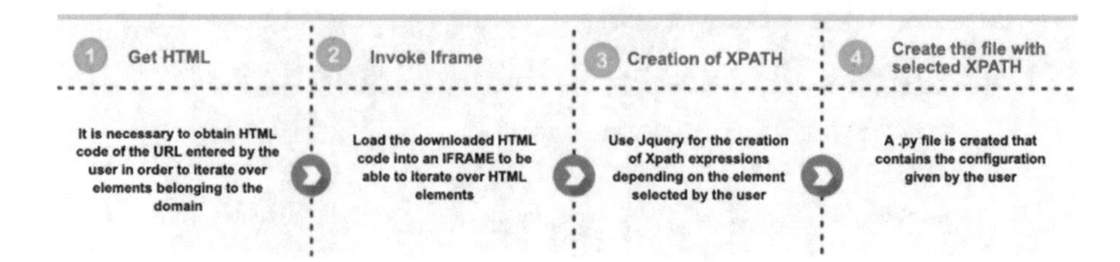

Fig. 2. Flow of the solution - Data collection

Developing a script to form XPATH (expressions that traverse an XML): Perform an algorithm capable of forming XPath expressions taking into account attributes such as class.

To determine a standard format for data collection: The optimal solution to collect data is to follow the scheme given by Scrapy where it proposes:

- Container block searching
- Selection of items
- Paging selection if necessary

3.2 Modeling Vocabularies

For the conceptualization of the data collected, the user has the main role, in which there are several activities to have a modeling such as: URI validation, vocabulary searching, classes, and properties.

To reach the stage of consensus vocabularies, it is necessary to classify the metadata or determine the names of the classes, once the parameters are established, it begins to consensus on the vocabularies, classes and properties described in the following section begin to be agreed upon.

Consensus of Vocabularies: To carry out searches of vocabularies coupled to the classes or subclasses of the collected data, a catalog of vocabularies called LOV (Linked Open Vocabularies) is used, which has an API that allows to consult about vocabularies, classes and terms, in this present work, this API is used since it gathers key aspects to structure RDF.

Create Model: Once the established vocabularies on the previous activity to be used have been selected, they need the intervention and knowledge of the user to make an optimal model according to their needs, which will have a friendly interface for the user to interact with the application. in such a way that it has a work area to build a formal model, where the user can assign the information collected and assign appropriately.

3.3 Generation of RDF

With the formal model designed by the user, necessary fields are established for the generation of RDF data, this process is carried out with the PHP easyRDF library which receives parameters from the previous activities such as vocabularies, classes, properties, attributes, data collected.

4 Web Architecture

For the design of the web application (Distributed Architecture) a coupled architecture is built, the same that covers general aspects of a client server architecture. This architecture has a flow where clients access to the services provided by the server through remote procedure calls using the HTTP protocol, Fig. 3 shows the architecture of the web application.

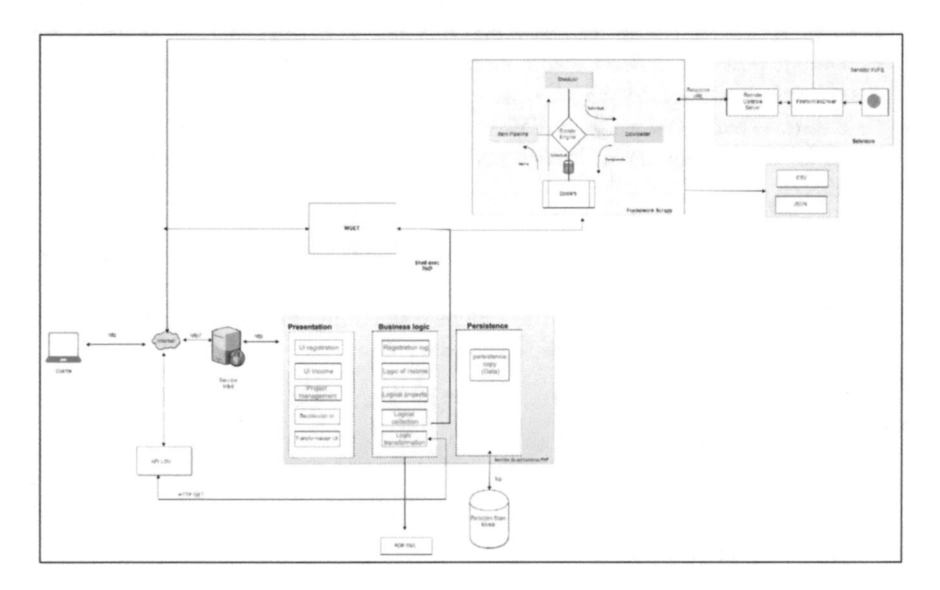

Fig. 3. Web architecture of the application

The flow of the architecture is done through requests to a web server where it hosts the different functionalities of the web page.

PHP Application Server: The server is implemented in PHP, it has a three-layer structure that allows to work separately the improvements and updates of the components, besides being an apt architecture when developing a client/server application, then details each of them:

Presentation Layer. It is responsible for presenting the information to the user and managing their interactions, it includes five user interface components such as: registration and entry that is essential within a web application that provides a service, in addition to these components it presents a management component of projects, likewise interfaces for data collection and transformation.

Business Logic Layer. The logic of the application is implemented, where a specific file is found for each presentation component, in this layer it is required to use the Shell and Scrapy framework for data collection, it is integrated by making a call to the function " exec ", which simulates the behavior of a Shell, it is also convenient to use Selenium that helps render web pages that load the content dynamically.

Data Collection Module. Within the data collection module it is necessary to integrate technologies for the fulfillment of the objectives, the following reference is made to the use of the technologies used in this module:

- **Wget**: It is used to download web pages with the necessary content to view the web page.
- **Framework Scrapy:** *It makes use of spiders, items, download and Scrapy's own engineering to collect structured data from any web page, just as any tool has limitations and one of them is to render dynamic pages that is why Coupling with Selenium helps to render pages resulting in structured data.*
- **Selenium WebDriver** nside an XVFB server: It executes a controller that makes it possible for the browser to be automatically managed in memory through a web driver.

Data Collection Module. Within this module it is necessary to use APIs that provide a search service for ontological vocabularies that exist within the web, using the LOV API through HTTP GET call to obtain results, in addition to using EasyRDF, which is a PHP library for generate RDF serialization.

Persistence Layer. It is directly related to the business logic layer to make requests to the database.

4.1 Data Collection Module

URL management

For the necessary configurations, the management of the URL entered by the user is done in this section. Wget is used as it is a free tool used to download content from web servers. Figure 4 shows the source code to run WGET.

The proposed solution when using WGET is limited to the download of static pages; which does not cover the dynamic content is therefore an additional solution was made able to give the user a section in which he can load an HTML file.

Development of XPATH

For the interactive management, we used our own jquery methods where through events such as.mouseover (),.mouseout (),.click ()it was able to obtain a nice interface for the user, in Fig. 5 we show the fragment of code that helps the travel of the html document.

```
1 #Bajar paginas web con wget(imagenes,estilos)
2 $cmd="wget  --no-directories \
3       --recursive \
4       --page-requisites \
5       --html-extension \
6       --convert-links\
7       --restrict-file-names=windows \
8       --domains hola.org --no-parent\
9       - P '$proyecto/scrapy/scrapy/' \
10      --user-agent='Mozilla/5.0 \
11        (Macintosh; Intel Mac OS X 10.9; rv:32.0) \
12        Gecko/20100101 Firefox/32.0' "." "'.$url.'"';
```

Fig. 4. Wget Command - Download Web Pages

```
1  //Iteracion con el DOM
2   $(inframeDoc).mouseover(function (event) {
3     class_valido[0]=$(event.target).attr('class');
4     if(class_valido!=""){
5       longi=class_valido[0].length;
6       var ps=class_valido[0].slice(-1)
7     }
8     if(ps==" "){
9       $(event.target).addClass('clase_espacio');
10    }
11    $(event.target).addClass('outline-element');
12    }).mouseout(function (event) {
13      $(event.target).removeClass('outline-element');
14    }).click(function (event) {
15        //Proceso Xpath
16     });
```

Fig. 5. Code for document path

4.2 Scrapy Data Collection Algorithm

For the elaboration of the data collection algorithm, a general structure was determined for any project where it is necessary to consider that the Xpath expressions must be generated correctly, in the development of this section the use of the Selenium technology is determined that helps in the automation of the dynamic pages and the management of the pagination building a.py file where four sections are established.

Libraries and items

Libraries are established to integrate Scrapy with Selenium allowing the fusion of the two powerful tools, it also contains a class allowing defining items to collect data in Fig. 6 illustrates the libraries that each.py file contains.

Pagination

It contains Selenium methods that help to track the paging "self.driver.find_element_by_link_text ()" allowing the paging search for the text of the link, in Fig. 7 the code fragment that allows paging is shown, it has the capacity to extract links from the

```
1  //Librerias e items
2  # coding=utf-8
3  import scrapy
4  from selenium import webdriver
5  from selenium.webdriver.common.keys import Keys
6  import time
7  from selenium.common.exceptions import NoSuchElementException
8  from pyvirtualdisplay import Display
9  display = Display(visible=0, size=(800, 600))
10 display.start()
11 class product_spiderItem(scrapy.Item):
12     item1=scrapy.Field()
13     ...
14     pass
```

Fig. 6. Libraries and items to integrate Scrapy and Selenium

```
1  //Manejo de paginacion
2  self.count = self.count + 1
3  link = self.driver.find_element_by_link_text(str(paginacion))
4  try:
5     if(self.count < self.COUNT_MAX):
6        link.click()
7        print self.driver.current_url
8        page_number += 1
9     else:
10       break
11 except:
12    break
13 def closed(self, reason):
14 display.stop()
15 self.driver.quit()
16 self.driver.close()
```

Fig. 7. Handling of paging

searched text, use is made of functions that allow the browser to be clicked, when the condition is false a cut is made within the Selenium Web Driver to stop executing this process.

Development of the Data Collection Algorithm (CRAWLER)

Unlike the previous algorithm changes the configuration logic of the project since a solution is made by obtaining the links and stored within an array to be traversed later, this solution is optimal so that in the beginning the rules of the Scrape are defined. Figure 8 shows the code fragment necessary to perform the Crawler.

Development of the mechanism to generate RDF

To generate RDF from a CSV file it is necessary to make a map together with libraries capable of creating graphs from input, in this section a sequence of instructions

```
1  //Solucion Crawler
2  while True:
3      #Almacenamiento de links
4      linklist = []
5      for link in self.driver.find_elements_by_xpath():
6          linklist.append(link.get_attribute('href'))
7      #Recorriendo los links y extrayendo informacion
8      for link in linklist:
9          self.driver.get(link)
10         print self.driver.current_url
11         sel = scrapy.Selector(text=self.driver.page_source)
12         item = product_spiderItem()
13         item['item1']=sel.xpath().extract()
14         item['intem2']=sel.xpath().extract()
15         yield item
16         self.driver.back()
```

Fig. 8. General configuration (Crawler)

Table 2. EasyRDF methods

Method	Description
public object resource ($uri)	It creates a resource to be stored within a graph, it is used to later make a description of each resource given by the CSV
public integer set (string $resource, string $property, mixed $value)	It establishes literal values, dates and number taking into account the representations established to generate RDF
EasyRdf_Literal_Date ($valor)	EasyRdf_Literal_Date ($valor)
EasyRdf_Literal_Integer ($valor)	It gives the necessary format to represent a whole type value

was developed that serves to build RDF serializations; within the PHP server there is a file called "rdf.php" that allows the construction of different serializations in Table 2, it mentions the methods used given by EasyRDF.

4.3 Results

A CSV is established with the list of people with different variables, taking into account the existing vocabularies within the Semantic Web; currently there is a vocabulary called "FOAF" that helps describe people and their activities, the application in addition to perform the reuse of vocabularies can be extended to use the vocabulary of the user, the searching for vocabularies are requests to the LOV catalog which facilitates the search of terms by the user.

Name: foaf:name,
Surname: foafLastName,

Age:foaf:age,
Address:voabulary_proper,
Date of birth: schemaBirtdate,
Phone:foaf:phone.

The searching of vocabularies is done as shown in Fig. 9 where the key words defined by an expert are entered into the application. Afterwards, the data was entered into the application and joint mapping with the established values was made.

Seleccion de terminos

http://example.org/ | Generar RDF

Nombre	Buscar Termino	foaf:name	Uri
Apellido	Buscar Termino	foaf:lastName	Tipo de dato
edad	Buscar Termino	foaf:age	Int
direccion	Buscar Termino	http://example.org/direccion	Tipo de dato
Fecha de nacimeinto	Buscar Termino	schema:birthDate	Date
Telefono	Buscar Termino	foaf:phone	Tipo de dato
Pais	Buscar Termino	place:Country	Tipo de dato

Fig. 9. Code for crawler

The results obtained within this stage are the serializations that the user can export in Fig. 10 shows the output of the collection process.

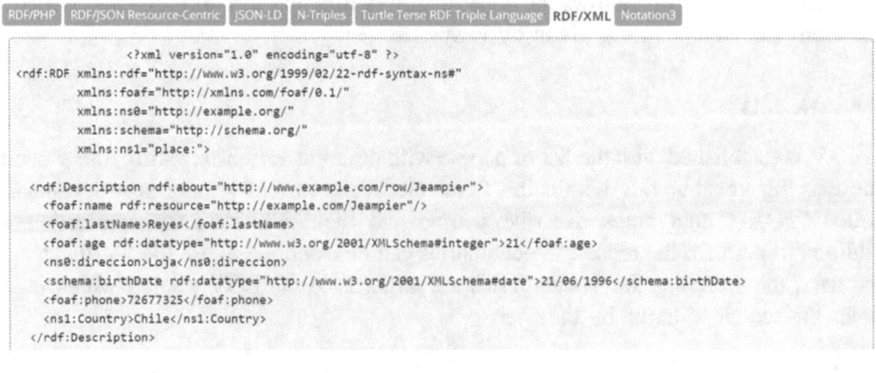

RDF/PHP | RDF/JSON Resource-Centric | JSON-LD | N-Triples | Turtle Terse RDF Triple Language | RDF/XML | Notation3


```
            <?xml version="1.0" encoding="utf-8" ?>
<rdf:RDF xmlns:rdf="http://www.w3.org/1999/02/22-rdf-syntax-ns#"
        xmlns:foaf="http://xmlns.com/foaf/0.1/"
        xmlns:ns0="http://example.org/"
        xmlns:schema="http://schema.org/"
        xmlns:ns1="place:">

  <rdf:Description rdf:about="http://www.example.com/row/Jeampier">
    <foaf:name rdf:resource="http://example.com/Jeampier"/>
    <foaf:lastName>Reyes</foaf:lastName>
    <foaf:age rdf:datatype="http://www.w3.org/2001/XMLSchema#integer">21</foaf:age>
    <ns0:direccion>Loja</ns0:direccion>
    <schema:birthDate rdf:datatype="http://www.w3.org/2001/XMLSchema#date">21/06/1996</schema:birthDate>
    <foaf:phone>72677325</foaf:phone>
    <ns1:Country>Chile</ns1:Country>
  </rdf:Description>
```


Fig. 10. Collection process

5 Conclusions

- The integration of tools such as Scrapy and Selenium allowed in a more precise way the development of an algorithm capable of extracting data from different web pages so it has very powerful functions that help the analysis of an HTML.
- The use of Jquery technology to assemble XPATH expressions by analyzing the DOM of an HTML allowed an analysis of the elements found taking into account upper and lower levels of each element.
- Most of web pages do not use standards established by W3C, making it difficult to have a reliable process when collecting data.
- To have a good data collection it is necessary to have user supervision, since many web pages contain errors within their HTML.

References

1. Bernal D, Castro A (2014) Web Semántica, más de una decada de su aparición, pp 61–69
2. Fabricio M, (2016) Catalogación de Recursos Educativos utilizando Vocabulario Controlado a partir de Ontologías
3. Scrapinghub (2018) Scarpy. https://scrapy.org/. Accessed 01 May 2018
4. López Poveda Y (2017) Análisis y diseño de un sistema para la extracción, análisis y comparación de precios de tiendas en la ciudad de Bogotá. Univiersidad Distrital Francisco José de Caldas
5. Leotta M, Clerissi D, Ricca F, Spadaro C (2013) Improving test suites maintainability with the page object pattern: An industrial case study. In: Proceedings - IEEE 6th Software Testing, Verification and Validation Workshops, ICSTW 2013, pp 108–113
6. SeleniumHQ (2018) Selenium Web Driver. https://www.seleniumhq.org/projects/webdriver/. Accessed 01 May 2018
7. Selenium (2017) Selenium-WebDriver, India
8. Gojare S, Joshi R, Gaigaware D (2015) Analysis and design of selenium webdriver automation testing framework. Procedia Comput. Sci. 50:341–346
9. Avasarala S (2014) Selenium WebDriver Guide. Birmingham
10. Import.io (2018) Import.io Extract. https://www.import.io/builder/data-extraction/. Accessed 06 May 2018
11. Mitchell R (2018) Web Scraping with Python, vol. I, p 239

Use of Apache Flume in the Big Data Environment for Processing and Evaluation of the Data Quality of the Twitter Social Network

Gladys-Alicia Tenesaca-Luna(✉), Diego Imba, María-Belén Mora-Arciniegas, Verónica Segarra-Faggioni, and Ramiro Leonardo Ramírez-Coronel

Departamento de Ciencias de la Computación y Electrónica, Universidad Técnica Particular de Loja, San Cayetano Alto y Marcelino Champagnat S/N, Loja, Ecuador
{gtenesaca, djimba, mbmora, vasegarra, rlramirez}@utpl.edu.ec

Abstract. The present work uses Hadoop as the core processing in the Big Data environment. There are several open sources tools from the Hadoop ecosystem that facilitate the processing of enormous volumes of data. In this paper, we have worked with Apache Flume and Apache Hive tools for the study case of the 2017 presidential elections in Ecuador. The analysis of data generated from Twitter social network focuses mainly in the first round of balloting of Ecuador's 2017 presidential election. These generated data have been obtained, stored, processed and analyzed to comply with the characteristics of the information that is considered Big Data. The selected tools have been evaluated in their architecture, installation, and use. Finally, the data have been evaluated under certain quality criteria or dimensions.

Keywords: Big Data · Twitter · Apache Flume

1 Introduction

In recent years, and due to the advances in information technology, the exchange of information has gone to a great growth which is evidenced in large volumes of data, it is necessary to take into consideration tools and storage media that allow collecting all this information for processing.

The human being has become a dynamic actor that contributes to the generation of a large amount of information using technological devices that have allowed generation, transfer, and exchange of data. With the passing of time, it has become necessary to assign a name to a great deal of information of resulting data for its analysis, use, value, and contribution that it provides a person, company, organization, etc. In this work, we have documented information regarding Big Data environments and tools like Apache Flume and Hive.

M. Botto-Tobar et al. (Eds.): TICEC 2018, AISC 884, pp. 314–326, 2019.
https://doi.org/10.1007/978-3-030-02828-2_23

The selected tools have been evaluated in their architecture, installation, and use in order to design a methodology that allows processing data from a social network source, as well as to evaluate the quality of extracted data. Finally, the case study is focused on the analysis and interpretation of data extracted from the Twitter social network which are about the 2017 presidential elections in Ecuador.

2 Theoretical Foundation

The objective of this bibliographical review is to find scientific papers that define the context of the implementation of the Big Data architecture, analysis and processing of large volumes of data extracted from Twitter, and indeed an evaluation of the data generated. Therefore, the following questions are raised what motivates this research, giving, as a result, a broader set of works.

2.1 Why Is Big Data Important?

Big Data refers to the collection of data that are generated daily. According to [1], the term "Big Data" is a deluge of information that is generated by digital devices which record and practically dictate the personal and professional life patterns of each one.

Therefore, patterns, trends, and associations are revealed by massive data processing. In our case of study, this information can become accurate about the Twitter[1] social network, in order to make interpretations of the 2017 presidential elections in Ecuador. Considering that Twitter (see footnote 1) is an important source of data that provides information in real time. The characteristics of a Big Data described in [2, 3] are: "volume" is related to the massive amount of data that is produced continuously, in addition, the massive data processing and storage is always complex. "Variety" refers to types of data like structured, unstructured and semi-structured data, it is important to mention that structured data is always of a fixed format. Finally, the last feature of a Big Data is "velocity" that refers to the speed of generating data in real-time. Table 1 shows the challenge and the impact to be considered by each of the Big Data dimension. Although there are some other new dimensions proposed by several researchers.

Therefore, the importance of the massive data is reflected in the ability to transform them into new information and then analyze and extract knowledge that it will help to make better decisions.

The authors [4] argue that the implementation of technological improvements enable the acquisition of data and allow discovering the needs and points of improvement for companies. On the other hand, learning from massive data allows researchers, analysts, and organizations to improve their operations [5, 6]. Therefore, the importance of Big Data is not only focused on the technological challenges involved but also on the way of the interaction with information has changed. This is how the Big Data analysis enters the scene using different techniques to extract information.

---

[1] https://twitter.com/.

Table 1. Dimension, challenge, and impact of Big Data.

Dimension	Challenge	Impact
Volume	How is data collected?	Decision making
Variety	How to analyze heterogeneous data?	Analysis
Velocity	What are the parameters to minimize latency in collecting data, analysis and decision making?	Business processes

2.2 What Methodology for Extracting Big Data?

As explained in the previous section, Big Data consists of structured and unstructured data, which includes video, audio, data generated by electronic devices. In our case of study, Twitter provides a large amount of data because of the online interaction of millions of Twitter users. The extraction of relevant information from massive data has several challenges. However, different platforms have been developed for data processing is effective.

In order to Big Data management, there are a series of steps to go through: acquisition, extraction, preprocessing, and data analysis. Figure 1 shows the mentioned stages:

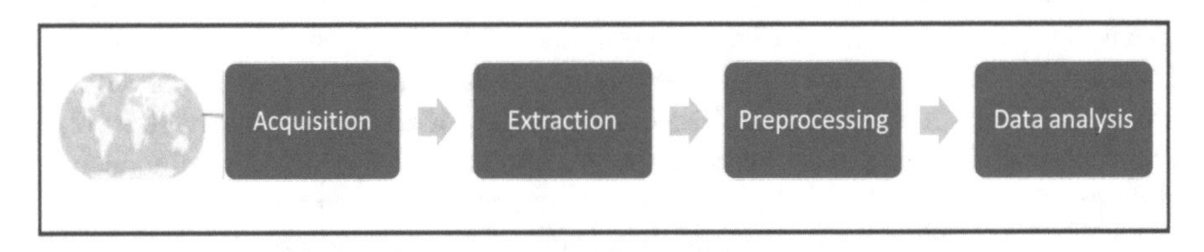

Fig. 1. Stages of Big Data processing

Acquisition: The data is collected from internal or external data sources, for example ERP and social networks respectively. As mentioned in [7], the data is obtained from social networks servers. It is important to mention that data is unstructured when it is obtained from external sources.

Extraction: Obtaining and extracting metadata also requires technical aspects. The aim at this stage is to extract the incompatible data to transform them into a readable format to reduce computational overload and perform data analysis in Big Data.

Preprocessing: This stage is designed to clean and remove some kind of variability in the input data. In this way, the data is provided with an organized structure that allows for quick access to data.

Data Analysis: The aim at this stage is to discover useful information, this process is facilitated with the help of technology.

2.3 How Can the Hadoop Ecosystem Be Used?

Hadoop[2] is an open source framework for processing data, it provides massive storage for any type of data on a distributed computational architecture. In addition, Hadoop is considered one of the best tools for analyzing data from Twitter.

Twitter data is instant data updates [5, 8, 9]. The authors of [9] emphasize that Hadoop is designed to solve problems such as processing and analyzing massive data, which is known as Big Data.

The Hadoop ecosystem is composed of several projects that offer functionalities such as storage, query, indexing, transfer, transmission and messaging [10].

In addition, Hadoop comprises an ecosystem that performs parallel processing in the same data stored in the HDFS nodes (Hadoop Distributed File System) using its parallel programming paradigm, namely Map Reduce [3, 7].

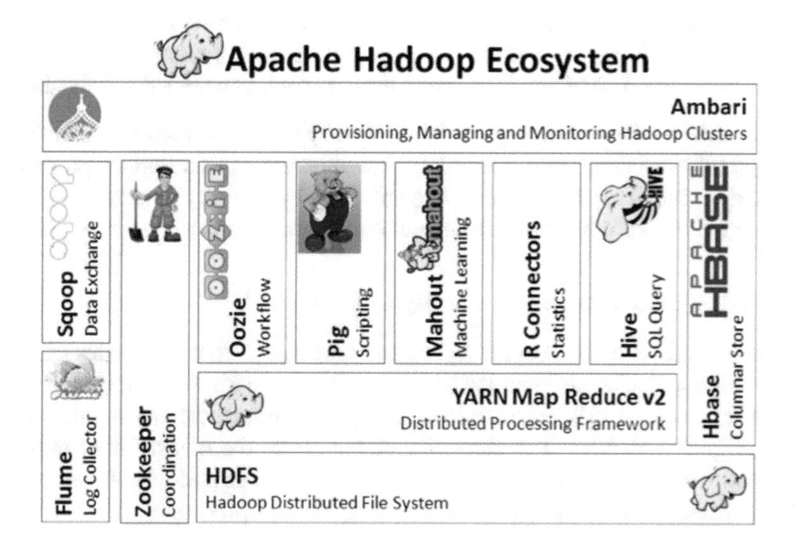

Fig. 2. Hadoop ecosystem

According to [11], the components of the Hadoop ecosystem are HDFS, Ambari, HBase, Hive, Sqoop, Pig, ZooKeeper, Mahout, Lucene/Solr, Avro, Oozie, and Flume (see Fig. 2). We describe the tools reviewed for our case study as follows:

Apache Flume: It is a framework based on streaming data streams to collect, aggregate and transfer large amounts of data. That is, it is flexible for obtaining and processing large volumes of unstructured and semi-structured data from social networks, facilitating the acquisition of information required by the user.

[2] https://hadoop.apache.org/.

Therefore, Flume is an efficient and reliable distributed service. Twitter data is semi-structured, and when collected through Flume, it is given a structure that allows knowing the information of a Tweet (e.g. date, id, text, etc.) [10].

Apache Hive: It is a data storage framework for consulting and managing large data sets stored in distributed Hadoop (HDFS) file systems.

Also, it is responsible for the process of extracting the bits from all the files that are generated in HBase. When the performance is not correct, Hive makes use of the MapReduce. Hive is recommended for those developers who are familiar with SQL, considering that Hive Query Language (HiveQL) is the primary data processing method explained in [12].

Sqoop: It is a tool for the treatment of structured data. It is based on command lines, which allows transferring large volumes of data between Hadoop and storage systems (MySQL or Oracle), which includes the Hadoop Distributed File System (HDFS), Apache Hive and Apache HBase. That is, Sqoop allows transfer data between Hadoop and relational databases, in addition, to import data from a relational database management system into HDFS and then export the data back into a relational database [12].

Pig: It facilitates the processing of large volumes of data present in Hadoop. In addition, it helps Hadoop users spend less time on creating MapReduce programs and focus more on data analysis. The advantage of using Pig is that you only need to write a few lines of code, so it reduces the general development and testing time [12].

2.4 What Is the Evaluation of the Big Data Quality?

The quality of the data is one of the most important aspects that companies take care of before obtaining results based on the analysis of the large information sets. It does not only refer to data that lack errors, this is only part of the whole environment that this concept of quality implies. The extraction of data from the thousands of information sources that are daily obtained or downloaded by experts is part of the various quality problems faced by companies prior to the analysis and processing of large volumes of data.

The thousands of data that are generated daily is one of the great challenges that we can face when talking about the evaluation of data quality because the data comes from different heterogeneous sources, which means that they are in countless numbers of formats, besides coming from several types of applications and domains [13].

In the context of Big Data, the evaluation of the quality of the data must be done before the Big Data analysis, this will provide veracity and confidence in the quality of the data that will be processed in the company. The large size of the volume of data requires certain mechanisms and strategies to evaluate the quality of the data quickly and efficiently, taking into account that such evaluation would be too expensive if applied to the entire complete data set, for this reason, one of the strategies applied is to separate sub-sets the data to evaluate them [14].

2.5 Related Work

The era of social networks generates a huge amount of data, which is why many authors consider it an opportunity to analyze this data. Some works are focused on the analysis of data from Twitter, investigations like [5, 7, 9, 15] consider that Hadoop is one of the best options of tools for analysis of Twitter data which are known as Big data. In [7] it focuses on exploring the social network Twitter to monitor Earth events, incidents, medical illnesses, user trends to make future decisions. In this study, Spark is implemented in the upper part of the Hadoop ecosystem to perform a real-time analysis.

Similarly, the work [15] is focused on the use of social networks for the detection of spatiotemporal events related to logistics and planning.

Specifically, it uses big data and artificial intelligence platforms, including Hadoop, Spark, and Tableau, to study Twitter data about London.

We can identify that these works have not explored other tools that make up the Hadoop ecosystem such as Apache Flume, Apache Hive, Pig or Sqoop.

3 Methodology

The proposal is framed in the implementation of a Big Data environment by configuring the Apache Flume tool that allows the process of downloading and processing a volume of data from the Twitter social network, performing an evaluation of the quality of the data extracted. The appropriate cycle for Big Data processing [16] is composed of 4 stages: Collection, Processing, Analysis, and Visualization. During the cycle, the data quality evaluation activity is considered, in which certain criteria and dimensions are applied to evaluate the data generated. The same that can be adapted for social data sources as shown in Fig. 3.

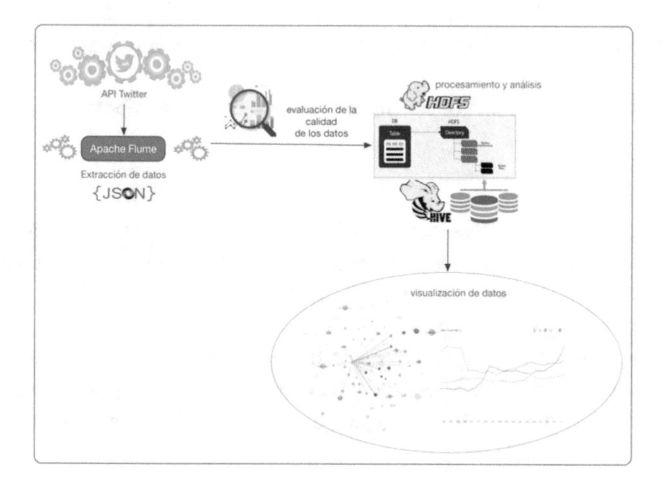

Fig. 3. Big Data processing cycle with Apache Flume

3.1 Data Collection

Data collection refers to the process of gathering, collecting information from data sources to process. Social networks such as Twitter, Facebook, Instagram among others are sources of inexhaustible data [17], have the following characteristics, among them:

1. It allows combining text with other multimedia resources in the same message.
2. It has an agile and simple dissemination mechanism (tweet, retweet, response, and mention)
3. It facilitates the creation of own channels for each product, service or campaign through tags or hashtags.

One of the tools that allow the collection of information on the ecosystem of Hadoop is Flume, a tool that obtains large volumes of unstructured and semi-structured data from various social networks quickly, which facilitates the acquisition of personalized information by the user. In addition, it is a highly flexible tool for data acquisition purposes [18]. Apache Flume transmits data through Streaming, which facilitates the origin of the data is configurable allowing logs monitoring and downloading information from different social networks, as well as taking information from e-mails, among others.

3.2 Data Processing

The data processing is done in Apache Hive, prior to the installation of Hadoop, which allows the processing of the data and through consultations the information is obtained for its corresponding analysis.

Apache Hive has become a Hadoop project that provides easier ways to create, query and manage large volumes of distributed data that are contained in the form of relational tables. Hive provides a type of query language derived from SQL which is called HiveQL or HQL. HiveQL is designed based on MapReduce in order to take advantage of the features for the processing of large amounts of data that are stored in Hadoop.

The processing and result of the information will not be given in real time [19].

3.3 Data Analysis

The analysis of the data is classified as confirmation analysis and exploratory analysis. The first is done under a deductive approach, which investigates the cause of the phenomenon that is assumed in advance. It is based on a hypothesis and after the analysis, the hypothesis is tested or refuted to give definitive answers.

In contrast, the exploratory analysis is done under an inductive approach that is closely linked to data mining. There are no hypotheses or assumptions since the data is explored through analysis to develop an understanding of the cause of the phenomenon.

3.4 Evaluation of the Quality of the Data

The evaluation of the quality of the data was focused on some quality dimensions, such as integrity and consistency. This activity is important to perform due to the high impact of the erroneous data at the time of the results of the Big Data analysis.

There are some previous criteria to consider in order to perform the evaluation; one of them is the data and types of data with which we are going to find, normally the data in the companies are quite heterogeneous if we focus on the sources, when talking about the social network data we find structured data and not structured from the various user publications.

For the process of quality assessment, a common framework characterized by several quality dimensions has been chosen, which allows us to compare these dimensions in different types of data. The mentioned frame is based on a classification in groups of dimensions according to the similarity of the data, each dimension is characterized by (Table 2):

Table 2. Dimensions of evaluation de quality

Dimension	Description
Accuracy	Validity and precision focus on the adherence of the data to a given reality
Integrity	Relevance and relevance refer to the ability to represent the relevant aspects of reality according to the domain
Redundancy	Minimization and conciseness refer to the ability to represent the reality of a domain with the minimum use of resources
Reliability	Clarity and simplicity refer to the ease of understanding of data by users
Accessibility	Availability are related to the user's ability to access personal data, culture, physical status/functions and available technologies
Consistency	Cohesion and coherence refer to the capacity of the data to comply without contradictions to all the properties of the reality of interest

4 Case Study

The proposal is framed in the implementation of a Big Data environment by configuring the Apache Flume tool that allows the process to be carried out.

4.1 Data Collection

The first case study focuses on downloading data related to the first round of balloting of Ecuador's 2017 from the Twitter social network, it is the social network that provides the most facilities with the least restrictions when obtaining information. Searching for topics by keywords helps a lot when establishing information patterns and exploration systems. Therefore, the information to obtain is related to each one of the presidential candidates using their Twitter account, these are:

- Lenín Moreno - @Lenin
- Guillermo Lasso - @LassoGuillermo

- Cynthia Viteri - @CynthiaViteri6
- Paco Moncayo - @PacoMoncayo
- Abdalá Bucaram - @daloes10
- Iván Espinel - @IvanEspinelM
- Washington Pezántez - @pesanteztwof
- Patricio Zuquilanda - @ZuquilandaDuque

Figure 4 shows a sample of the data extracted from Twitter with Apache Flume, this data is presented in JSON format to be evaluated under the quality dimensions defined and later processed in Apache Hive.

Fig. 4. Data download from Apache Flume

4.2 Data Quality Assessment

The process of evaluating the data extracted in Apache Flume is done based on the established dimensions versus the conventional errors that are presented in the data after the discharge with Apache Flume. To carry out this process, a search was made of the most relevant errors found when downloading data from the social network, then a classification based on the dimensions of accuracy, redundancy, integrity, readability was made., accessibility and consistency that the data can possess and finally an evaluation comparison is made. Table 3 presents the evaluation performed.

4.3 Data Processing

Start Hive

After Hadoop starts, Hive starts in the virtual machine executing the following command in the system terminal (Fig. 5):

Fig. 5. Start Apache Hive

Table 3. Evaluation of quality dimensions vs. conventional problems.

Quality problems	Dimension					
	Accuracy	Redundancy	Integrity	Reliability	Accessibility	Consistency
Duplicity of data						x
Distortion in the information extracted	x		x	x		
Incorrect data, error in data entry	x	x	x		x	x
Irrelevant data		x	x			x
Outdated data	x		x			
Information privacy				x	x	x
Incorrect data types					x	x
Lack of restrictions on the integrity of information	x	x	x	x	x	x

Export Data from HDFS to Hive

The process of managing the information of Twitter is done in several steps in Hive: the creation of the database, creation of a table, export of HDFS data to the created table, execution of the query scripts on Hive, obtaining results and data.

4.4 Data Analysis

By having the information in a Hive table called "data_primera_vuelta" we proceed to execute the queries that allow us to give us information on trends of the first round of balloting of Ecuador's 2017 according to the consultation sentences that we want to make. The information search was carried out in 5 queries, which have shown the following information:

- Total records in first round of balloting of Ecuador's 2017 (Fig. 6).

```
hive> select count(*) from data_primera_vuelta;
```

Fig. 6. Sentence for total records in first round

El resultado de la consulta es el siguiente (Figs. 7 and 8):

```
Total MapReduce CPU Time Spent: 1 minutes 29 seconds 380 msec
OK
131512320 TOTAL
Time taken: 304.719 seconds, Fetched: 1 row(s)
hive>
```

Fig. 7. Result of the total of records in first round

```
hive> select '@Lenin fue mencionado' as nombre, count(id) as cantidad from data_primera_vuelta where text like '%@Lenin%'
> union
> select '@CynthiaViteri6 fue mencionado' as nombre, count(id) as cantidad from data_primera_vuelta where text like '%@CynthiaViteri6%'
> union
> select '@daloes10 fue mencionado' as nombre, count(id) as cantidad from data_primera_vuelta where text like '%@daloes10%'
> union
> select '@ZuquilandaDuque fue mencionado' as nombre, count(id) as cantidad from data_primera_vuelta where text like '%@ZuquilandaDuque%'
> union
> select '@pesanteztwof fue mencionado' as nombre, count(id) as cantidad from data_primera_vuelta where text like '%@pesanteztwof%'
> union
> select '@LassoGuillermo fue mencionado' as nombre, count(id) as cantidad from data_primera_vuelta where text like '%@LassoGuillermo%'
> union
> select '@PacoMoncayo fue mencionado' as nombre, count(id) as cantidad from data_primera_vuelta where text like '%@PacoMoncayo%'
> union
> select '@IvanEspinelM fue mencionado' as nombre, count(id) as cantidad from data_primera_vuelta where text like '%@IvanEspinelM%';
```

Fig. 8. Processing of queries

4.5 Data Visualization

Results of the Case Study: Ecuador's 2017 Presidential Elections

When obtaining the results after using the Flume and Hive tools in the case study, it is necessary to analyze and interpret the data thrown. This is done as follows.

Total Records in the First Round

The first round of balloting of Ecuador's 2017 began on January 3rd, 2017, and ended on the day of the elections on February 19th, 2017. During all that time, the social network Twitter was constantly active, from which a total of 131512320 twitters that were generated at different times of the day in the entire Ecuadorian territory were downloaded with the Flume tool. Of the downloaded twitters, a total of 28055040 tweets relate to the candidates of the first electoral round. This relationship is shown in the following table:

The Fig. 9 shows the percentage relationship of Table 4:

Table 4. Total data downloaded from Twitter - 1st round

Description	Amount
Tweets about other topics	103457280
Tweets about candidate	28055040
TOTAL	131512320

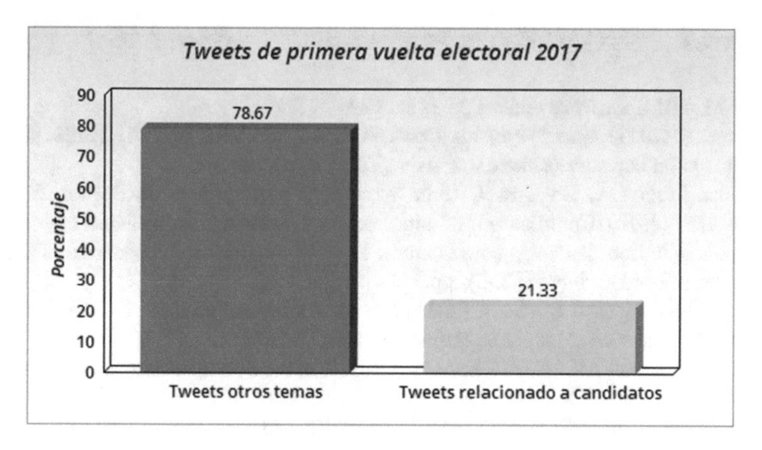

Fig. 9. Percentage of election results Ecuador 2017

Interpretation: During the period of the of balloting of Ecuador's 2017, the publications made by several people in Ecuador covered a total of 131512320 tweets, which represents 100% of the information obtained. Of that percentage, 78.67% (103457280) are tweets related to various topics other than politics and 21.33% (28055040) of the tweets talked about the presidential candidates who were part of the first round of elections 2017. Many of the people have used the social network to make their comments known by some candidate, and this is a good starting point to exploit the information using Hadoop.

5 Conclusions

- Carrying out an investigation at the corporate level, it has been established that in business organizations that handle Big Data, the main tools of the Hadoop ecosystem that are used are: Flume, Hive, These tools they make a significant contribution to an organization, providing value so that decision-making is smarter, faster and makes a difference.
- The use of the usability prototype has made it possible to efficiently compare the structured and unstructured data generated during the first round of the 2017 elections.
- The evaluation of data quality is one of the current challenges within Big Data since it ensures that the volume of data to be analyzed and processed is consistent and legible, guaranteeing that the results are reliable and suitable for making the right decisions within organizations or companies, this being one of the activities that today are proposed as a new alternative to research and work.
- As future work, a validation of data can be proposed with other tools of the Big Data ecosystem that allow to validate in different levels of veracity of the forecasts and results obtained in the processing of a volume of data.

References

1. Riffat M (2014) Big data—not a panacea. ISACA 3:19–21
2. Dasoriya R (2017) A review of big data analytics over cloud. In: 2017 IEEE international conference on consumer electronics-Asia (ICCE-Asia), pp 1–6
3. Tenesaca Luna GA, Chicaiza J, Mora Arciniegas MB, Torres JP, Segarra-Faggioni V, Vinan MS (2016) Contribution of big data in E-leaming. A methodology to process academic data from heterogeneous sources. In: 35th international conference of the chilean computer science society (SCCC), pp 1–12
4. Mayer-Schönberger V, Cukier K (2013) Big data: a revolution that will transform how we live, work, and think. Houghton Mifflin Harcourt, Boston
5. Sehgal D, Agarwal AK (2016) Sentiment analysis of big data applications using Twitter data with the help of HADOOP framework. In: 2016 international conference system modeling & advancement in research trends (SMART), pp 251–255
6. Sabar NR, Yi X, Song A (2018) A bi-objective hyper-heuristic support vector machines for big data cyber-security. IEEE Access 6:10421–10431
7. Mazhar Rathore M, Ahmad A, Paul A, Hong W-H, Seo H (2017) Advanced computing model for geosocial media using big data analytics. Multimed Tools Appl 76(23):24767–24787
8. Soche López S (2016) Metodología para el modelamiento de datos basado en big data, enfocados al consumo de tráfico (voz-datos) generado por los clientes. Universidad Militar Nueva Granada, Bogota, Colombia, p 17
9. Sehgal D, Agarwal AK (2018) Real-time sentiment analysis of big data applications using Twitter data with hadoop framework. In: Soft computing: theories and applications, pp 765–772
10. Vohra D (2016) Introduction. In: Practical hadoop ecosystem. Apress, Berkeley, pp 3–162
11. Hurwitz J, Nugent A, Halper F, Kaufman M (2013) Big data for dummies. Wiley, New Jersey
12. Bhardwaj A, Vanraj, Kumar A, Narayan Y, Kumar P (2016) Big data emerging technologies: a CaseStudy with analyzing Twitter data using Apache Hive. In: 2015 2nd international conference on recent advances in engineering & computational science, RAECS 2015, December 2016
13. Batini C, Rula A, Scannapieco M, Viscusi G (2015) From data quality to big data quality. J Database Manag 26(1):60–82
14. Taleb I, El Kassabi HT, Serhani MA, Dssouli R, Bouhaddioui C (2016) Big data quality: a quality dimensions evaluation. In: 2016 international IEEE conferences on ubiquitous intelligence & computing, advanced and trusted computing, scalable computing and communications, cloud and big data computing, internet of people, and smart world congress (UIC/ATC/ScalCom/CBDCom/IoP/SmartWorld), pp 759–765
15. Suma S, Mehmood R, Albugami N, Katib I, Albeshri A (2017) Enabling next generation logistics and planning for smarter societies. Procedia Comput Sci 109:1122–1127
16. Tenesaca-Luna GA, Chicaiza J, Mora-Arciniegas M-B, Ureña-Torres J-P, Faggioni AS, Santiago M, Ludeña V (2016) Contribution of big data in e-learning. A methodology to process academic data from heterogeneous sources
17. Logicalis (2015) Redes sociales como fuentes de datos: el caso de Twitter. https://blog.es.logicalis.com/analytics/redes-sociales-como-fuentes-de-datos-el-caso-de-tweeter. Accessed 24 Mar 2018
18. The Apache Software Foundation (2017) Apache Flume. https://flume.apache.org/. Accessed 16 Apr 2018
19. Thusoo A, Sen Sarma J, Jain N, Shao Z, Chakka P, Anthony S, Liu H, Wyckoff P, Murthy R (2009) Hive - a warehousing solution over a map-reduce framework. Sort 2:1626–1629

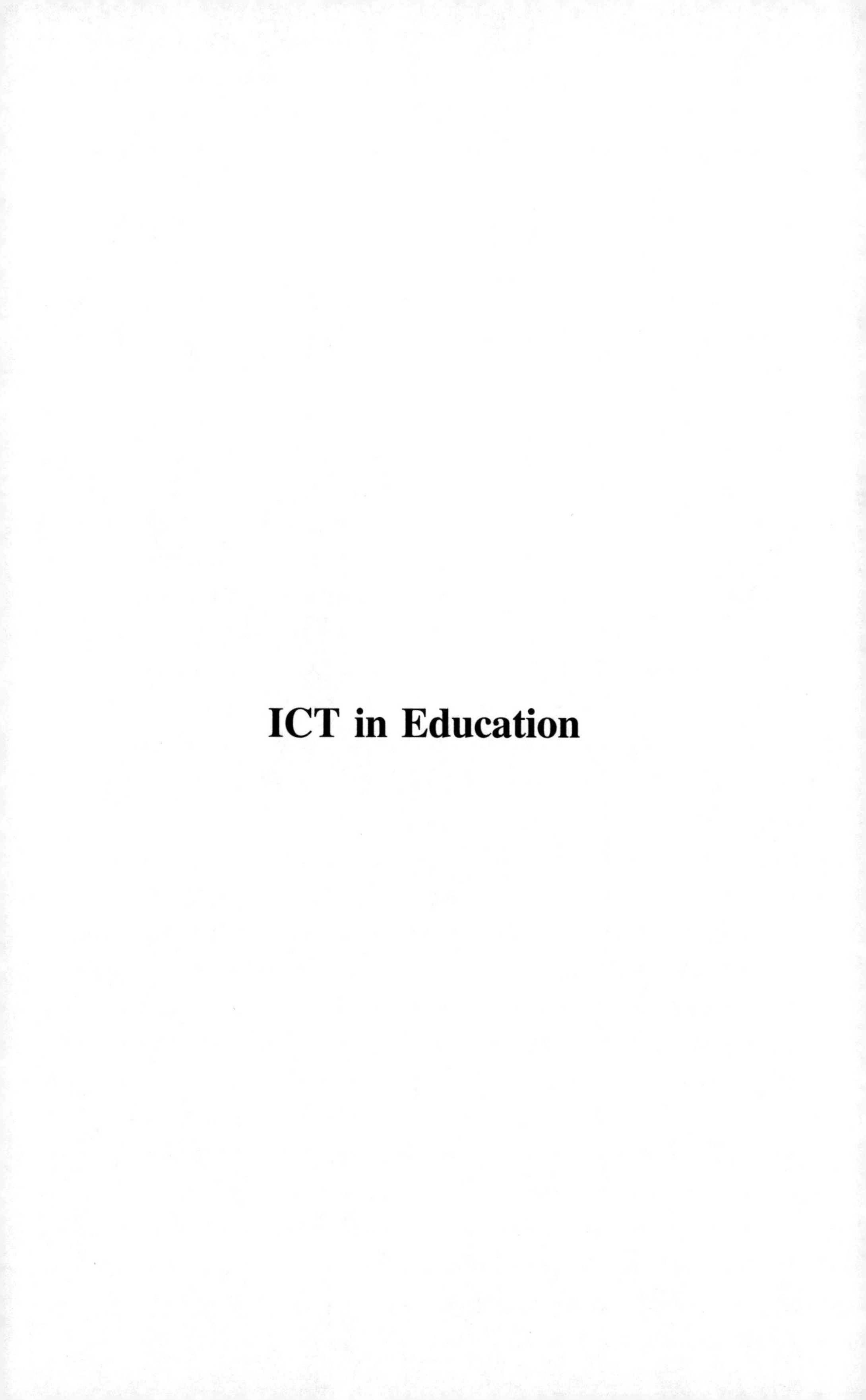

ICT in Education

Sophomore Students' Acceptance of Social Media for Managing Georeferenced Data in a Socially-Enhanced Collaborative Learning Process

Erika Lozada-Martínez(✉), Félix Fernández-Peña, and Pilar Urrutia-Urrutia

Universidad Técnica de Ambato, Ambato 180103, Ecuador
{elozada7423,fo.fernandez,elsapurrutia}@uta.edu.ec

Abstract. Social media has been influencing different initiatives of collaborative learning during the last decade. Nevertheless, the lack of studies on the acceptance of georeferenced data in social media as well as the relevance of specific features in education has been identified as a missing point in this research area. In this scenario, this paper evaluates the technology acceptance of a Facebook group and a georeferenced wiki as learning tools. The study was carried out with the participation of 250 sophomore students of software engineering at the Universidad Técnica de Ambato. A controlled experiment was designed based on the Technology Acceptance Model. After using the evaluated tools for eigth weeks, two surveys were conducted and 52 participants were interviewed. The results show that both, the perceived ease of use and the perceived usefulness of the analyzed tools have a positive impact on the behavioral intention of using social media in learning processes. No significant differences were found but both tools had positive acceptance among students. Moreover, the interviewed students have the opinion that an hybrid model of socially-enhanced collaborative learning, based on Facebook and wikis of georeferenced data, has a positive impact on collaborative learning.

Keywords: Social media · TAM · Education · e-Learning

1 Introduction

In the last few years, the impact of social media on the academia has been remarkable, being actually used for supporting the activities of both, learners and learning facilitators [1–4]. These tools are been considered even more effective than the traditional methods of collaborative learning [5–7]. In this scenario, social media let students to communicate and interact, among them and with facilitators, and they also promote autonomous work and self-training [2,3,7]. This aims to turn the model where teachers teach into a model where students learn [1,8,9].

M. Botto-Tobar et al. (Eds.): TICEC 2018, AISC 884, pp. 329–344, 2019.
https://doi.org/10.1007/978-3-030-02828-2_24

Social media-based collaborative learning embraces the use of Facebook, Twitter, blogs, forums and wikis [8,9]. Connolly et al. evaluated the readiness that teachers and students of Australia have for using Facebook as a vocational learning tool [7]. Meanwhile, Moore et al. studied the activity, the actions, interactions and publications of the members in a Facebook group created with educational purposes [10].

Several studies have been focused on using Facebook groups [2,7,10–12], which is the most common alternative of social media in education [13]. Nicolai et al. studied the performance of a Facebook group for educational purposes; their research took place with the participation of medicine students in Munich [2]. Castro and González-Palta used a Facebook group named Face-Critic to improve the critical thinking of psychology students [11]. Reyes-Garcés et al. implemented Sigma, an application that renders in the Facebook canvas and extends the functionalities of a Facebook group, demonstrating that the Facebook platform is flexible enough for adapting the user experience to a specific context [14].

On the other hand, wikis have had a broad use in collaborative learning [15–17]. Particularly in education, He and Yang studied the effectiveness of wikis for facilitating lesson planning and collaboration between workgroups [15]. Wang et al. proposed a wiki where students and teachers collaborate and cooperate in the design and implementation of real time lab experiments, improving also the user experience and reducing time and space limitations [16]. Zander et al. developed WIKL, a wiki platform on which students can freely access and create content related to subjects of interest through their learning process [17]. So it is obvious that wikis are also used in the academia. [15–17].

Meanwhile, the management of georeferenced data in a socially-enhanced collaborative wiki has been considered relevant [18] but, based on the literature review, the benefits and acceptance of its use in education have not been assessed. Furthermore, the lack of studies on the relevance of specific features of social media tools, such as the type of data they support, has been identified as a missing point in this research area [8,19].

As extending a Facebook group has not proven to be accepted by students as a better option for collaborative learning [14], we decided to extend Dokuwiki for managing georeferenced data and testing its acceptance in education; the resultant wiki was named LearnersWiki.

In this paper, we compare the technology acceptance of LearnersWiki against a Facebook group named LearnersGroup. Both tools were used for supporting collaborative learning about graphs as a data structure. Graphs were chosen because collaborative learning about graphs have had good results [20–22] and because this data structure is usually learned by sophomore students, for whom it has been proved that social media have a positive influence in education [13]. It is not our goal to determine in which case social media allows to have the best performance of students, but to assess the acceptance of social media for managing georeferenced data in a socially-enhanced collaborative learning process. Table 1 shows a comparison between the features supported by LearnersGroup and LearnersWiki. Both tools support, in someway, sharing and rating content.

Meanwhile, LearnersGroup supports online dialog whilst LearnersWiki implements collaborative edition of georeferenced content. In this context, this study aims to answer the following research question:

- Has LearnersWiki better acceptance than LearnersGroup as a collaborative learning tool for software engineering sophomore students in Ambato?

Table 1. Features supported by LearnersGroup and LearnersWiki.

Features	LearnersGroup	LearnersWiki
Share content	Posts	Pages
Collaborative edition	No	Yes
Online dialogue	Yes	No
Rate useful content	Likes	1-5 star rating
Georeferenced content	No	Yes

This paper is structured as follows. In Sect. 2, the technology acceptance model and supporting technologies of the study are introduced. In Sect. 3, the experiment setting is described. In Sect. 4, the obtained results are discussed. Finally, in Sect. 5, conclusions and future work are presented.

2 Materials

2.1 Technology Acceptance Model

The Technology Acceptance Model (TAM), developed by Davis [23], lets understand the relationships between attitudes, intentions and internal beliefs of users [12,23]. Based on these aspects, the model has the capacity to evaluate the acceptance of technology, including information systems.
This model is composed by four factors which are: (i) perceived ease of use, (ii) perceived usefulness, (iii) attitude towards using and (iv) behavioral intention to use. The relationships between them, which are depicted in Fig. 1, allow us to predict the reasons why a technology is accepted or not, explain those reasons and determine corrective actions [1,19,24–27].

TAM is widely used in research because this model is simple, robust, influential and applicable [1,19,24–27]. Dumpit and Fernández studied the use of social media in higher education with TAM [1]. Efiloğlu Kurt and Tingöy analyzed the acceptance and use of a virtual learning environment in Turkey and The United Kingdom applying TAM [24]. Chai and Fan studied relationships between social media and creativity by establishing correlations between TAM factors [25]. Wamba extended TAM for evaluating the use of social media in the workspace across multiple countries [26]. In Ecuador, Ramirez-Anormaliza et al. evaluated the acceptance and use of e-learning systems in a public university [27].

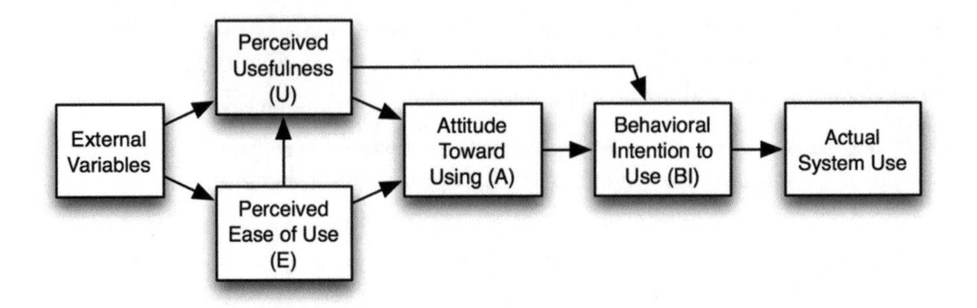

Fig. 1. Technology acceptance model.

From these studies [1,24–27] we understand the wide scope of application of TAM. We conclude that TAM is valid for quantifying the level of acceptance of both tools under study. In this study, based on TAM and the design of experiment of Efiloğlu Kurt and Tingöy [24], the following hypotheses were tested in the context of social media use by software engineering sophomore students at the Universidad Técnica de Ambato.

- **H1**: The perceived usefulness has a positive impact on the behavioral intention to use LearnersGroup and LearnersWiki.
- **H2**: The perceived ease of use has a positive impact on the behavioral intention to use LearnersGroup and LearnersWiki.
- **H3**: There is a significant difference in perceived ease of use between LearnersGroup and LearnersWiki.
- **H4**: There is a significant difference in perceived usefulness between LearnersGroup and LearnersWiki.
- **H5**: There is a significative difference in behavioral intention to use between Learnersgroup and LearnersWiki.

2.2 Facebook Groups

LearnersGroup's virtual space, which interface is depicted in Fig. 2, was created in the Facebook platform for allowing students to debate about trees and graphs data structures. Participants in the experiment used it for sharing useful content, discussing topics of interest through dialogues and reacting by using social actions. With the analysis of its specific features, shown in Fig. 1, our hypothesis was that:

- **H6**: Online dialogue has a positive impact on the behavioral intention to use LearnersGroup.

2.3 LearnersWiki

LearnersWiki, which interface is depicted in Fig. 3, was implemented by extending the open source wiki system *Dokuwiki*. It has the ability to manage georeferenced data and rating metadata. Learners can use LearnersWiki to collaborate

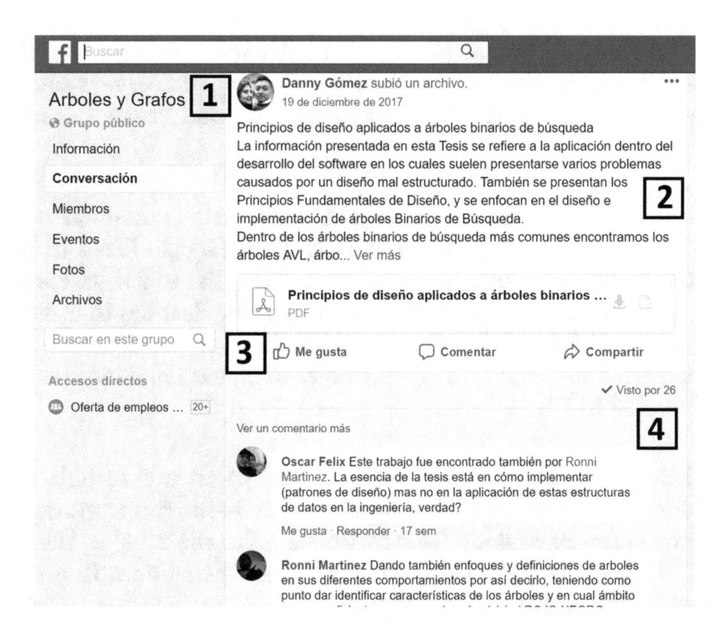

Fig. 2. LearnersGroup interface. (1) Declaration of the topic of the Facebook group. (2) Example of published content in LearnersGroup. (3) Available social actions. (4) Example of online dialogue.

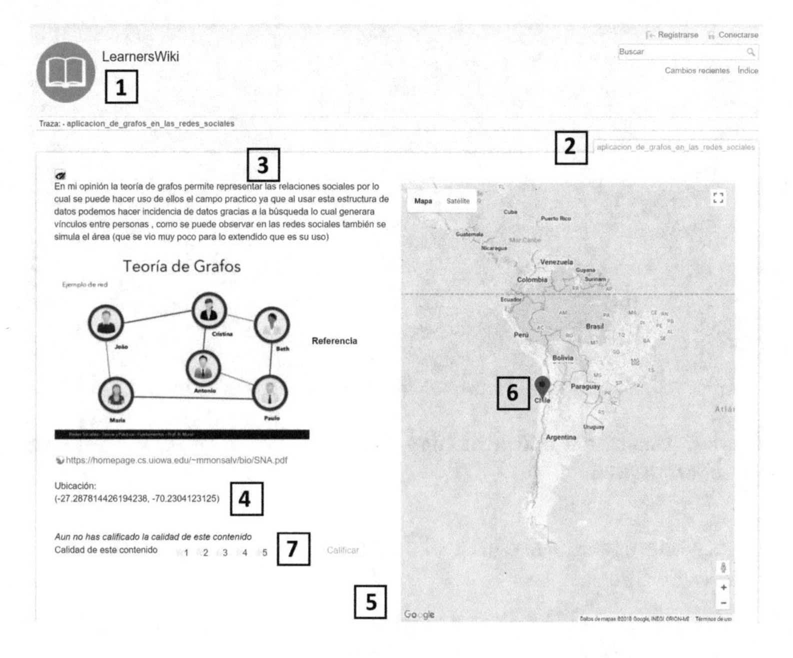

Fig. 3. LearnersWiki interface. (1) Name and Logo of the tool. (2) Page title. (3) Page content. (4) Coordinates of the location where the content was generated. (5) Map component. (6) Spot marker of a location. (7) Rating of the content quality.

in creating, representing and acquiring knowledge and also for rating the quality of wiki pages' content. Students were able to create, edit, review and rate wiki pages with content concerning the implementation and use of trees and graph data structures.

Maps Manager.- The map manager uses the location coordinates associated to content and makes an asynchronous request to Google Maps to render the map component. The information provided is presented to the user and the corresponding spot marker is displayed on the map. The decision to use the Google Maps API is based on the functions and methods it provides for annotating points through markers and highlighting areas of educational interest using heat maps, as depicted in Fig. 4.

Rating Manager.- The feature of rating wiki pages was included in LearnerWiki. The rating manager allows to visualize a 5-star rating system with the aim to improve user experience. This feature is quite the same as the like action in LearnersGroup, which evaluates the quality of the content in a qualitatiive scale of level of enjoyability of the content.

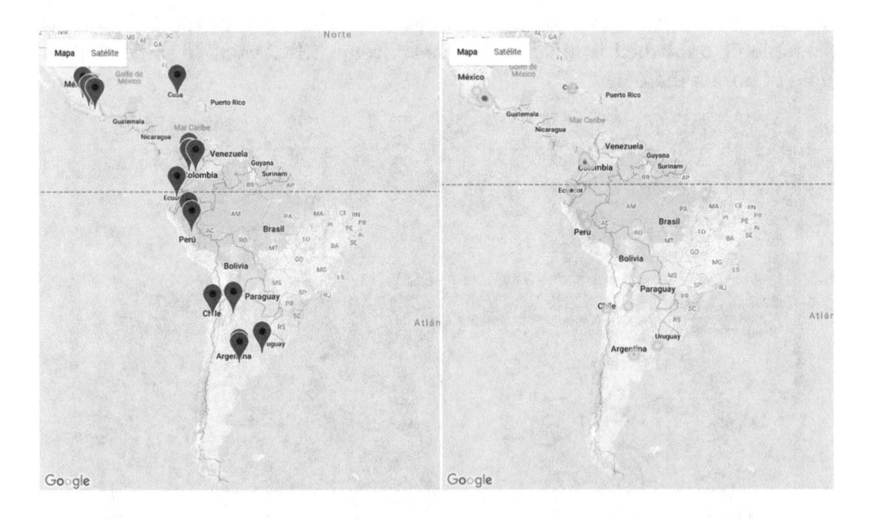

Fig. 4. Marker annotation and heat map functionalities that show where content of LearnersWiki originated

Analysing the specific features of LearnersWiki, shown in the Table 1, our hypotheses were that:

- **H7**: Collaborative edition has a positive impact on the behavioral intention to use LearnersWiki.
- **H8**: Georeferencing has a positive impact on the behavioral intention to use LearnersWiki.
- **H9**: LearnersWiki is a sound social media tool for collaborative learning.

3 Experiment Design

3.1 Population and Sample Size

The study sample size was 250 software engineering sophomore students from the Universidad Técnica de Ambato, of a total of 632 students of this area. Those 250 students voluntarily responded to the call for the study, made through university-sponsored email accounts. The size of the sample is consistent with other studies in the area of social media on education [1,8,24,28]. The Universidad Técnica de Ambato was chosen for the study because it is a well known public academic institution and also because this institution gave us the facilities for carrying out this experimental evaluation.

3.2 Survey Design

In order to quantify the level of technological acceptance, the information gathering method, as in other studies conducted with TAM [1,16,24,25,27], was carried out by using questionnaires. Two questionnaires were designed, one for LearnersGroup and the other one for LearnersWiki. Both of them were organized in five sections:

1. Demographic information.
2. Attitude toward using.
3. Perceived ease of use.
4. Perceived usefulness.
5. Behavioral intention to use.

In both questionnaires, survey items were included to evaluate each TAM factor; those items are detailed in Table 2. Following the proposal of Junco [13], a Likert scale was used with values from 1 to 5 as in: (1) Completely disagree, (2) Disagree, (3) Neutral, (4) Agree, and (5) Completely agree.

Both surveys were hosted in the free survey-hosting website Google Forms. Links to surveys were sent to participants through their university-sponsored email accounts. Meanwhile, as part of the qualitative evaluation of the technology acceptance of the tools, a formal and voluntary interview was conducted. The following questions were used:

1. What is your opinion about the possibility of georeferencing content in LearnersWiki?
2. How do you use interactions in LearnersGroup to rate content?
3. What is your opinion about the possibility of rating wiki pages content?
4. What is your opinion regarding the possibility of participating in an online dialogue about a publication in LearnersGroup?
5. What is your opinion about using LearnersGroup and LearnersWiki for Education?

Table 2. Items included in the technological acceptance surveys.

Factors	Item	Item ID	
		LearnersGroup	LearnersWiki
ATU	Contribution as information consumer	G1	L1
	Contribution as information provider	G2	L2
PEU	Ease to share content	G3	L3
	Ease of collaborative editing	—	L4
	Ease of creating online dialogs	G4	—
	Ease of georeferencing	—	L5
PU	Usefulness for collaborative learning	G5	L6
	Interest generation in the content	G6	L7
	Ease to determine useful content	G7	L8
	Collaborative editing facilitates learning	—	L9
	Online dialogue facilitates learning	G8	—
	Georeferencing facilitates learning	—	L10
BIU	Behavioral intention to use the tool for collaborative learning	G9	L11

Note: ATU Attitude toward using, PEU Perceived ease of use, PU Perceived usefulness, BIU Behavioral intention to use.

3.3 Methodology

The experiment was executed in two phases. In the first phase, the students made use of LearnersGroup and LearnersWiki. The 250 students participating in the experiment were separated in two independent groups of the same size. The first group of participants, 90 male and 35 female, were added as members of LearnersGroup. The second group of participants, 90 male and 35 female, were enrolled as users of LearnersWiki. In the second phase of the experiment, the evaluation of the technology acceptance level of both tools was carried out. One survey was applied to each group and a total of 52 participants were interviewed.

4 Results and Discussion

Statistical tests in this research were carried out using IBM SPSS ver. 23.

4.1 Demographic Information

Table 3 shows the most significant demographic data of students that participated in the experiment. The same amount of males and females was included in each group and both groups were mostly below the age of twenty.

Previous experience with the use of wikis and Facebook Groups was not a requirement for participating in the experiment. Nevertheless, 245 of the students which means a 98% of the participants in the study had previous experience with wikis, and the 100% have used Facebook before.

Table 3. Demographics of participants in the experiment.

Characteristic	*Quantity*		*Percentage*	
	LearnersGroup	LearnersWiki	LearnersGroup	LearnersWiki
Age				
Less than 20	110	110	44%	44%
Between 20 and 30	15	15	6%	6%
Gender				
Male	90	90	36%	36%
Female	35	35	14%	14%

4.2 Reliability Analysis

Cronbach Alpha was used to determine the reliability of the designed questionnaires. A pilot test was conducted with 25 students, corresponding to 10% of the sample size. These students did not participated in the evaluation of the tools. The Cronbach Alpha coefficients obtained were 0.787 and 0.880 for the surveys evaluating the technological acceptance of a LearnersGroup and of LearnersWiki, respectively. The obtained values indicated that both surveys had a high level of reliability and gave way to the application in the experiment.

4.3 Data Analysis

The dataset was downloaded from Google Forms as a .csv file. The initial screening showed that all responses were completed without anomalies. So, the final sample size was 250 (N = 250).

Following the proposals of [1,24,29,30], our data analysis is based on correlations between variables, the comparison of mean values and standard deviations, as part of a descriptive statistic, and a t-test for comparing the values of actual use of each proposal. A correlation analysis between the variables corresponding to survey items was made using Kendall's Tau-b. In Table 4, the correlation coefficients for the variables of LearnersGroup survey are shown.
According to students' answers, determining useful content and facilitating the creation of online dialogs have a significant relation with the fact that LearnersGroup is considered useful for collaborative learning and that it is able to generate interest for the content. Table 5 shows correlation coefficients between variables of the LearnersWiki survey. From moderate to significant correlation between variables L9 and L4, L6 and L7, we can affirm that LearnersWiki's capacity for generating interest for the content has a significant relation with its ease of use and with the perceived usefulness of learners. The hypothesis H1 is accepted according to the results shown in Tables 4 and 5. These results corroborate that the perceived usefulness of learners, corresponding to the items G5 to G7 and L6, L7 and L9, has a positive impact on the behavioral intention of

Table 4. Correlation between items of the LearnersGroup's questionnaire.

Items	G1	G2	G3	G4	G5	G6	G7	G8	G9
G1	1,000								
G2	,415*	1,000							
G3	-,411*	-,220	1,000						
G4	,129	,116	0,000	1,000					
G5	,299	,233	-,008	,508**	1,000				
G6	,286	,333	,008	,390*	,718**	1,000			
G7	,188	,045	,207	,215	,484**	,519**	1,000		
G8	-,030	,065	,142	,146	-,015	,121	,132	1,000	
G9	,131	,194	-,007	,501**	,698**	,546**	,451*	,064	1,000

Table 5. Correlation between items of the LearnersWiki's questionnaire.

Items	L1	L2	L3	L4	L5	L6	L7	L8	L9	L10	L11
L1	1,000										
L2	,256	1,000									
L3	,239	,249	1,000								
L4	,145	,035	,080	1,000							
L5	,307	,181	,312	,267	1,000						
L6	,088	,294	,079	,597**	,254	1,000					
L7	,377*	430*	,271	,420*	,371*	,564**	1,000				
L8	,276	,011	,068	,060	-,116	-,156	,021	1,000			
L9	,233	,257	,317	,367*	,266	,348	,458*	,101	1,000		
L10	-,016	,274	,245	,160	-,006	,085	,072	,270	,274	1,000	
L11	,285	,401*	,025	,406*	,401*	,627**	,587**	0,000	,401*	,134	1,000

using LearnersGroup and LearnersWiki, which correspond to the items G9 and L11, respectively.

The results shown in Tables 4 and 5 also prove that the items G4, L4 and L5, related to the perceived ease of use, have a positive impact on the values of items G9 and L11, which represent the behavioral intention to use LearnersGroup and LearnersWiki. Thus, the hypothesis H2 is accepted.

A T-test was run for comparing differences between TAM factors for each tool. The results are listed in Table 6. The p-value for the ease of use, the perceived usefulness and the behavioral intention to use between a LearnersGroup and a LearnersWiki are of 0.53, 0.33 and 0.93, respectively. None of these values is lower than 0.05. Therefore, hypothesis H3, H4 and H5 are rejected. Related to specific features, in Table 4, it is observed that there is a significant relation between the items of facility for online dialog creation and the intention to use, G4 and G9. These results allow to accept hypothesis H6. In Table 5 it is identified a significant relation between items L4, L9 and L11. These results indicate that

Table 6. Independent samples test.

		Levene's test for equality of variances		t-test for equality of means				
		F	Sig	t	df	Sig.	Mean Difference	Std. Error Difference
ATU	Equal variances assumed	1,573	,21	1,215	98	,227	,240	,197
	Equal variances not assumed			1,215	96,48	,227	,240	,197
PEU	Equal variances assumed	,399	,53	-,084	123	,933	-,013	,159
	Equal variances not assumed			-,082	94,78	,935	-,013	,163
PU	Equal variances assumed	,961	,33	-,786	223	,433	-,092	,117
	Equal variances not assumed			-,790	216	,430	-,092	,116
BIU	Equal variances assumed	,007	,93	,141	48	,889	,040	,284
	Equal variances not assumed			,141	47,99	,889	,040	,284

Note: ATU Attitude toward using, PEU Perceived ease of use, PU Perceived usefulness, BIU Behavioral intention to use.

the collaborative edition functionality has a positive impact on the intention to use. So, hypothesis H7 is accepted too.

The results of descriptive statistics that were carried out, which are depicted in Table 7, and the results of the T-test that was run, depicted in Table 6, show that the evaluation of both, LearnersGroup and LearnersWiki, is positive.

Table 7. Group statistics of TAM factors.

Factor	Tool	N	Mean	Std. Deviation	Std. Error Mean
ATU	G	250	3,62	,923	,131
	L	250	3,38	1,048	,148
PEU	G	250	4,36	,942	,133
	L	375	4,37	,818	,094
PU	G	500	4,06	,851	,085
	L	625	4,15	,890	,080
BIU	G	125	3,92	,997	,199
	L	125	3,88	1,013	,203

Note: ATU Attitude toward using, PEU Perceived ease of use, PU Perceived usefulness, BIU Behavioral intention to use.

For both tools, the average score for measuring TAM factors took values between 3.38 and 4.37, with standard deviation values in a range from 0.018 to 1.043. This result indicates that both tools were technologically accepted by participants in the experiment. Factors with the highest means were the perceived ease of use and the perceived usefulness with values between 4.06 and 4.15 for LearnersGroup and LearnersWiki, respectively. This result shows that both tools were considered easy to use and useful.
The most significant answers during the interview are depicted in Tables 8, 9, 10, 11 and 12.

Table 8. Students' answers to question 1 of the interview.

What is your opinion about the possibility of georeferencing content in LearnersWiki?	
Student 13	The spot in the map is good, you can see where the content originates
Student 24	Georeferencing is easy, you just have to drag the pointer and you can place the information origin in the Map
Student 31	It is something additional that allows LearnersWiki and can not be done in Facebook

Table 9. Students' answers to question 2 of the interview.

How do you use interactions in LearnersGroup to rate content?	
Student 29	I use Like to show that I checked a publication
Student 41	For me, a Like is a neutral opinion about a content
Student 50	When I really read a post, because it is about something I am interested in, I put another reaction different from Like, I love it, for example

Table 10. Students' answers to question 3 of the interview.

What is your opinion about the possibility of rating wiki pages content?	
Student 16	A way to rate with numbers is simple to use
Student 27	It is easier to use a rating system than using Likes or another reaction to rate content

Table 11. Students' answers to question 4 of the interview.

What is your opinion regarding the possibility of participating in an online dialogue about a publication in LearnersGroup?	
Student 38	Conversation in social media is an effective dialogue method if a person finds something interesting

Table 12. Students' answers to question 5 of the interview.

What is your opinion about using LearnersGroup and LearnersWiki for Education?	
Student 8	LearnersWiki is suitable for education, it requires more organization and more dedication
Student 12	In LearnersGroup we only look for information to share, many times we do not even read or review the content
Student 19	LearnersWiki was very good because the information could be found more easily with the index and through the map. My recommendation would be to connect it to Facebook so created pages can be shared
Student 45	LearnersGroup is better because I receive homework notifications, that should be included in LearnersWiki

These comments support the acceptance of hypothesis H6 and H7. Besides, from these opinions, it is possible to deduct that LearnersWiki is a valid social media tool for collaborative learning. Therefore, hypothesis H9 is also accepted. During the interviews, students also made the recommendation of creating an hybrid socially-enhanced collaborative learning environment by integrating the use of LearnersWiki together with LearnersGroup. This idea enlight the future development of social media-based collaborative learning.
Concerning hypothesis H8, it is shown in Table 5 that no significant correlation is observed between variables L10 and L11. However, in Tables 8 and 12, positive opinions of students regarding the usefulness of georeferentiation are shown. In this research, no analysis of the content provided by participants was carried out. Nevertheless, by mean of the management of georeferenced data that

Table 13. Hypotheses summary.

Hypothesis	Corroborrated	Tool	Detail	Items	p-val
H1	YES	Survey	Tables 4 and 5	G5 to G7 in relation to G9 and L6,L7,L9 in relation to L11	<0,05
H2	YES	Survey	Tables 4 and 5	G4 in relation to G9. L4 and L5 in relation to L11	<0,05
H3	NO	Survey	Table 6	PEU	0.53
H4	NO	Survey	Table 6	PU	0.33
H5	NO	Survey	Table 4	BIU	0.93
H6	YES	Survey	Table 5	L4 and L9 in relation to L11	<0,05
H7	YES	Interview	Tables 8, 9, 10, 11 and 12	—	—
H8	YES	Interview	Tables 8 and 12	—	—
H9	YES	Data analysis	Figure 4	—	—

LearnersWiki provides, in Fig. 4, it is obvious that this data allows to show the levels of concentration of the origin of content. This fact gives added value of the tool and its usefulness for the participants in the experiment. This way, hypothesis H8 and H9 are considered accepted.

Additionally, results of Tables 4, 5, 7, 6, 8, 9, 10, 11 and 12 show that there is no significant difference regarding the technology acceptance level as collaborative learning tool when comparing LearnersGroup with LearnersWiki. A summary of hypotheses corroboration can be observed in Table 13.

5 Conclusions and Future Work

The obtained results show that the perceived ease of use and perceived usefulness have a positive impact on the behavioral intention to use social media in learning processes. No significant difference were found regarding the perceived ease of use, perceived usefulness and intention to use LearnersGroup and LearnersWiki. On the other hand, it was proven that the online dialogue, the collaborative edition and the content georeferentiation features influenced the behavioral intention to use the tools of the participants. Furthermore, location metadata represent an added value because when any shared content is associated to a map location, learners can better assimilate the theoretic knowledge. TAM made possible to determine that LearnersGroup and LearnersWiki are social media tools accepted by sophomore students. It was also proven that there is no significant difference with respect to the acceptance of both tools. Nevertheless, it is important to note that the obtained results in the study cannot be generalized because all students who participated in the experiment were software engineering sophomores of the Universidad Técnica de Ambato. As future work, more experiments with the participation of students from different fields and different universities are going to be carried out for corroborating the results in a broader scenario. In addition, an hybrid tool is proposed to be designed, in which Facebook Groups' features lay down combined with those features of a wiki for managing georeferenced data.

References

1. Dumpit DZ, Fernandez CJ (2017) Analysis of the use of social media in higher education institutions (HEIs) using the technology acceptance model. Int J Educ Technol High Educ 14(1)
2. Nicolai L, Schmidbauer M, Gradell M, Ferch S, Antón S, Hoppe B, Pander T, Borch PV, Pinilla S, Fischer M, Dimitriadis K (2017) Facebook groups as a powerful and dynamic tool in medical education: mixed-method study. J Med Internet Res 19(12)
3. Gan B, Menkhoff T, Smith R (2015) Enhancing students' learning process through interactive digital media: new opportunities for collaborative learning. Comput Hum Behav 51:652–663
4. Manca S, Ranieri M (2016) Facebook and the others. Potentials and obstacles of Social Media for teaching in higher education. Comput Educ 95:216–230

5. Cole D, Rengasamy E, Batchelor S, Pope C, Riley S, Cunningham AM (2017) Using social media to support small group learning. BMC Med Educ 17(1):1–7
6. Bazelais P, Doleck T, Lemay DJ (2018) Investigating the predictive power of TAM: a case study of CEGEP students' intentions to use online learning technologies. Educ Inf Technol 23(1):93–111
7. Connolly T, Willis J, Lloyd M (2018) Studies in Continuing Education Evaluating teacher and learner readiness to use Facebook in an Australian vocational setting. Stud Contin Educ 1–15
8. Zheng B, Niiya M, Warschauer M (2015) Wikis and collaborative learning in higher education. Technol Pedagog Educ 24(3):357–374
9. Bordel B, Mareca MP (2017) Using wikis in the higher education: The case of Wikipedia [Empleo de wikis en la docencia universitaria:el caso de la Wikipedia]
10. Moore-Russo D, Radosta M, Martin K, Hamilton S (2017) Content in context: analyzing interactions in a graduate-level academic Facebook group. Int J Educ Technol High Educ 14(1)
11. Castro PJ, González-Palta IN (2016) Percepción de estudiantes de psicología sobre el uso de facebook para desarrollar pensamiento crítico. Formacion Universitaria 9(1):45–56
12. Teo T, Doleck T, Bazelais P (2017) The role of attachment in Facebook usage: a study of Canadian college students. Interact Learn Environ 4820(April):1–17
13. Junco R (2015) Student class standing, Facebook use, and academic performance. J Appl Dev Psychol 36:18–29
14. Reyes-Garcés E, Fernández-Peña F, Pérez-Nata W, Urrutia-Urrutia P (2018) App sigma and facebook groups: quantitative evaluation of the usability and technology acceptance by software engineering students of universidad técnica de ambato. Formación Universitaria 11(5)
15. He W, Yang L (2016) Using wikis in team collaboration: a media capability perspective. Inf Manag 53(7):846–856
16. Wang N, Chen X, Lan Q, Song G, Parsaei HR, Ho SC (2017) A novel wiki-based remote laboratory platform for engineering education. IEEE Trans Learn Technol 10(3):331–341
17. Zander L, Tanneberger T, Peukert J, Mensah GA (2017) The implementation of WikL - An educational wiki supporting collaborative learning in engineering university courses. In: Proceedings of the ASME turbo expo, vol 6. American Society of Mechanical Engineers (ASME)
18. Lozada-Martínez E, Fernández-Peña F, Naranjo-Avalos H, Urrutia-Urrutia P (2018) Social knowledge harvesting based on a geographically-annotated wiki. case study: history and culture of ambato. In: 2018 International Conference on eDemocracy eGovernment (ICEDEG), pp 229–234
19. Kear K, Donelan H, Williams J (2014) Using wikis for online group projects: student and tutor perspectives. Int Rev Res Open Distance Learn 15(4):70–90
20. Takači D, Stankov G, Milanovic I (2015) Efficiency of learning environment using GeoGebra when calculus contents are learned in collaborative groups. Comput Educ 82:421–431
21. Lam HL, Tan RR, Aviso KB (2016) Implementation of P-graph modules in undergraduate chemical engineering degree programs: experiences in Malaysia and the Philippines. J Clean Prod 136:254–265
22. Promentilla MAB, Lucas RIG, Aviso KB, Tan RR (2017) Problem-based learning of process systems engineering and process integration concepts with metacognitive strategies: the case of P-graphs for polygeneration systems. Appl Therm Eng 127:1317–1325

23. Davis FD (1989) Perceived usefulness, perceived ease of use, and user acceptance of information technology. MIS Q 13(3):319–340
24. Efilolu Kurt Ö, Tingöy Ö (2017) The acceptance and use of a virtual learning environment in higher education: an empirical study in Turkey, and the UK. Int J Educ Technol High Educ 14(1)
25. Chai JX, Fan KK (2018) Constructing creativity: social media and creative expression in design education. Eurasia J Math Sci Technol Educ 14(1):33–43
26. Wamba SF (2018) Social media use in the workspace: applying an extension of the technology acceptance model across multiple countries. Springer
27. Ramirez-Anormaliza R, Sabaté F, Llinàs-Audet X, Lordan O (2017) Aceptación y uso de los sistemas e-learning por estudiantes de grado de Ecuador: El caso de una universidad estatal. Intang Cap 13(3):548–581
28. De Wever B, Hämäläinen R, Voet M, Gielen M (2015) A wiki task for first-year university students: the effect of scripting students' collaboration. Internet High Educ 25:37–44
29. Lisha C, Goh CF, Yifan S, Rasli A (2017) Integrating guanxi into technology acceptance: an empirical investigation of WeChat. Telemat Inform 34(7):1125–1142
30. Altanopoulou P, Tselios N (2017) Assessing acceptance toward wiki technology in the context of higher education. Int Rev Res Open Distance Learn 18(6):127–149

Engineering, Industry, and Construction with ICT Support

Random Sub-sampling Cross Validation for Empirical Correlation Between Heart Rate Variability, Biochemical and Anthropometrics Parameters

Erika Severeyn[1(✉)], Jesús Velásquez[2], Héctor Herrera[1], and Sara Wong[3]

[1] Departamento de Tecnología de Procesos Biológicos y Bioquímicos, Universidad Simón Bolívar, Valle de Sartenejas-Baruta, Caracas 89000, Venezuela
severeynerika@usb.ve

[2] Departamento de Termodinámica y Transferencia de Calor, Universidad Simón Bolívar, Valle de Sartenejas-Baruta, Caracas 89000, Venezuela

[3] Departamento de Ingeniería Eléctrica, Electrónica y Telecomunicaciones, Universidad de Cuenca, Av. 12 de Abril y Agustín Cueva, 010201 Cuenca, Ecuador

Abstract. According to National Cholesterol Education Program-Adult Treatment Panel III, metabolic syndrome (MS) is a condition characterized by: Dyslipidemia, abdominal obesity, high levels in fasting glucose and arterial hypertension. Studies have explored indexes using dimensional analysis (DA) formed by anthropometric, biochemical and heart rate variability parameters for the diagnosis of MS. The dimensionless numbers made from DA have the capability to manage them as a mathematical functionality; therefore it is possible to relate them, even when the parameters used are not connected. The aim of this work is to find a polynomial equation using as variables two dimensionless numbers designed from anthropometrical and biochemical (π_{IS}) parameters and from heart rate variability (π_{HRV}) parameters. A fitting using a parametrical random sub-sampling cross validation (RSV) was performed using as an objective function the least squares method. A database of 40 subjects (25 control subjects and 15 subjects with MS) was employed. The polynomial parameters that best fit the database used correspond to a polynomial of order eight. The RSV substantially improves the adjustment of the polynomial compared to the application of the least squares method only (0.6678 vs. 0.3255). The polynomial relationship between π_{IS} and π_{HRV} allows the possibility to determine biochemical and anthropometric variables from heart rate variability parameters. Due to the limited number of subjects in the database used, it is necessary to repeat this methodology in a more extensive database to determine a more general polynomial that can be used with any type of population.

Keywords: Random sub-sampling cross validation · Empirical correlation
Metabolic syndrome

1 Introduction

According to National Cholesterol Education Program-Adult Treatment Panel III (NCEP-ATPIII) [1], metabolic syndrome (MS) is a condition characterized by:

M. Botto-Tobar et al. (Eds.): TICEC 2018, AISC 884, pp. 347–357, 2019.
https://doi.org/10.1007/978-3-030-02828-2_25

Dyslipidemia, abdominal obesity, high levels in fasting glucose and arterial hypertension [1]. The early diagnosis of MS is very important to prevent the development of diabetes mellitus, insulin resistance and cardiovascular diseases.

The NCEP-ATPIII criterion is the most used for diagnosis of MS. There are studies that had been explored other variables to MS diagnosis. For instance, the HOMA-IR had been used to characterize the MS subjects [2]. Also waist circumference (WC) and body mass index (BMI) have been applied to determine MS diagnosis cut-off points [3, 4]. On the other hand, investigations have been revealed that heart rate variability parameters can discriminate between diabetic subjects and control [5] and significant differences were also found between subjects with MS and control in the high normalized frequencies [6].

There are numerous methodologies for the design of indexes that can diagnose metabolic diseases such as metabolic syndrome and insulin resistance. Among these methods is the random sub-sampling cross validation (RSV). RSV has been used to perform parametric optimizations for the improvement of the index detection capacity in the diagnosis of insulin resistance [7, 8]. The receiver operating characteristic (ROC) curves had been used to evaluate the index detection performance [9, 10]. Regression analysis had been applied to observe the nature of the relationships between variables by coefficient of correlation (R^2) [7].

On the other hand, dimensional analysis (DA) is another method that had been used to find indexes that can diagnosis MS. DA offers a method for reducing complex physical problems to the simplest form in order to obtain a quantitative solution [11]. The DA resumes the concept of similarity. In physical terms, similarity refers to some equivalence between two phenomena that are actually different. Mathematically, similarity refers to a transformation of variables that leads to a reduction in the number of independent variables that specify the problem. Dimensional analysis addresses both these views. Its main utility derives from its ability to succinct the functional form of physical relationships [12]. The DA is possible due to the π Vaschy-Buckingham theorem, which gives a methodology to obtain dimensionless numbers (DN) from physics variables [12].

The DA had been used in aerodynamics, hydraulics, ship design [13, 14], astrophysics [15], heat and mass transfer [16], biology [17] and even economics [18]. Also the DA was used with heart rate variability [9], anthropometrical and biochemical variables [10] to design indexes for MS diagnosis. The dimensionless numbers (DNs) made from dimensional analysis have the capability to manage them as a mathematical functionality; therefore it is possible to relate them, even when the parameters used are not connected. In fluids, heat transfer and thermodynamics fields, usefully correlations have been reached by experimentally finding empirical relationships between DNs.

The aim of this work is to find a polynomial equation using as variables two dimensionless expressions: One designed from anthropometrical and biochemical parameters (π_{IS}) and the second design from heart rate variability parameters (π_{HRV}) [9, 10]. For this a fitting using RSV was performed in a database of 40 subjects. In the next section the materials and method used will be explained. In Sects. 3 and 4, results and discussion will be presented. And finally, in Sect. 5, the conclusions and future works proposals will be presented.

2 Materials and Methods

2.1 Database

In 2009, 40 adult males between 18 and 45 years old without diabetes were enrolled in the DID-USB project called: Electrocardiographic study of cardiac neuropathy in the metabolic syndrome [19]. Each subject underwent high density lipoprotein cholesterol, triglycerides, blood pressure, anthropometric measures (height, weight, waist circumference, BMI), fasting glucose and electrocardiographic signals. The Table 1 shows the characteristics of the database used. MS subjects were diagnosed according to NCEP-ATPIII criteria (Table 2) [1].

Table 1. Anthropometric, biochemical and heart rate variability parameter of database.

Parameters		Total database (n = 40)	Control[b] (n = 25)	MS subjects according to NCEP-ATPIII (n = 15)	*p*-value[a]
Demographic	Age (years)	30.88 ± 7.18[c] [18.00–47.00][d] [28.65-33.10][e]	30.24 ± 7.80 [18.00–47.00] [27.18–33.30]	31.93 ± 6.13 [22.00–44.00] [28.83–35.04]	0.99
Anthropometric	Waist Circumference (cm)	90.89 ± 22.48 [66.60–161.00] [83.92–97.86]	77,71 ± 10,27 [66,60–106,00] [73,69–81,74]	112.85 ± 20.00 [77,00–161.00] [102.73–122.98]	<0.001
	BMI (Kg/m^2)	26.52 ± 7.60 [18.17–48.53] [24.16–28.87]	22.10 ± 3.42 [18.17–33.01 [20.76–23.44	33.88 ± 6.92 [22.11–48.53] [30.38–37.39]	<0.001
Biochemical	High Density Lipoprotein Cholesterol (HDL) (mg/dL)	45.88 ± 9.02 [32.00–70.00] [43.08–48.67]	48.12 ± 9.18 [32.00–70.00] [44.52–51.72]	42.13 ± 7.63 [32.00–57.00] [38.27–46.00]	0.20
	Triglycerides (TG) (mg/dL)	131.93 ± 97.40 [36.00–448.00] [101.74–162.11]	74.56 ± 41.93 [36.00–219.00] [58.12–91.00]	227.53 ± 87.99 [124.00–448.00] [183.01–272.06]	<0.001
	Fasting Glucose (mg/dL)	96.15 ± 11.05 [75.00–119.00] [92.73–99.57]	90.56 ± 8.81 [75.00–107.00] [87.11–94.01]	105.47 ± 7.61 [92.00–119.00] [101.61–109.32]	<0.001
Clinical	Systolic Blood Pressure (mmHg)	122.23 ± 12.95 [100.00–147.00] [118.21–126.24]	114.08 ± 7.80 [100.00–130.00] [111.02–117.14]	135.80 ± 6.85 [118.00–147.00] [132.33–139.27]	<0.001
	Diastolic Blood Pressure (mmHg)	77.73 ± 11.02 [60.00–103.00] [74.31–81.14]	72.40 ± 8.27 [60.00–90.00] [69.16–75.64]	86.60 ± 9.26 [70.00–103.00] [81.91–91.29]	<0.001
Heart Rate Variability	RMSSD (ms)	53.82 ± 38.36 [13.00–179.65] [41.93–65.70]	64.14 ± 46.91 [17.98–179.65] [44.08–84.21]	41.81 ± 20.31 [13.00–76.04] [31.53–52.09]	0.65
	HF × 10^5 (ms^2)	6.35 ± 10.02 [0.32–46.59] [3.25–9.46]	8.42 ± 12.90 [0.32–46.59] [2.90–13.93]	3.58 ± 3.25 [0.44–11.91] [1.93–5.22]	0.95

[a]Statistically significant difference if p-value < 0.05.

[b]Control subjects are those who do not belong to MS subjects.

[c]Average and standard deviation.

[d]Maximum and minimum value.

[e]95% confidence interval.

Table 2. Metabolic syndrome diagnosis criteria according to NCEP-ATPIII.

Measure	Categorical cut-points
WC	Men ≥ 90 cm Women ≥ 80 cm
TG	≥150 mg/dL
HDL	Men < 40 mg/dL Women < 50 mg/dL
Blood Pressure (Systolic/Diastolic)	≥130/ ≥ 85 mmHg
Fasting Glucose	≥100 mg/dL

2.2 Dimensionless Numbers

The dimensionless numbers used in this work were designed in [9, 10]. In [9] were proposed DNs that can be used to diagnose subjects with MS, IR and obesity using five HRV parameters: RR interval, RMSSD (root mean square of the successive differences), SD, HF (high frequencies) and LF (low frequencies). For this purpose, seven DNs, designed from the π Vaschy-Buckingham theorem, were assessed using ROC curves. The DN built with the variables: HF and RMSSD; obtained the best performance as classifier of MS, presenting an area under the ROC curve greater than 0.70.

In [10] were proposed DNs that can discriminate subjects with MS using biochemical (HDL, triglycerides) and anthropometric variables (weight, height, waist circumference). Three DNs were constructed and their capability as classifier was assessed using ROC curves. The DN constructed with the variables: Abdominal circumference, triglycerides, weight and height; was the one that obtained a better performance as classifier of MS, presenting an area under the ROC curve of 0.86. In Eqs. 1 and 2 are shown the two DNs chosen from [9, 10].

$$\pi_{HRV} = \frac{HF}{RMSSD^2} \tag{1}$$

$$\pi_{IS} = \frac{100\ W\ H}{WC^4\ TG} \tag{2}$$

Where HF should be entered in ms^2, RMSSD in ms, weight (W) in Kg, height (H) in meters, WC in meters and TG in mg/dL.

2.3 Empirical Correlation

In dimensional analysis the finding of correlation between two or more DNs that describe two similar phenomena is very common [20]. The most used procedure to find a functional relation between two variables is the regression analysis by least squares method.

In other words, the idea is to find a polynomial equation (Eq. 1) of degree n with the best fit with the experimental data used.

$$\pi_{IS} = a_0 + a_1\pi_{HRV} + a_2\pi_{HRV}^2 + \cdots + a_n\pi_{HRV}^n \tag{3}$$

Where, $a_0, a_1, a_2, \ldots a_n$ are the parameters of the polynomial equation.
With Eqs. 1 and 3 it is possible to estimate the TG (TG_C) by substitution of Eq. 2 in 3:

$$TG_c = \frac{100\ W\ H}{WC^4\left(a_0 + a_1\pi_{HRV} + a_2\pi_{HRV}^2 + \cdots + a_n\pi_{HRV}^n\right)} \quad (4)$$

In this work, a RSV method was performed to find the best fit polynomial equation. The RSV consist in randomly separates data in: training data and validation data. In each

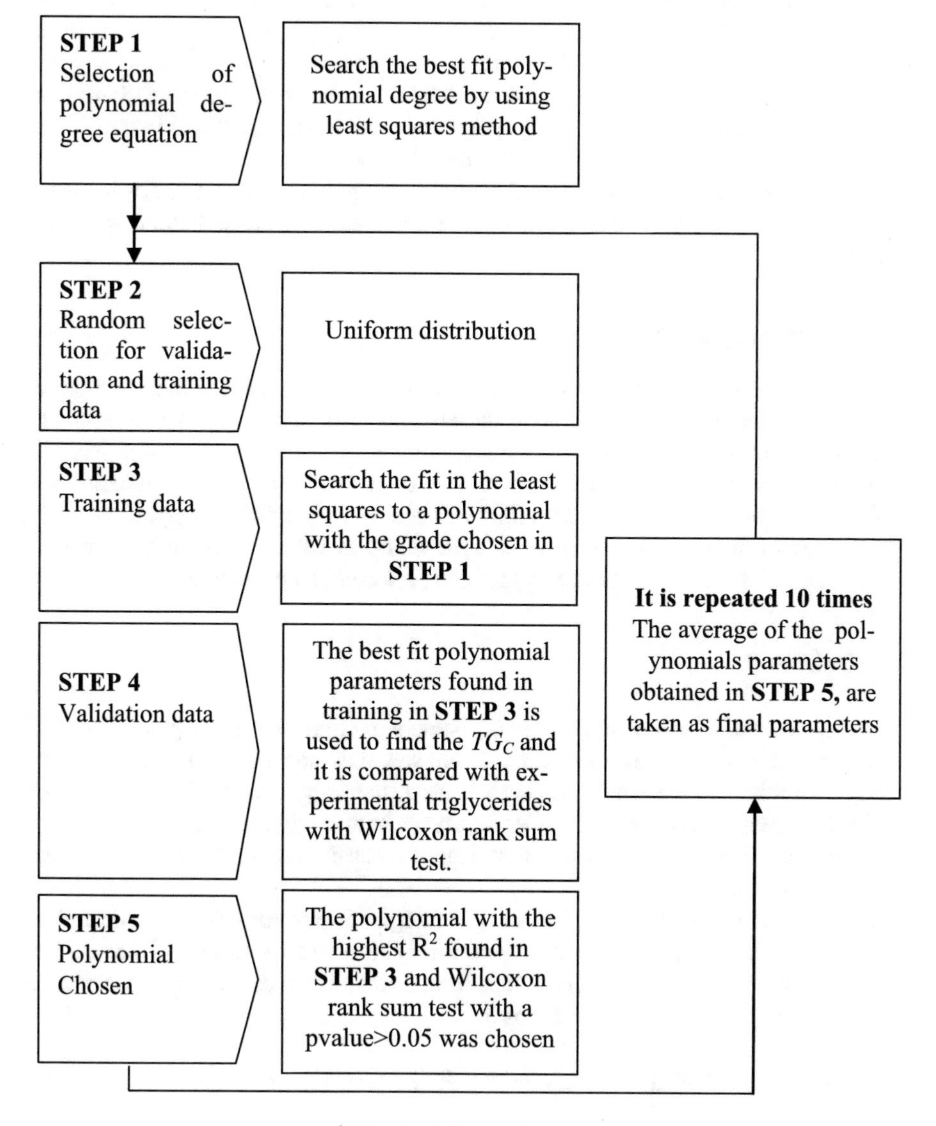

Fig. 1. Scheme of RSV.

separation, classifier algorithm is re-trained with correspondent data and validated with remaining data. Results are later averaged [7, 8].

In Fig. 1 a scheme of RSV applied in this study is showed. In step 1, the degree of the polynomial with the best fit is chosen by testing (with all the subjects of the database) polynomials with grades from 1 to 9, and the degree that best fitting presents (R^2 closest to 1) was selected. In step 2, a random selection of training and validation data is made using a uniform distribution (75% of data was taken for training and remaining data for validation in each population). In step 3, training sample is used to search the polynomial parameters ($a_0, a_1, a_2, \ldots a_n$) with the best fitting using the least squares method. In step 4, the best fitting polynomial parameters found in training in step 2, is used to calculate the triglycerides (Eq. 4) and it is compared with experimental triglycerides with the Wilcoxon rank sum test. In step 5, the polynomial with the higher R^2 found in step 2 and Wilcoxon rank sum test with a p value > 0.05 found in step 5, is chosen. Steps from 2 to 5 are performed ten times. Finally, polynomial parameters obtained in the 10 iterations were averaged to acquire final polynomial.

The Wilcoxon rank sum test is used to choose the polynomial that can estimate the triglycerides with no significant differences between the TG_C and experimental triglycerides.

2.4 Statistical Analysis

All data were processed using the MATLAB program version R2015a. Non-paired samples were handled with a different distribution than normal, so to determine the differences between groups of two, the non-parametric Wilcoxon rank sum test was used, and between groups of three or more were used the non-parametric statistical test of Kruskal Wallis, a p-value less than 0.05 was considered statistically significant [21]. The data in the text and in the tables are presented as values of mean and standard deviation, minimum and maximum values and 95% confidence interval.

3 Results

Table 3 shows the R^2 calculated for the adjustment of curves by least squares for polynomials of different degrees, these results correspond to step 1 explained in the methodology. In this step the whole database is used, the polynomials of degree 8 and 9 obtained the best R^2, however, as the R^2 difference between the polynomial of degree 9 and degree 8 is 0.0001, the polynomial of degree 8 was chosen for the RSV since it has fewer parameters.

In Eq. 5 can be observed the polynomial of degree 8 used for RSV. The parameters and coefficient of determination are presented in Table 4 with the mean and the standard deviation. These results corresponds steps from 2 to 5. The mean a standard deviation is the results of the ten iterations of step 5.

$$\pi_{IS} = a_0 + a_1\pi_{HRV} + a_2\pi_{HRV}^2 + \cdots + a_6\pi_{HRV}^6 + a_7\pi_{HRV}^7 + a_8\pi_{HRV}^8 \tag{5}$$

Table 3. R^2 for each polynomial degree in Step 1 without RSV.

Polynomial degree	Coefficient of determination (R^2)
1	0.2556
2	0.2622
3	0.2973
4	0.3085
5	0.3105
6	0.3105
7	0.3110
8	0.3159
9	0.3160

Table 4. Parameters and coefficient of determination of resultant polynomial fitting.

Polynomial Parameters	Mean and standard deviation
a_8	$1.060\ 10^{-12} \pm 2.973\ 10^{-12}$
a_7	$-1.763\ 10^{-9} \pm 4.323\ 10^{-9}$
a_6	$1.245\ 10^{-6} \pm 2.714\ 10^{-6}$
a_5	$-4.886\ 10^{-4} \pm 9.604\ 10^{-4}$
a_4	$1.166\ 10^{-1} \pm 2.094\ 10^{-1}$
a_3	$-1.731\ 10^{1} \pm 2\ .878\ 10^{1}$
a_2	$1.562\ 10^{3} \pm 2.437\ 10^{3}$
a_1	$-7.831\ 10^{4} \pm 1.161\ 10^{5}$
a_0	$1.670\ 10^{6} \pm 2.385\ 10^{6}$
R^2	0.6678 ± 0.0131

Figure 2 shows the fitting curve. The green points correspond to the control subjects, the red points correspond to the subjects with MS. The black line corresponds to the curve adjusted without the application of the RSV and the blue line the curve adjusted by RSV. Figure 3 shows the box plot for calculated triglycerides (TG_C) from Eq. 4 and experimental triglycerides.

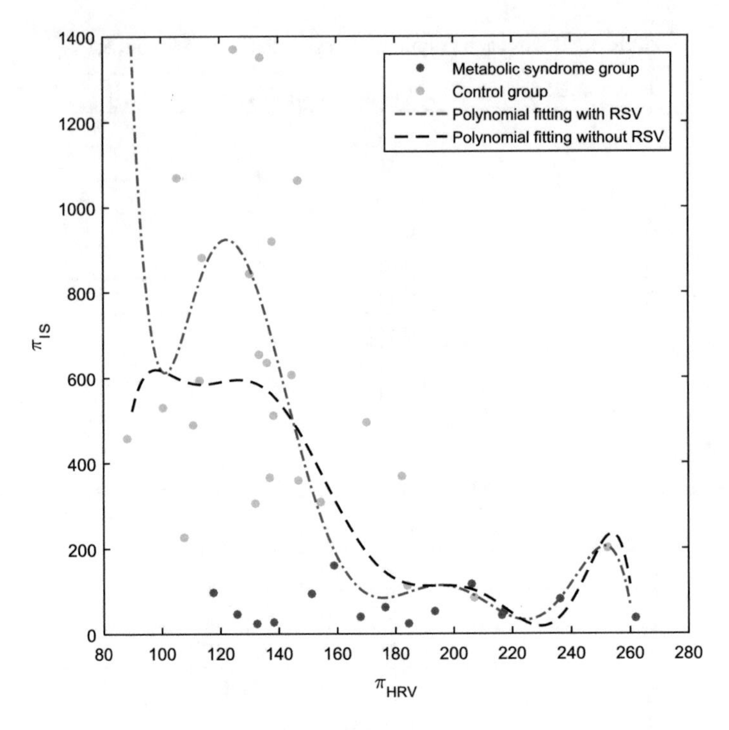

Fig. 2. Database scatter plot (green points) and polynomial fit (red points).

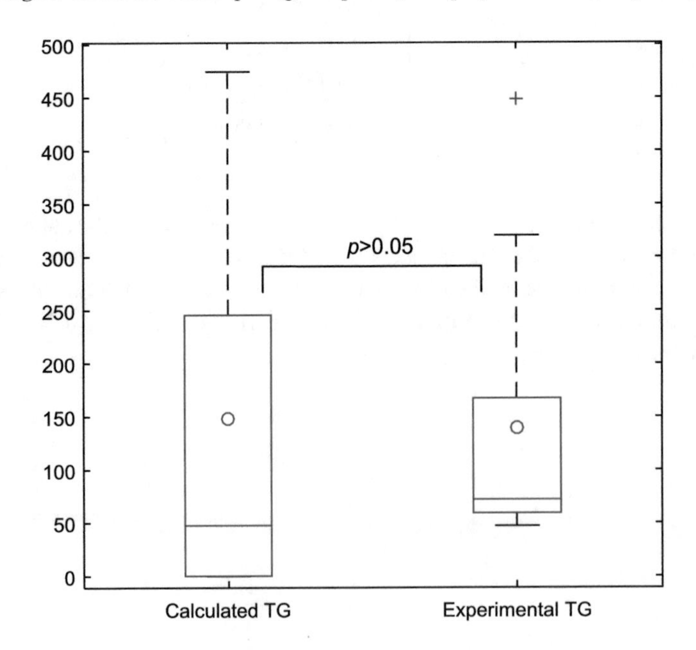

Fig. 3. Boxplot of calculated triglycerides and experimental triglycerides, and mean (red point).

4 Discussion

In Table 3, the parameters that best fit the database used correspond to a polynomial of order eight. In Table 4, it can be seen that the RSV substantially improves the adjustment of the polynomial compared to the adjustment without the RSV (0.6678 vs. 0.3159). This corroborates the works [7, 8], where the RSV was used for the parametric adjustment of a proposed index for the diagnosis of insulin resistance. In both studies, the application of RSV resulted in an improvement in the detection capacity of the proposed index.

Figure 2 shows the equations adjusted to the data. It can be verified that in the polynomial adjusted without RSV, the control subjects with a $\pi_{IS} > 600$ (40% of the control subjects) are away from the curve; therefore they are not represented in it. This group has the particularity of having lowest values of fasting glucose (<97 mg/dL), triglycerides (<56 mg/dL), blood pressure (Diastolic blood pressure <80 mmHg, systolic blood pressure <118 mmHg), waist circumference (<79.1 cm) and presenting normal values of BMI (<21). This suggests that the adjusted curve by least squared method may not represent subjects who have not at least one of the diagnostic criteria for the MS. On the contrary, in the adjusted curve by RSV, all the control subjects are near to the curve, hence all control subjects are represented in it and the determination coefficient obtains an improve value for this polynomial compare to polynomial fitting by least squared method.

In the same sense, the MS subjects with $\pi_{HRV} < 150$ (33% of the subjects with MS), are placed away from both of adjusted curves (With and without RSV). It suggests a part of MS group is not represented in the curves, and it could be the reason that the adjusted curve by RSV has not presented a better fitting. On the other hand, the group of subjects with MS represented in both curves, are those who had the highest values of triglycerides (>167 mg/dL), fasting glucose (>102 mg/dL), blood pressure (Diastolic blood pressure > 80 mmHg, systolic blood pressure > 136 mmHg) and waist circumference (>103 cm), even 27% of MS group present morbid obesity (BMI > 40). This suggests that the curves found (with or without RSV) may represent only subjects with a marked MS.

Figure 3 shows that the calculated and experimental triglycerides averages do not present significant differences ($p > 0.05$), corroborating that the polynomial equation according to RSV can be used to find anthropometric and biochemical values from heart rate variability parameters, and vice verse.

5 Conclusions

In this investigation, two dimensionless numbers were used (π_{IS} and π_{HRV}) and correlated by the design of a polynomial. For this, a random sub-sampling cross validation was applied to find a polynomial correlation with the best fit to a database of 40 subjects. The database used had as variables: biochemical, anthropometric and heart rate variability parameters.

The findings confirm that the random sub-sampling cross validation method improves the parametric adjustment compared to the application of the least squares method only. All this corroborates the works [7, 8] where the use of the RSV method improved the detection capacity of the proposed indexes for the diagnosis of MS.

The polynomial between the π_{IS} and π_{HRV} found in this work can represents control subjects and subjects with marked MS. Also, the polynomial adjust with RSV allow the possibility of being able to determine biochemical and anthropometric variables from of heart rate variability parameters and vice verse. Due to the limited number of subjects in the database used, it is necessary to repeat this methodology in a more extensive database and thus be able to determine a more general polynomial that can be used with any type of population.

References

1. Grundy SM, Cleeman JI, Daniels SR, Donato KA, Eckel RH, Franklin BA, Spertus JA (2005) Diagnosis and management of the metabolic syndrome: an American Heart Association/ National Heart, Lung, and Blood Institute scientific statement. Circulation 112(17):2735–2752
2. Motamed N, Miresmail SJH, Rabiee B, Keyvani H, Farahani B, Maadi M, Zamani F (2016) Optimal cutoff points for HOMA-IR and QUICKI in the diagnosis of metabolic syndrome and non-alcoholic fatty liver disease: a population based study. J Diabetes Complications 30(2):269–274
3. Worachartcheewan A, Dansethakul P, Nantasenamat C, Pidetcha P, y Prachayasittikul V (2012) Determining the optimal cutoff points for waist circumference and body mass index for identification of metabolic abnormalities and metabolic syndrome in urban Thai population. Diabetes Res Clin Pract 98(2):e16–e21
4. Gozashti MH, Najmeasadat F, Mohadeseh S, Najafipour H (2013) Determination of most suitable cut off point of waist circumference for diagnosis of metabolic syndrome in Kerman. Diabetes Metab Syndr Clin Res Rev 8(1):8–12
5. Quintero L, Wong S, Parra R, Cruz J, Antepara N, Almeida D, Ng F, Passariello G (2007) Stress ECG and laboratory database for the assessment of diabetic cardiovascular autonomic neuropathy. In: 29th annual international conference IEEE EMBS, Lyon, Francia, pp 4339–4342
6. Severeyn E, Wong S, Passariello G, Cevallos JL, Almeida D (2012) Methodology for the study of metabolic syndrome by heart rate variability and insulin sensitivity. Revista Brasileira de Engenharia Biomédica 28(3):272–277
7. Velásquez J, Wong S, Encalada L, Herrera H, Severeyn E (2015) Lipid-anthropometric index optimization for insulin sensitivity estimation. In: 11th international symposium on medical information processing and analysis, vol 9681, p 96810R). International Society for Optics and Photonics, December 2015
8. Velásquez J, Severeyn E, Herrera H, Encalada L, Wong S (2017) Anthropometric index for insulin sensitivity assessment in older adults from Ecuadorian highlands. In: 12th international symposium on medical information processing and analysis, vol 10160, p 101600S. International Society for Optics and Photonics, January 2017
9. Velásquez J, Severeyn E, Herrera H, Astudillo-Salinas F, Wong S (2017) Dimensional analysis of heart rate variability parameters for metabolic dysfunctions diagnosis. In: 2017 IEEE Ecuador technical chapters meeting (ETCM), pp 1–6. IEEE, October 2017

10. Velásquez J, Herrera H, Encalada L, Wong S, Severeyn E (2017) Análisis dimensional de variables antropométricas y bioquímicas para diagnosticar el síndrome metabólico. Maskana 8:57–67
11. Bridgman PW (1969) Dimensional analysis. In: Encyclopaedia Britannica (Wm. Haley, Editor-in-Chief), vol 7, pp 439–449. Encyclopaedia Britannica, Chicago
12. Buckingham E (1914) On physically similar systems; illustrations of the use of dimensional analysis. Phys Rev 4:345–376
13. Sedov LI (1959) Similarity and dimensional analysis in mechanics. Academic Press, New York
14. Baker WE, Westine PS, Dodge FT (1973) Similarity methods in engineering dynamics. Hayden, Rochelle Park
15. Kurth R (1972) Dimensional analysis and group theory in astrophysics. Pergamon Press, Oxford
16. Lokarnik M (1991) Dimensional analysis and scale-up in chemical engineering. Springer Verlag, Berlin
17. McMahon TA, Bonner JT (1983) On size and life. Scientific American Library, New York
18. De Jong FJ (1967) Dimensional analysis for economists. North Holland, Amsterdam
19. Severeyn E, Wong S, Herrera H, Altuve M (2015) Anthropometric measurements for assessing insulin sensitivity on patients with metabolic syndrome, sedentaries and marathoners. In: 2015 37th annual international conference of the IEEE engineering in medicine and biology society (EMBC), pp 4423–4426. IEEE, August 2015
20. Esfahani JA, Majdi H, Barati E (2014) Analytical two-dimensional analysis of the transport phenomena occurring during convective drying: apple slices. J Food Eng 123:87–93
21. Marusteri M, Bacarea V (2010) Comparing groups for statistical differences: how to choose the right statistical test? Biochemia Medica 20(1):15–32

Robotic Arm Manipulation Under IEC 61499 and ROS-based Compatible Control Scheme

Carlos A. Garcia[1], Gustavo Salinas[1], Victor M. Perez[1], Franklin Salazar L.[1], and Marcelo V. Garcia[1,2](✉)

[1] Universidad Tecnica de Ambato, UTA, 180103 Ambato, Ecuador
{ca.garcia,leonidasgsalinas,victormperez,fw.salazar,mv.garcia}@uta.edu.ec
[2] University of Basque Country, UPV/EHU, 48013 Bilbao, Spain
mgarcia294@ehu.eus

Abstract. Currently, the generation of traditional and straight-line production systems are being replaced with the objective of generating flexible and modular production systems. To achieve this goal, one of the main alternatives used by enterprises is the integration of robotic systems in the production cells. This robotic integration, indirectly allows robots to be compatible with the concepts of Smart Factories and Industry 4.0, where the high processing and communication capacities of the devices allows better and faster data exchange among the devices conforming the system. IEC-61499 proposes an architecture for the generation of distributed and modular production systems under Industry 4.0 paradigm. This paper proposes the development of an based IEC-61499 agile architecture implementation using a low-cost controller as Raspberry Pi 3B for control of KUKA youBot™ industrial robotic arm in a Pick and Place process.

Keywords: Robotic manipulator · IEC-61499 · Industry 4.0
Robot Operating System (ROS)

1 Introduction

IEC-61131 is the automation oriented standard destined to norm the integration of Programmable Logic Controllers (PLCs) in industrial processes. It is based on the use of a centralized architecture for the develop of its systems. Throughout the years of its usage, it successfully accomplished to generate the rigid production structures, characteristic of the previous industrial paradigm. However, the current industry development is aimed towards a structure more flexible and modular. In recent years, the concepts of Smart Factories and Industry 4.0, are slowly being implemented in enterprises worldwide. They are built over the conception of incorporating improved industrial communications and devices with more processing characteristics into the production systems. This allows to all

M. Botto-Tobar et al. (Eds.): TICEC 2018, AISC 884, pp. 358–371, 2019.
https://doi.org/10.1007/978-3-030-02828-2_26

the devices involved in the process to share real time data and information about their operations and easily group them into flexible production modules [1].

The main objective of an automated production system is to transfer human abilities and dexterities to machines. Since the first introduction of automated systems into manufacturing processes, a high improve in performance and quality was obtained. The combination of this success with the dexterity needs brought by the creation of newer and more complex production tasks, was the base point of the development of industrial robots [2]. Their operation is based on the joint work of three devices: the operative element (being robotic arms the most used representatives), the CPU where the control programming is stored and executed, and the power source. Until recent years, the PLC was chosen by many developers and programmers as control device following IEC-61131 guidelines [3]. However, with the advances of technology and the requirements of the new industry structure, current robots tend to incorporate CPUs running embedded software for their control.

IEC-61499 was developed in order to overcome the previous standard flaws while facing the new industrial paradigm. It is based on the creation of generic Function Blocks (FBs) and their implementation in modular applications for the obtaining of distributed control schemes. Similarly, to the IEC-61131's requirement for the use of a PLC to control an automated system, the main physical requirement of IEC-61499 to be implemented is the use of a controller capable of running an embedded software. Having the standard's requirement to use an embedded controller and the new generation industrial robots' characteristic to operate under one, this paper proposes the generation of an IEC-61499 compatible control system directed to the operation of a robotic arm.

The layout of the paper is as follows: Sect. 2 presents a general overview of the IEC-61499 standard, Sect. 3 presents the point of view of some related works used as starting point for the present investigation, Sect. 4 illustrates the task used as case of study, Sect. 5 details the functions developed under IEC-61499, Sect. 6 describes the final system implementation together with the obtained tests results and finally, some conclusions and ongoing work are discussed in Sect. 7.

2 Standard IEC-61499

The new standard for automation IEC-61499, is designed to generate flexible, modular, and distributed control systems which can meet the requirements of:

1. Portability.- Software tool's capacity to accept, interpret, and correctly execute systems developed by other software tools.
2. Interoperability.- Capacity of multi-brand integration in the same system.
3. Configurability.- Capability of easy modification of the devices configuration and the software components.

It specifies a generic architecture for distributed process, using as base the FB model introduced for the first time in its predecessor IEC-61131 [4] and adding event handling mechanisms. It defines the FB type as the basic unit for encapsulating and reusing programming algorithms rated to: mathematical operations, specific operation functions, communication functions, etc. As shown in Fig. 1, the FB structure is composed by a head which manages the event flow and a body in charge of the data flow management. This structural scheme is adopted to provide an improved synchronization between the data transfer and the program execution. The development of IEC-61499 compatible FBs is done in two stages. The first one is the External Structure Generation with the use of a desktop engineering environment, able to provide an extensible development framework for modelling distributed control applications under IEC-61499. After the external case is build, the second stage related to the Programming Integration begins. Once a FB structure is generated, the source files containing the programming related to its behavior are generated as well. In the standard guidelines, it is specified that this source files can be written in one of the IEC-61131 programming languages and additionally in high level programming languages such as Java, C++, Phyton, etc. [5].

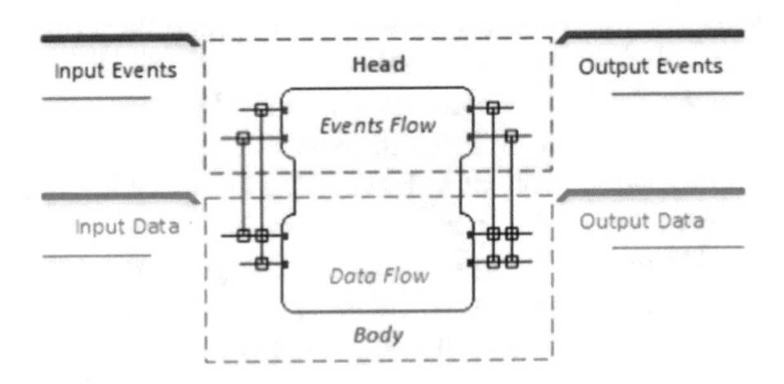

Fig. 1. FB structure elements.

As previously described, the FB is the IEC-61499 fundamental unit, out of which a distributed control systems are built. With the use of the desktop development framework, FBs are interconnected for the generation of applications. Finally, distributed control systems are generated using the same engineering tool to relate the built applications to the devices of a specified control network and then downloading into them. For an embedded controller to be able to execute the system operations, it is needed the installation of a Runtime Environment in its operative system which can run the algorithms specified in the system FBs according IEC-61499 guidelines.

3 Related Works

Due to their efficiency and high performance, robots are the perfect alternative for the execution of repetitive tasks which need a high level of precision or present a high safety risk for a person's integrity. Industrial robots have many applications, being the most representative: spot welding, spray painting, electronic testing, metrology, assembly and machining of mechanical parts. A practical example of this applications is presented in [6], where a robotic automation system for steel bean assembly in building constructions is generated. Similarly, in [7] robots are used for the grinding and polishing process of aspherical mirrors. In the work presented in [8] a more complex architecture is given with the implementation of a pick-and-sort application based on the joint work of real time vision and motion planning, allowing a robotic arm to pick objects from a moving conveyor belt.

With the increasing industrial needs for more flexible and modular systems, robot's development is also being directed towards the coordinated work of multiple devices. An example of this is shown in [9], where processes of: welding, transporting and polishing; are automated with the integration of robotic systems capable of sharing information among themselves. This is mainly done using the concepts of Smart Factories and Industry 4.0, which are characterized by the high efficiency and easy customization given by the improve of interconnections between devices. The work proposed in [10], shows the benefits of this easy data exchange through the system with the generation of a general methodology for mobile robots navigation. However, the improve in communications not only benefits the control elements in the system. The authors in [11] propose an interaction solution between robots and humans, with the generation of a virtual environment which can access to the production system data and replicate accurately the robots behavior in the physical world.

Many researchers in recent years, have contributed to IEC-61499 industrial scalability growth. In [12] the authors present the integration of a distributed control system in an industrial batch process, using embedded development boards as controllers. Similarly, in [13] an automation architecture of Cyber Physical Processes (CPPs) based on UML for industrial processes is proposed. The high integration level of the new standard as result of the use of embedded devices is appreciated in these works. In the list of advantages that allowed the accelerated development of the IEC-61499 there is also the capacity of generating generic functions using high-level programming languages. An example of this is shown in [14], where an advanced fuzzy control programming is integrated in the FBs codification to control a CPPS using an embedded board as controller.

The evolution of robot programming has gone from the setting of sequence individual actions of the actuators, to the referenced joint movement in order to generate smother and more precisely actions. Therefore, the programing involved in the operation of a robot has also gotten more complex. One of the most implemented alternatives by many researchers is the kinematic controller. In [15] the authors present a controller based on the kinematics of a SCARA robot in order to integrate it in a distributed winding process. Additionally, in [16] one of the investigations presents a switching algorithm between kinematic control strategies for the operation of a robotic arm. These algorithms are generally developed in high level programming, making them fully compatible with IEC-61499. As example, the authors in [17] introduce a robotic control initiative compatible with the new standard. And in [18] a robotic cell management with the use of an IEC-61499/ROS architecture is created, showing this way the success at integrating the management and control of robotic elements with the new standard for the creation of modular and flexible production lines.

4 Study Case

One if the most common tasks at manufacturing processes is the positioning of objects. The speed and precision required for the placing of parts in continuous assembly lines goes beyond human capacities. Thus, the implementation of industrial robots in the process is needed for an enterprise to be competitive in the current market. KUKA is one of the main representatives at industrial robot's manufacturing worldwide. Their product KUKA youBot™ shown in Fig. 2, is a small versatile robot destined for development and educational purposes. It combines a mobile omnidirectional platform enclosing a CPU working under an embedded software and a robotic arm with five Degrees of Freedom (DOF) using a two- finger gripper as final actuator. Both of its components can work as one entity by configuring the arm to work as a slave of the platform's CPU, or separately by controlling the arm with another Ethernet compatible processing unit.

The current works presents a proposal based on the use of the youBot's robotic manipulator under the control of the embedded development card Raspberry PI 3B, for the accomplishment of a "pick and place" task. The final control application shown in Fig. 3 is generated with the integration of two groups of FBs under IEC-61499 FORTE™ runtime. The first set encapsulates the manipulation of the youBot's drivers, allowing the data write and read of the manipulator's physical resources. The second set includes a Point to Point control algorithm based on the kinematic model of the robotic arm. The development board Raspberry Pi 3B is used as controller for the following investigation. Previous works

Fig. 2. KUKA youBot™.

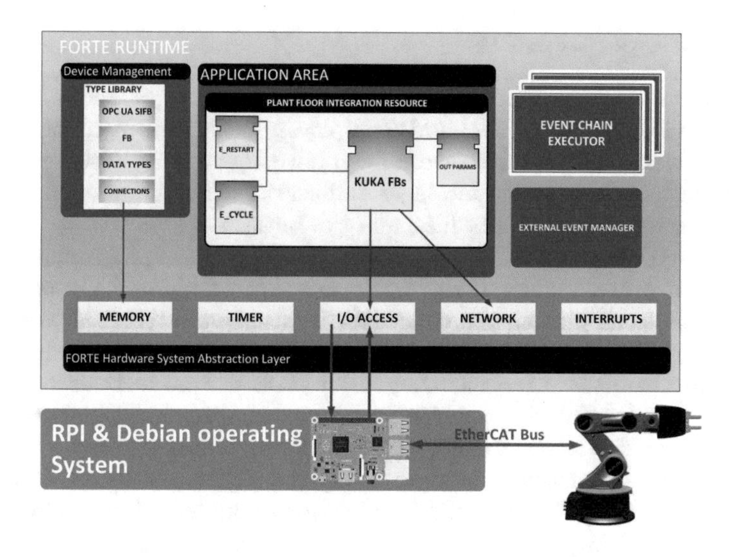

Fig. 3. Final operation architecture scheme.

such as [12–14] demonstrate the success at generating low-cost CPPS solutions for industrial processes. This is because the fact that despite being created for educational purposes, the capability of operating with an embedded software and the physical characteristics presented in Table 1, make not only the Raspberry devices but virtually every development embedded board, totally compatible with IEC-61499 standard.

Table 1. Raspberry Pi 3B physical characteristics.

Physical characteristics	Description
CPU	ARM 1176JZFS a 700 MHz
RAM	512 MB
GPIO's	8xGPIO, SPI, I2C, UART
Interfaces	2 × USB 2.0; Ethernet 10/100; WiFi

5 Service Interface Function Blocks (SIFBs) Proposal

For the creation of the following FBs, the ICE-61499 engineer environment 4DIAC-IDE is used to design and implement the external structure of the blocks. With the help of this program, the input and output tunnels used for the exchange of data and events are added based on the final service that the block gives to the application. In a lower design-level, each FB is generated with its respective .CPP and .H source files. The algorithms related to the individual operations of the functions are incorporated to the programming with the use of the C++ software editor ECLIPSE-IDE™.

5.1 Robot Operation FBs

These FBs uses Robot Operating System (ROS) messages under IEC 61499 runtime like FORTE to control and manipulate Kuka youBot robot. ROS is a component-based software framework for robotic applications. ROS nodes communicate with each other via messages and services. The formats of these messages are defined in special text files, and can be seen as standardized datatypes in the ROS environment. ROS has a set of tools that make programming easier.

The build process under FORTE is also able to automatically transfer the message definitions into source code. ROS Software is organized in packages, which are grouped thematically into stacks. FBs programmed under FORTE are built to incorporate the KUKA youBot ROS drivers in their codification, allowing the interaction between the functions and the robot logical resources.

The youBot driver stack contains C++ classes which grant to applications low-level access by means of two parts: (1) the EtherCAT driver and (2) the KUKA youBot API. The EtherCAT driver incorporates a Source Open EtherCAT Master (SOEM) operation structure in the processing device, making it capable to communicate in the lowest level and drive all of the actuators of the youBot. Using the operations established by the first stack layer as operations starting point, as shown in Fig. 4, the youBot API is a group of C++ classes developed to provide additional functionality related to platform kinematics, arm kinematics and joint data manipulation. In summary, these FBs send ROS messages as commands to reads proprioceptive sensors measurements like odometry and joint angles of the arm and the wheels. FORTE provides a API wrapper to mapping between the ROS messages and KUKA youBot API method calls.

Using the description of previous paragraph, two FBs are designed and development to write and read the joint angles at KUKA youBot^{TM}. First FB is YouBotArm_READ FB presented in Fig. 5. This FB includes the following events:

- **INIT (input)**.- Trigger event of the initial configuration tasks execution.
- **REQ (input)**.- Trigger event of the operations corresponding to a normal work cycle.
- **EMR (input)**.- Trigger event of the ending of any operation currently being executed.
- **INITO (output)**.- Flag event generated once the initial operations are finished.
- **CNF (output)**.- Flag event generated with the end of a normal work cycle.
- **ERR (output)**.- Flag event triggered with the "emergency shutdown".

The YouBotArm_READ FB is designed to operate with the following data input and outputs:

- **STATUS (STRING, output)**.- Returns as a string value the text corresponding to the state of the FB, being the options: initialized, operating, and stopped.
- **JR1 to JR5 (REAL, outputs)**.- Returns the angle values of the joints at the moment of the read operation as formatted a real type data.

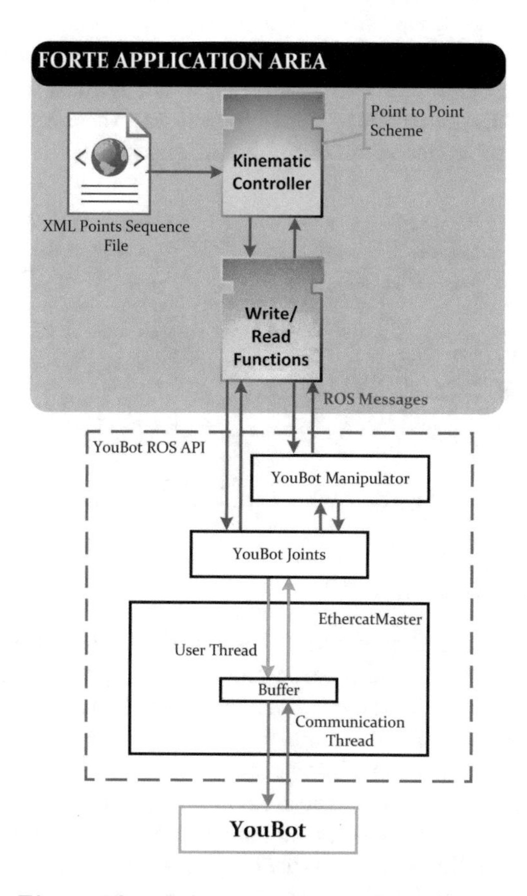

Fig. 4. FORTE communication with ROS API.

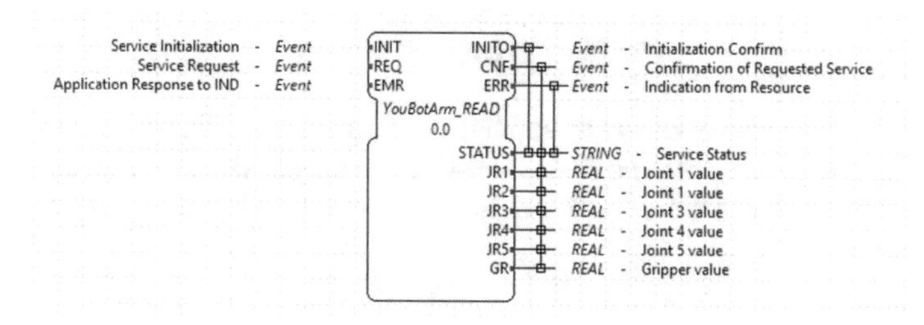

Fig. 5. YouBotArm_READ FB.

The second FB designed is YouBotArm_WRITE FB is used to write joint angles data to Robot API. This FB provides events like INIT, REQ, EMR INITO, CNF and ERR (see Fig. 6) as well as the following output data:

- **GR (REAL, output)**.- Returns the gripper opening length value at the moment of the read operation formatted as a real type value.

- **JOINT_1 to JOINT_5 (REAL, inputs)**.- Real type values of the desired joint angles, used as reference to manipulate the actuators.
- **GRIPPER (REAL, input)**.- Real type value of the desired gripper opening length, used as reference to manipulate the actuator.

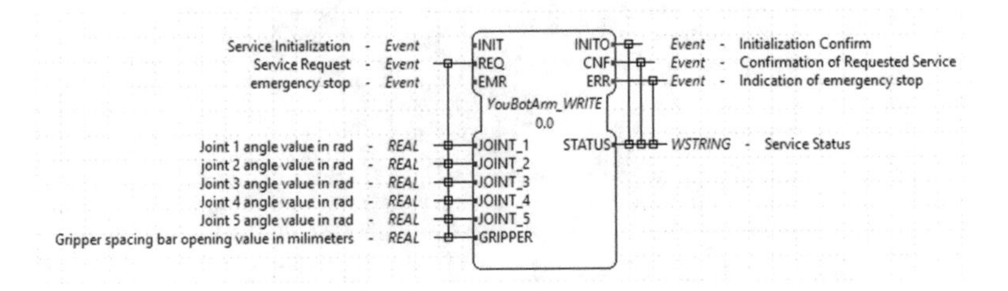

Fig. 6. YouBotArm_WRITE FB.

5.2 Kinematic Controller FB

The operation of this FB is centered in the execution of a Point to Point control scheme based on the youBot kinematic model. The representation of this model as function of the arm configuration gives in return the location of the effector **h**. It also can be described with its operational coordinates represented as function of the robotic arm general coordinates:

The operation of this FB is centered in the execution of a Point to Point control scheme based on the youBot kinematic model. The representation of this model as function of the arm configuration gives in return the location of the effector h. It also can be described with its operational coordinates represented as function of the robotic arm general coordinates:

$$f : N_b \rightarrow \mu; (q_b) \rightarrow h = f(q_b) \tag{1}$$

Where N_b represents the arm spatial configuration. Therefore, the instantaneous kinematic model returns the derivative of the gripper location as a function of the derivative of the arm configuration:

$$\dot{h}(t) = J(q)v(t) \tag{2}$$

Where $\dot{h} = [\dot{h}_1 \quad \dot{h}_2 \quad \dots \quad \dot{h}_n]^T$ is the vector which identifies the velocity of the final effector, and $v = [v_1 \quad v_2 \quad \dots \quad v_m]^T$ is the vector which contains the arm joints velocities. The dimension of the v vector is directly related to the robotic arm, being the value of m equal to the number of links conforming it. The matrix:

$$J(q) = \frac{\partial f(q)}{\partial q} T(q) \tag{3}$$

Is the Jacobian matrix that defines the linear mapping formed between the velocity vectors of the arm $v(t)$ and the robot's gripper velocity $\dot{h}(t)$. $T(q)$ is defined as the transformation matrix which relates the robotic arm velocities $v(t)$ with the joint velocities $\dot{q}(t)$, having that:

$$\dot{q}(t) = T(q)v(t) \tag{4}$$

The Point to Point controller is designed to stabilize with a given orientation the youBot final effector to a specified point, where $h_d = [h_{dpos}^T \quad h_{dpor}^T]^T$. Therefore, the position and speed required for stability can be defined as:

$$h_d(t, s, h) = h_d \tag{5}$$

$$v_{h_d}(t, s, h) = \dot{h}_d \equiv 0 \tag{6}$$

The control task involving reaching the specified points is defined as:

$$v_{ref}(t) = f(\widetilde{h}(t)) \tag{7}$$

Where $\tilde{h}(t) = h_d - h(t)$, resulting in $lim_{t\to\infty}\tilde{h}(t) = 0$.

The Controller_PtP FB shown in Fig. 7 is designed to read a XML file in order to obtain the reference points for the execution of its operations. Following the Robot Operation FBs execution structure, this block is designed to work with the input events: INIT, REQ and EMR; and to generate the output events: INITO, CNF and ERR. The data inputs and outputs which the function receives and generates are:

- **XML_FILE (STRING, input)**.- Path of the XML file containing the control reference points.
- **J1_SENSOR to J5_SENSOR (REAL, inputs)**.- Angle values of the joints at the moment of the read operation formatted as a real type data.
- **GR_SENSOR (REAL, input)**.- Gripper opening value at the moment of the read operation formatted as a real type data.
- **STATUS (STRING, output)**.- Returns as a string value the text corresponding to the state of the FB, being the options: initialized, operating, and stopped.
- **J1_ACTUATOR to J5_ACTUATOR (REAL, outputs)**.- Results of the controller operations representing the desired joint angles of the manipulator.

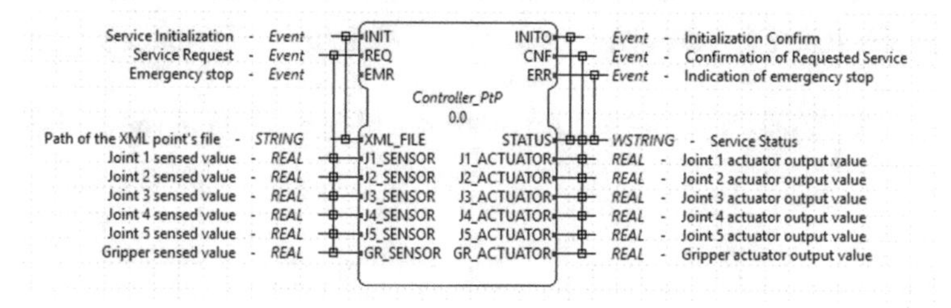

Fig. 7. Controller_PtP FB.

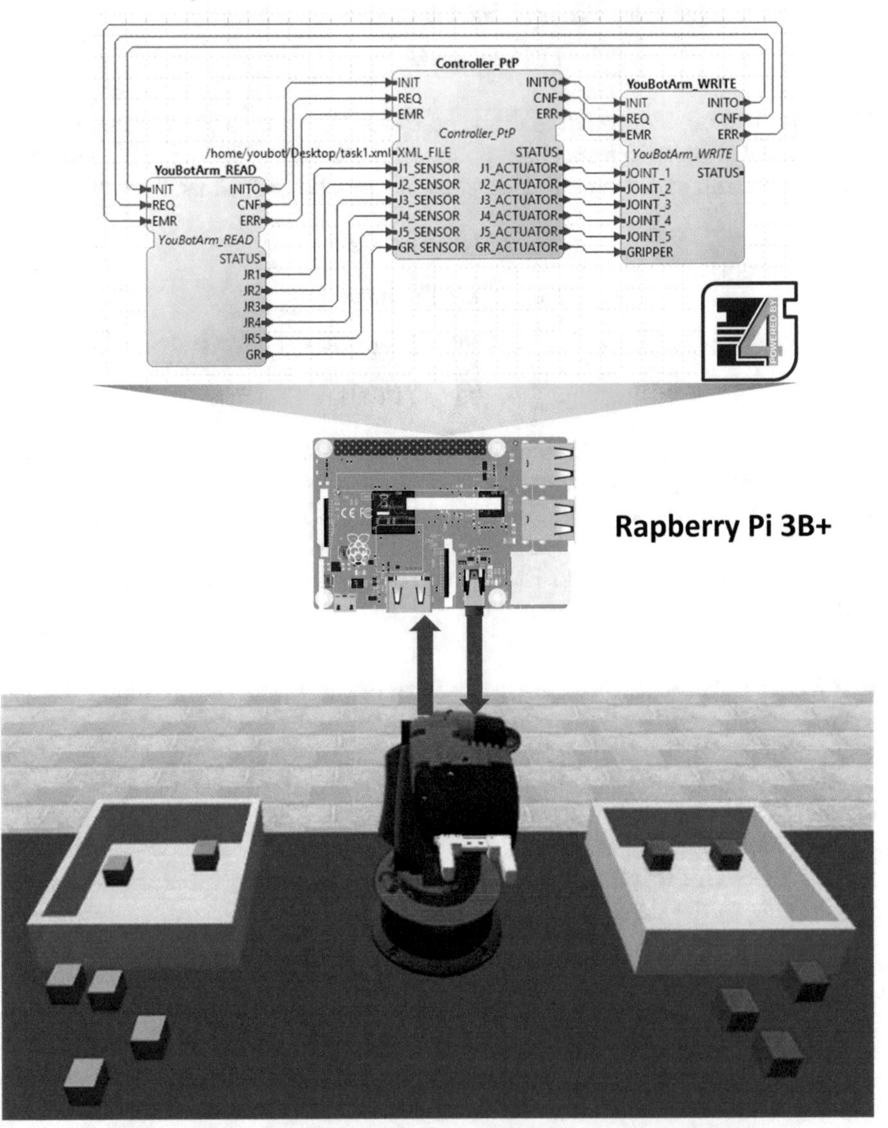

Fig. 8. (a) IEC 61499 Forte Control Application. (b) Simulation of the youBot robotic arm test scenario.

6 System Application Implementation and Results

For the accomplishment of the pick and place task, Fig. 8(a) shows the generated FBs distributed to fit the work loop corresponding to: read of the sensor's state and transmission of the resulting data to the control scheme (YouBotArm_READ FB), calculation and transmission of control actions based on the plant's current state (Controller_PtP FB), and interpretation of the resulting control actions into physical actions (YouBotArm_WRITE FB).

The logical and physical components of the system interact with each other as shown in Fig. 8(a). Once the system is build, it's design is downloaded into the Raspberry Pi, where the operations and processing are done using the FORTE runtime environment. On top of the logical structure of the system, the created FBs perform the corresponding transactions of data and events. The read and write FBs are the ones generating the interaction with the physical components of the youBot. This is done by means of the driver manipulation integrated in their programming. The functions given by the youBot API manage the data transaction among the logical hierarchies and allow the exchange of information between the Raspberry working as master and the youBot working as slave.

Additionally, to the already explained functionalities of the Controller_PtP FB, a part of its algorithm was designed to write the control errors resulting of its operation in a .TXT file. The performance tests done to the implemented system, used this function to export the resulting data and plot it with mathematical software. The general scenario of operations is based on the transportation of cubic elements between containers located in the arm's surroundings as presented in Fig. 8(b).

The XML file corresponding to the first task, incorporates seven points from the arm's operative space. After the first execution, the data displayed in Fig. 9(a) demonstrates a high level of error correction with the continuous error decrease. With the normal operation verified, a second test was implemented

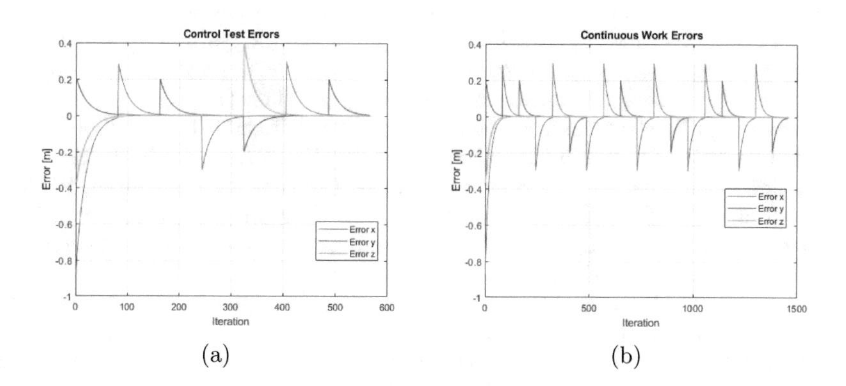

Fig. 9. Control errors resulting of (a) Test 1: XML file with seven points from the arm's operative space (b) Test 2: XML file with the points corresponding to a continuous operation

by entering a XML file with the sequence of points corresponding to a normal continuous execution. The results displayed in Fig. 9(b) shows a cyclical error correction, corresponding to the continuous move of the arm between containers.

7 Conclusions and Ongoing Work

The architecture proposed in this paper supports the scalability of Smart Factories where robotic systems are implemented to execute tasks with a high precision and speed. This proposal uses the IEC-61499 capacity of generating FBs using high-level programming languages such as C++, and creates functions which take advantage of the youBot open source drivers and use them to generate the joint work of KUKA's robotic manipulator and the embedded development board Raspberry Pi 3B. This way it is demonstrated the high compatibility between IEC-61499 standard and new generation robots working under an embedded software.

This work is implemented as the first stage of a research project and sets the foundations to the integration of the IEC-61499 standard with mobile robotic manipulators. The future lines of investigation will be focused on three main aspects: the generation of communication FBs using the MQTT protocol, the integration of image processing with Open CV libraries, and the analysis of an environment and determination of the best path to avoid obstacles.

References

1. Pedersen MR, Nalpantidis L, Andersen RS, Schou C et al (2016) Robot skills for manufacturing: from concept to industrial deployment. Robot Comput-Integr Manuf 37:282–291
2. Grau A, Indri M, Lo Bello L, Sauter T (2017) Industrial robotics in factory automation: from the early stage to the Internet of Things. In: IECON 2017 - 43rd annual conference on IEEE industrial electronics society, pp 6159–6164
3. Rossano GF, Martinez C, Hedelind M, Murphy S, Fuhlbrigge TA (2013) Easy robot programming concepts: an industrial perspective. In: 2013 IEEE international conference on automation science and engineering, pp 1119–1126
4. International Electrotechnical Comission, Standard IEC61131, vol 3 (2003)
5. Christensen JH, Strasser T, Valentini A, Vyatkin V, Zoitl A (2012) The IEC 61499 function block standard: overview of the second edition. In: ISA autom. Week
6. Chu B et al (2009) Robotic automation system for steel beam assembly in building construction. In: ICARA 2009 - proceedings 4th international conference on autonomous robots and agents, pp 38–43
7. Kim J, Lee W, Yang H, Lee Y: Real-time monitoring and control system of an industrial robot with 6 degrees of freedom for grinding and polishing of aspherical mirror. In: 2018 international conference on electronics, information, and communication (ICEIC), Busan, Republic of Korea, no 1, pp 6–9
8. Zhang Y, Li L, Ripperger M, Nicho J, Veeraraghavan M, Fumagalli, A (2018) Gilbreth: a conveyor-belt based pick-and-sort industrial robotics application. In: 2018 second IEEE international conference on robotic computing, pp 17–24

9. Borisov OI, Gromov VS, Kolyubin SA, Pyrkin AA, Bobtsov AA (2016) Human-free robotic automation of industrial operations. In: IECON 2016-42nd annual conference of the IEEE industrial electronics society. IEEE, pp. 6867–6872
10. Gonzalez AGC, Alves MVS, Viana GS, Carvalho LK, Basilio JC (2018) Supervisory control-based navigation architecture: a new framework for autonomous robots in industry 4.0 environments. IEEE Trans Ind Inform 14(4):1732–1743
11. Schuster K, Groß K, Vossen R, Richert A et al (2016) Preparing for industry 4.0-collaborative virtual learning environments in engineering education. In: Engineering education 4.0. Springer, Cham, pp 477–487
12. Garcia CA et al (2017) CPPS on low cost devices for batch process under IEC-61499 and ISA-88. In: Proceedings - 2017 IEEE 15th international conference on industrial informatics, INDIN 2017, pp 855–860
13. Castellanos EX, Garcia CA, Rosero C, Sanchez C, Garcia MV (2017) Enabling an automation architecture of CPPs based on UML combined with IEC-61499. In: 17th international conference on control, automation and systems, ICCAS, pp 471–476
14. Garcia CA et al (2017) Fuzzy control implementation in low cost CPPS devices. In: 2017 IEEE international conference on multisensor fusion and integration for intelligent systems, MFI, pp 162–167
15. Gerngroß M, Herrmann P, Westermaier C, Endisch C et al (2017) Highly flexible needle winding kinematics for traction stators based on a standard industrial robot. In: 2017 7th international electric drives production conference (EDPC). IEEE, pp 1–7
16. Tommaso L, Paolis D, Bourdot P, Mongelli A (2017) Augmented reality, virtual reality, and computer graphics: 4th international conference, AVR 2017, Ugento, Italy, June 12–15 2017, proceedings. Springer
17. Zoitl A, Strasser T, Valentini A (2010) Open source initiatives as basis for the establishment of new technologies in industrial automation: 4DIAC a case study. In: 2010 IEEE international symposium on IEEE industrial electronics (ISIE), pp 3817–3819
18. Iannacci N, Giussani M, Vicentini F, Tosatti LM (2016) Robotic cell work-flow management through an IEC 61499-ROS architecture. In: 2016 IEEE 21st international conference on emerging technologies and factory automation (ETFA). IEEE, pp 1–7

EDA and a Tailored Data Imputation Algorithm for Daily Ozone Concentrations

Ronald Gualán[1,2](✉), Víctor Saquicela[1], and Long Tran-Thanh[2]

[1] Department of Computer Science, University of Cuenca, Cuenca, Ecuador
{ronald.gualan,victor.saquicela}@ucuenca.edu.ec
[2] School of Electronics and Computer Science, University of Southampton, Southampton SO17 1BJ, UK
ltt08r@ecs.soton.ac.uk

Abstract. Air pollution is a critical environmental problem with detrimental effects on human health that is affecting all regions in the world, especially to low-income cities, where critical levels have been reached. Air pollution has a direct role in public health, climate change, and worldwide economy. Effective actions to mitigate air pollution, e.g. research and decision making, require of the availability of high resolution observations. This has motivated the emergence of new low-cost sensor technologies, which have the potential to provide high resolution data thanks to their accessible prices. However, since low-cost sensors are built with relatively low-cost materials, they tend to be unreliable. That is, measurements from low-cost sensors are prone to errors, gaps, bias and noise. All these problems need to be solved before the data can be used to support research or decision making. In this paper, we address the problem of data imputation on a daily air pollution data set with relatively small gaps. Our main contributions are: (1) an air pollution data set composed by several air pollution concentrations including criteria gases and thirteen meteorological covariates; and (2) a custom algorithm for data imputation of daily ozone concentrations based on a trend surface and a Gaussian Process. Data Visualization techniques were extensively used along this work, as they are useful tools for understanding the multidimensionality of point-referenced sensor data.

Keywords: Air pollution · Sensor data · Data imputation
Gaussian process

1 Introduction

Air pollution is a critical environmental problem with detrimental effects on human health. According to a report released the last year by the World Health Organization (WHO), more than 80% of people living in urban areas are exposed to air quality levels which exceed the World Health Organization limits. Air

M. Botto-Tobar et al. (Eds.): TICEC 2018, AISC 884, pp. 372–386, 2019.
https://doi.org/10.1007/978-3-030-02828-2_27

pollution contamination affects all regions of the world, however populations in low-income cities are the most impacted [24].

In order to mitigate air pollution, effective actions in the form of research studies and governmental decision, are required. Air pollution monitoring is the most important asset supporting such actions. Conventional approaches for monitoring based on expensive, complex, stationary equipment, are being replaced or complemented by The Next Generation Air Pollution Monitoring Systems (TNGAPMS) [20], which rely on low-cost, easy-to-use, portable air pollution sensors. Low price of the new sensor devices enable researchers to deploy a large number of these on field, meaning that data is collected at unprecedented spatial, temporal and contextual detail [21]. An additional effect, derived from the availability and commercialization of low-cost air pollution sensors, is the participation of non-expert public. Low-cost sensors have attracted such amount of attention from the general public, that community-led air quality sensing networks have emerged. This community-centered approach is referred to as participatory sensing or citizen science [3,10,20].

Although low-cost sensors offer advantages in terms of accessibility and volume of deployable devices, they also have a huge disadvantage: they are unreliable. This means that observations taken by low-cost sensors are prone to gaps, bias and noise. Gaps are caused by malfunction in the sensors. Bias might be caused by changes in the calibrated state of a sensor produced by environmental conditions or deterioration. All sensors present some level of noise. Low-cost sensors, however, are more susceptible to noise because they are built with low-quality electrical components. These problems should be addressed before data can be used to support research or decision making. However, effectively addressing them is a complex challenge that still is undergoing research. The presence of noise in sensor data has been cause of study for more than two decades in areas of sensor networks, mobile robotics, and machine learning [23].

Missing data is a common problem on legacy sensor network technologies, and arguably it is more serious on cheap sensor networks. Data completeness is crucial for modeling because even very short gaps preclude the calculation of important summary statistics [5]. To have a complete data set which can be used to support studies and decision taking, it is essential to use data imputation methods for optimally infilling missing data. As aforementioned, the context of this study is based on a new trend of sensor technologies based on cheap sensor equipment for capturing air pollution concentrations. However, since to date there is not a publicly available data set for such sensors, we decided to explore available legacy options. Thus, we built an air pollution data set composed by several air pollution concentrations including criteria gases and 13 meteorological covariates, which can be used as a benchmark in other spatio-temporal modeling projects; and we also developed and tested a custom algorithm for data imputation of daily ozone concentrations based on a trend surface and a Gaussian Process.

2 Background: Hierarchical Bayesian Models

In this section, we review the formulation of one of the most relevant spatio-temporal models: Hierarchical Bayesian models. The concepts and formulations in this section served as the basis for developing a custom data imputation algorithm described in Sect. 5.

Hierarchical Bayesian models are statistical artifacts expressed in multiple levels (hierarchical form) whose posterior probability parameters are estimated using the Bayesian framework [1]. The literature review regarding statistical modeling acknowledges that spatio-temporal associations are captured more effectively by models that build dependencies in different stages or hierarchies [7]. In the last years, Bayesian hierarchical models for point-referenced space-time data have become feasible thanks to recent advances in both statistical methodology and computation power [2].

According to [8] a hierarchical structure can be used to build Bayesian models, where three levels specify the distributions for data, process, and parameters, as follows:

$$\begin{array}{rl} \text{First} & [data|process, parameter] \\ \text{Second} & [process|parameter] \\ \text{Third} & [parameter] \end{array} \tag{1}$$

For example, for Gaussian Process (GP) [6], the first level determines the true underlying process, the spatio-temporal random effects are specified in the second level of the hierarchy, and the prior distribution of the parameters or hyper-parameters are introduced in the third level [2].

For the specification of a hierarchical model, a common basic structure is presented in different manuscripts [2,12,16], where the random variable representing the observed value of the point-referenced response is typically depicted as $Z(\mathbf{s}_i, t)$, with $i = 1, \ldots, n$ representing the sites $\mathbf{s}_i = (x_i, y_i)$, monitoring some variable (eg. air pollution concentrations) at times $t = 1, \ldots, T$. Space is considered as a continuous variable, while time is considered as a discrete variable. The formulation of the first level is given by:

$$Z(\mathbf{s}_i, t) = Y(\mathbf{s}_i, t) + \epsilon(\mathbf{s}_i, t), \quad \epsilon(\mathbf{s}_i, t) \sim N(0, {\sigma_\epsilon}^2) \tag{2}$$

where $Y(\mathbf{s}_i, t)$ is the true process and $\epsilon(\mathbf{s}_i, t)$ is the independent nugget effect absorbing micro-scale variability. The next stage of the hierarchy is defined as:

$$Y(\mathbf{s}_i, t) = \mu(\mathbf{s}_i, t) + \eta(\mathbf{s}_i, t) \tag{3}$$

where $\mu(\mathbf{s}_i, t)$ denotes the mean surface and $\eta(\mathbf{s}_i, t)$ is a space-time process. The mean surface and the space-time process can be defined in several ways, which will depend on the available covariates and space-time structures used. A simple mean surface model would be to use several covariates, as the independent Gaussian process (GP) model does:

$$\mu(\mathbf{s}_i, t) = \boldsymbol{\beta}\boldsymbol{x}(\boldsymbol{s}_i, t)$$

That is, the mean surface is calculated as a linear regression of p covariates (including the intercept) $\boldsymbol{x} = (1, x_1, \ldots, x_{p-1})$. Some of these covariates may vary in space and time. The regression coefficients are presented in vectorial form $\boldsymbol{\beta} = (\beta_1, \ldots, \beta_p)$. For the case of a Gaussian process, $Y(\boldsymbol{s}_i, t)$ denotes all the random effects [2].

3 Assembling the *EPA+NNRP(2016)* data set

This section describes the assembly of a real-life data set, meant to provide a test case for assessing spatio-temporal models. This data set was named *EPA+NNRP(2016)* because it combines the ozone concentration from the Air Data project from the U.S. Environmental Protection Agency (EPA) and the weather covariates taken from the NCEP/NCAR Reanalysis Project (NNRP). Exploratory Data Analysis (EDA) techniques were extensively used throughout the study, as they are essential for understanding the multiple dimensions of air pollution sensor data and its meteorological covariates. The analysis and plots presented in this report were implemented using the R language and statistical environment [14] and are available on line[1].

The Air Data project from the U.S. Environmental Protection Agency (EPA) collects air quality and weather measurements from more than 4000 outdoor monitors across the United States, Puerto Rico, and the Virgin Islands [22]. The data provided cover a time period from 1980 to 2017 in hourly, daily and annual aggregation. For this study, we downloaded the pre-generated daily files for the year 2016[2]. The available files are grouped in two categories: (1) Criteria Gases such as Ozone, Sulfur dioxide (SO2), Carbon monoxide (CO), and Nitrogen dioxide (NO2); and (2) Particulates, such as PM2.5 (particles with a diameter of 2.5 μm or less), PM10 (particles with a diameter of 10 μm or less) [22].

Since the geographical area covered by the EPA data set is too extensive for our experimental purposes, we reduced the domain to the state of California based on the amount of ozone monitors available for this state. From the air pollutants available from this data set, we chose ozone as the main target variable as it is one of the most studied pollutants in modeling studies [9,12,17]. From the 181 stations which registered measurements of ozone during the year 2016, 142 were selected because its proportion of ozone missing data did not exceed 10%. Figure 1 shows the location of the selected stations. The missing observations were assumed to be missing at random. The average proportion of missing data for the ozone stations was around 1.5% (range: 0–9.56). The mean distance between the selected ozone sites was 370.69 km (range: 4.5–1187.6).

Although the EPA data set contains stations monitoring air pollutant concentrations, not all the stations monitor all the variables. This can be seen in Fig. 1(b) where some stations only monitor ozone concentrations. The lack of variables in some stations and the missing data represent the biggest problems of this data set. Figure 2 presents the daily mean levels of ozone aggregated

[1] https://github.com/rgualan/soton-data-science-thesis.

[2] https://aqs.epa.gov/aqsweb/airdata/download_files.html.

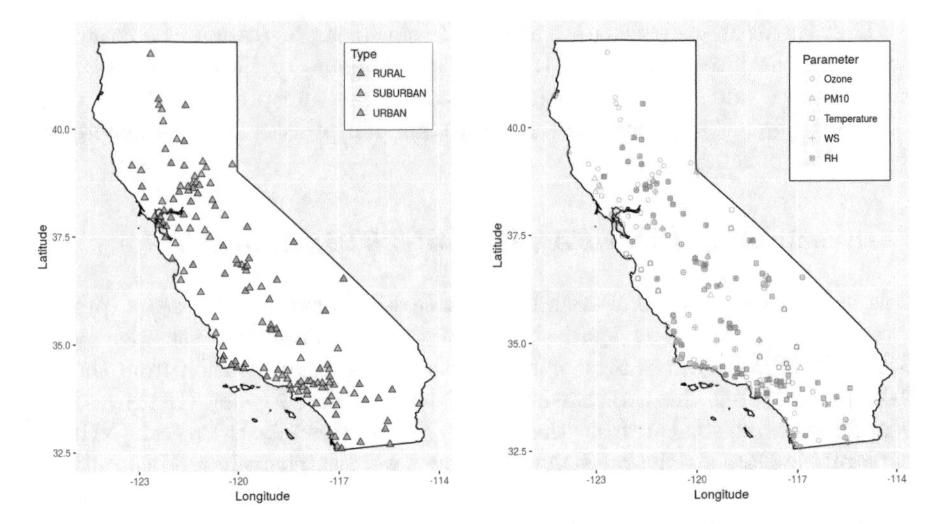

Fig. 1. Geographical area of the state of California including: (a) Location of the selected Air Data ozone monitors for the year 2016 by type of setting, and (b) Location of the Air Data monitors by the observed parameter.

across 50 randomly chosen monitoring sites from the 142 available ones, sorted by station code. The station code is built by three digits representing the county, a separator and four digits representing the site number. The daily values are displayed using a heat map. Overall, it can be seen that monitoring sites in the same county (i.e. stations having the same three first digits) seem to present slightly similar concentrations, which indicates spatial correlation. The bottom of the plot shows the daily mean values aggregated across the 142 stations, and a shaded area covering the interval between the 5 and 95 percentiles. The time series plot and the heat map (or Hovmöller diagram) show that the highest ozone levels of the year 2016 were registered between June and September, which is also the period of most variability between the stations.

The NCEP/NCAR Reanalysis Project (NNRP), chosen for the covariates, maintains more than 80 meteorological variables available at daily time resolution and $2.5^{\circ} \times 2.5^{\circ}$ space resolution (around 209 km), from the year 1948 to the present. This data set was selected for three reasons: (1) the availability of daily means for the year 2016, i.e. the year of study; (2) the spatial resolution is acceptable; and (3) reanalysis variables come from the interpolation of observations or are at least influenced by observations assimilated in a physical-based model [11]. From the range of available variables[3], we chose the following: Air Temperature at 2 m (K), Relative humidity (%), Wind speed (m/s), Precipitation (kg/m^2), Dew point (K), Water evaporation (kg/m^2), Ground Heat flux (W/m^2), Surface Geopotential Height (m), Tropopause Geopotential Height (m), Latent Heat Flux (W/m^2), Tropopause Pressure (Pa), Pressure at MSL (Pa),

[3] https://www.esrl.noaa.gov/psd/data/gridded/data.narr.monolevel.html.

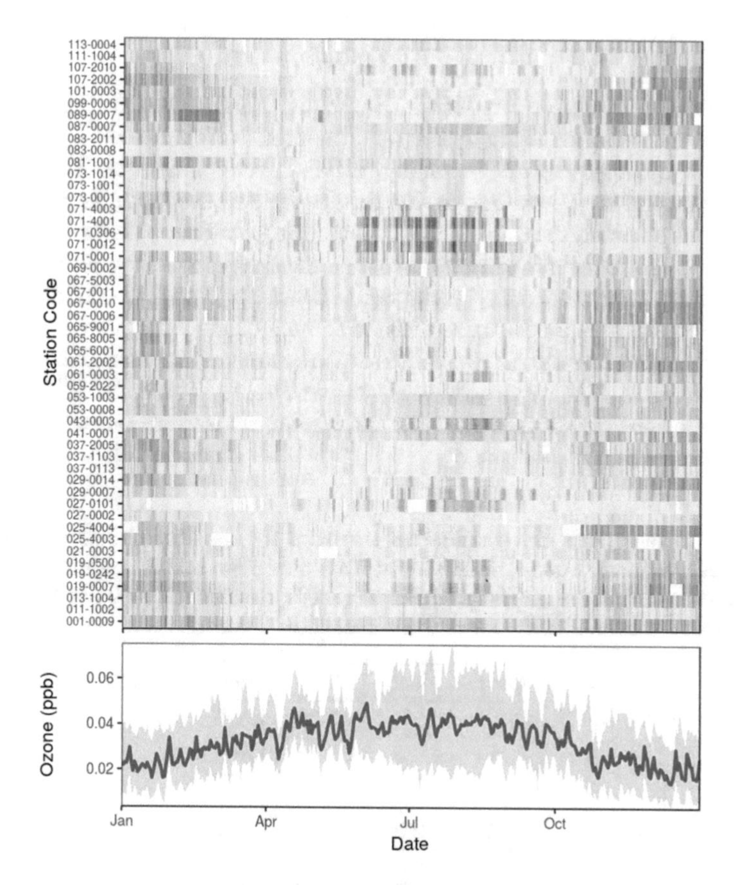

Fig. 2. Heat map or Hovmöller diagram for the daily ozone concentrations in parts per billion for 50 randomly chosen stations out of the 142 monitoring sites sorted by Station code ({county_code}-{site_num}) versus date. Low concentration levels are depicted in green while high concentrations are depicted in orange. Gaps are painted in white. The bottom panel shows the daily mean concentrations plus a shaded area covering the interval between the 5 and 95 percentiles.

and Vegetation (%). These variables were chosen because they have been used as predictors for air pollution concentration in previous studies [4,13,19,25], and because in a modeling application counting with a sufficient amount of covariates is important to achieve good precision levels. This is particularly true when modeling environmental phenomenons such as air pollution, which are influenced by hundred of variables. All the chosen meteorological variables are dynamical, i.e. change in space and time, except for Surface Geopotential Height and Vegetation, which are static in time. The data for the Reanalysis model are available in netCDF format, whereby predictors were extracted at the locations of the ozone stations using IDW interpolation, and hence predictors data are complete.

4 Data Imputation

As mentioned in the previous section, there are missing data in the ozone variable of the created data set, which should be imputed for assembling a complete data set for spatio-temporal modeling. IDW (Inverse Distance Weighted) was the first method to be evaluated for this task, because IDW is a simple yet efficient interpolation technique. Since IDW interpolation is a spatial interpolation method that does not take into account the time dimension, an independent model was created for each day of the year. In order to assess the validity of using IDW interpolation as an in-filling method for the ozone variable, we carried out a LOOCV (Leave-one-out Cross Validation) choosing as test stations only those with missing data. The results are presented in Table 1. The mean R^2 (coefficient of determination) of 0.61 and the min and maximum metrics indicate an overall poor performance.

Table 1. Leave-one-out cross-validation statistics for the IDW interpolation of the scaled ozone variable. The stations with at least 5 or more missing values were chosen as test stations. MSE: Mean Squared Error. RMSE: Root Mean Squared Error. MAE: Mean Absolute Error. BIAS: Bias. R^2: coefficient of determination

Metric	Mean	Min	Max
MSE	0.527	0.147	1.429
RMSE	0.693	0.384	1.195
MAE	0.572	0.302	1.116
BIAS	−0.061	−1.116	0.797
R^2	0.612	0.002	0.868

Three cases from the above experiment were examined in more detail (stations *021-0003*, *027-0101*, and *083-4003*) in order to analyze the problems and possible improvements. The two first cases correspond to the stations with the smallest and largest RMSE, while the latter belongs to the station which obtained the worst R^2. The geographical locations of these stations and their corresponding original and interpolated time series, are included in Figs. 3 and 4, respectively. By examining the geographical location and the closest neighbor stations, it was discovered that the station obtaining the smallest RMSE has 7 neighbor stations in a radius of 75 km, while the station with the worst RMSE has none in the same range. The third case also has seven neighbor stations in that range. However, despite of the relative small area surrounding the third case and its neighbor stations, the Elevation field of these stations varies considerably (between a range of 41–303 m), which is larger than the range of the first station (0–100 m). Thus, there is some evidence that the amount of neighbor stations and the variability in Elevation are relevant factors for the precision obtained by the IDW interpolation method.

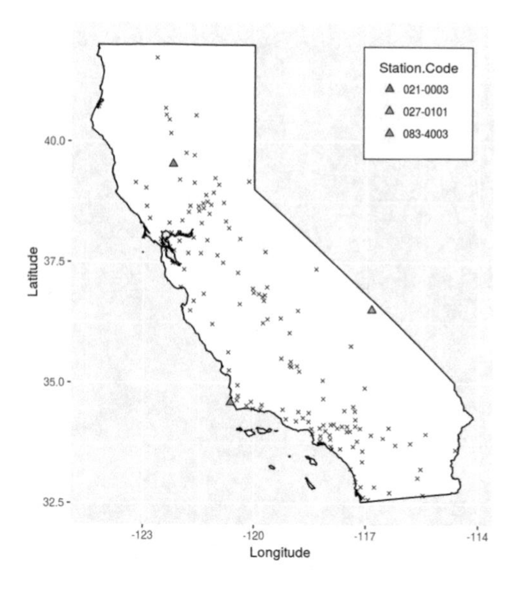

Fig. 3. Three selected cases for assessing the IDW interpolation experiment.

The time series plots included in Fig. 4, show that the first case obtains relatively acceptable predictions, the second case obtains biased predictions, and the third case obtains a wrong prediction which does not match the trend in the original time series. Linking these observations with those regarding the availability of neighbor stations, lead to three conclusions. First, the availability of close neighbor stations (with no significant differences in Elevation), can lead to acceptable prediction performance. Second, the lack of close neighbor stations leads to biased predictions. Third, high variability in the Elevation of the neighbor stations could produce very different predicted annual trends despite of the availability of close stations.

In addition to IDW interpolation, other three methods were tested: (1) GBM imputation based on Generalized Boosted Regression Modeling, (2) Gaussian process regression, and (3) Kernel K-nearest neighbors regression. For the former, the library *imputation* [15] was used. The GBM imputation method relies on two characteristics of decision trees: robustness against outliers and its skills to handle missing data by using surrogate splits [15]. The Gaussian process implies a spatio-temporal random effect which is used to model the space-time interactions [2]. Kernel k-nearest neighbors method replaces a missing value by finding and combining the k closest samples in the training set, usually by means of some combination weighed by the distances to the training set samples [18].

The additional imputation methods were tested in a subset of our data set. The subset includes 79 stations whose amount of missing data does not exceed 15 values (out of 366) on the variable ozone. One of the 79 stations was chosen for testing purposes, based on its location and relatively lack of close neighbor stations. This reduced data set was created to provide a more compact data

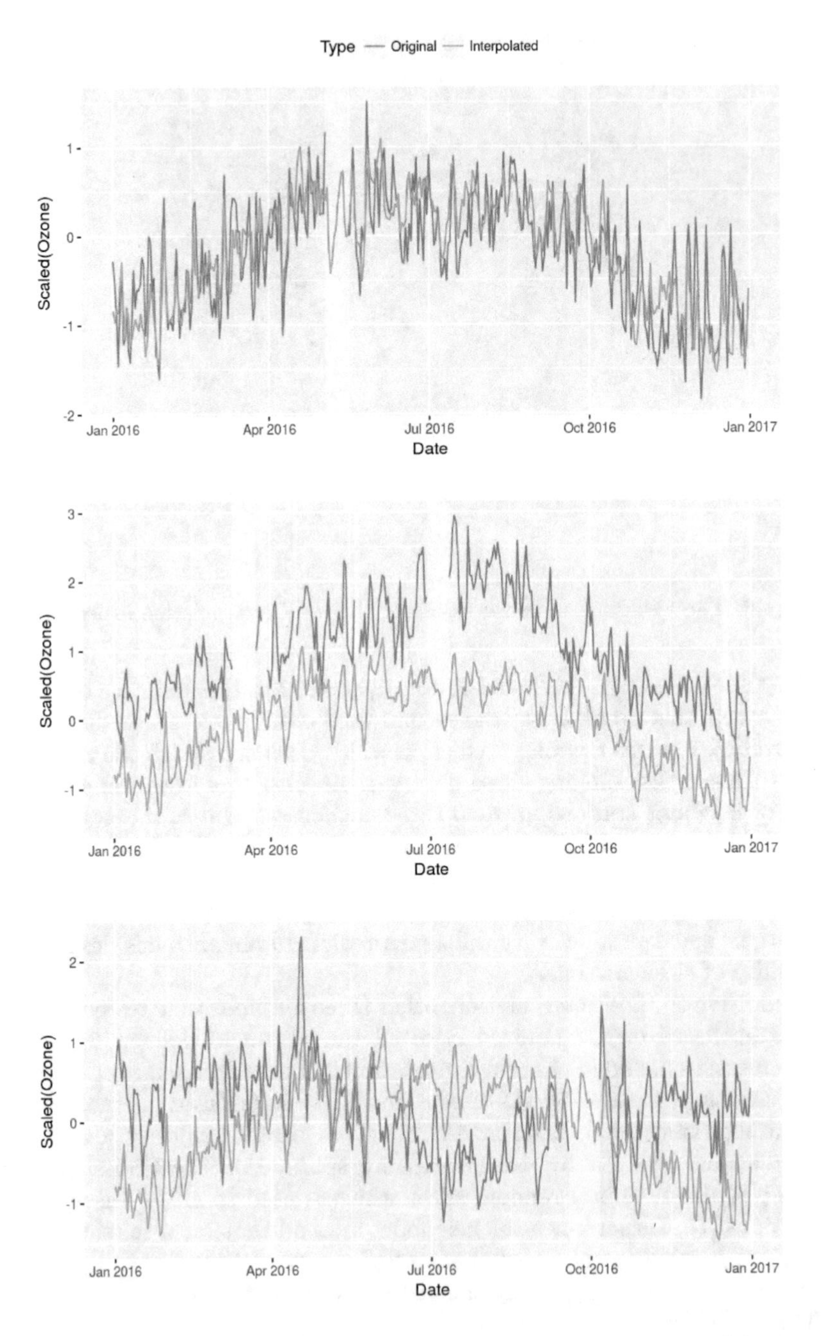

Fig. 4. Original and interpolated time series for the three chosen stations: (a) *021-0003*, (b) *027-0101*, and (c) *083-4003*. This stations correspond to the smallest RMSE, the highest RMSE, and the smallest R^2, in the LOO-CV of the IDW interpolation method evaluated as an option to infill missing values in the ozone variable of the EPA data set for the state of California. The highlighted yellow areas indicate zones where there are missing values.

set that can be efficiently handled by the costly GBM and GP models, and to provide an almost complete data set where there is plenty of space for injecting random missing values. Two flavors of missing values were introduced: one relatively large block of 15 contiguous missing values (contiguous missing data) and 10 random single missing values (i.e. sparse missing values). Figure 5 presents the original and modeled/imputed time series for the test station where the days with randomly injected missing values are highlighted in yellow. Table 2 summarizes some goodness-of-fit metrics estimated only on the injected missing days, as the GBM method produces replacements only for the missing values, and not a complete time series as the other methods. Figure 5 and Table 2 show that GBM, GP and KKNN produced biased predictions, similar to the IDW imputation model tested before. It is worth mentioning that the GP model was computationally expensive, reporting a runtime of approximately 6 min.

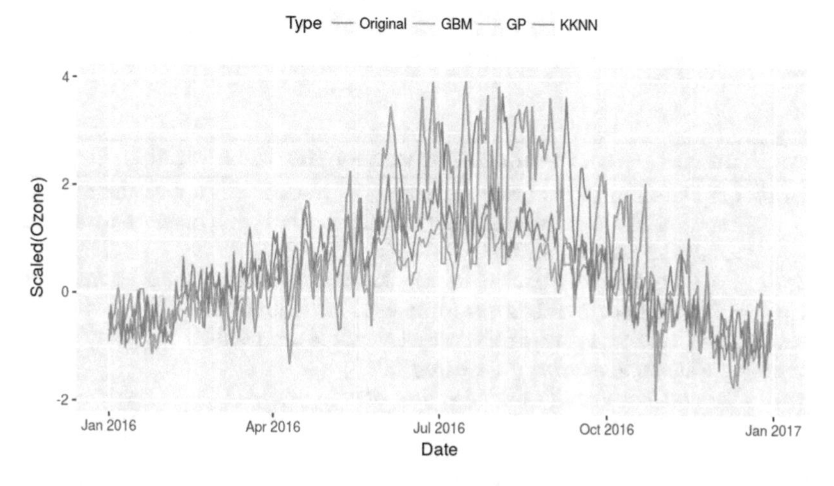

Fig. 5. Original time series and three imputation models based on GBM, Gaussian Processes, and Kernel KNN for a subset of the EPA data set formed by one selected test station (*107-0009*) and 78 semi-complete stations. The highlighted yellow areas indicate zones where there are missing values.

Table 2. Goodness-of-fit metrics predictions estimated from the GBM, GP and KKNN models for the station *107-0009*.

Model	MSE	RMSE	MAE	BIAS	R^2
GBM	2.626	1.621	1.351	−1.233	0.856
GP	1.948	1.396	1.186	−1.185	0.834
KKNN	1.511	1.229	1.002	−0.916	0.642

5 Custom Data Imputation

The imputation methods tried out in the previous section did not provide a precise enough solution for replacing the missing data in the ozone variable, because they provided biased predictions. Thus, we decided to apply a custom methodology which leverages the goodness of the aforementioned models by solving the bias deficiency. This methodology was inspired by the hierarchical models reviewed in Sect. 2, which model the observed variable as a mean surface (or trend) plus a space-time process. Our idea is to specify a polynomial linear trend as the mean surface and a Gaussian process as the space-time process. Our solution is based on three premises: (1) ozone has an annual trend which can be modeled using a polynomial linear model, (2) the ozone random variable can be modeled with an acceptable R^2 metric, and (3) each station should be modeled separately to achieve high accuracy. The first premise was observed in several of the time series plots included in the previous section. These figures show that there seems to be an annual trend underlying the ozone variable. The second premise is motivated by the relatively good R^2 metrics obtained

---

Algorithm 1. Custom in-filling approach for the ozone variable

---

input : A Data Frame A of size $n_{rows} \times n_{cols}$. Each row is a measurement with several features including Station.Code, Date, one or more covariates, and the target variable
input : A Data Frame S containing the location of each of the stations in A
input : The number of neighbors to be used for training the GP model: n
Result: The missing values of the target variable are in-filled
1st stage: Fill single missing gaps using MTB;
forall *Station.Code s with missing values* **do**
  ii ← findSingleGapIndices();
  forall *i in ii* **do**
    $a[i] \leftarrow (a[i-1] + a[i+1])/2$
  end
end
2nd stage: Fill remaining gaps using polynomial de-trending plus Gaussian process;
forall *Station.Code s with missing values* **do**
  $k \leftarrow$ getKneighbours(S,n);
  $B \leftarrow A[Station.Code == k]$;
  forall *Station.Code u in B* **do**
    Calculate polynomial trend: x_{trend};
    Calculate residuals after trend: x_{res};
  end
  model ← trainGP(residuals versus predictors);
  $\hat{x}_{res} \leftarrow$ predict(model, s)
  ii ← findMissingValuesIndices();
  $x[ii] \leftarrow x_{trend}[ii] + \hat{x}_r es[ii]$
end

---

by the models GBM and GP in the previous experiment. See Table 2. Only stations with missing data shall be modeled. Also, the observations of close stations are correlated. Therefore, in order to train a model to fit an station with location $\boldsymbol{s}_i$, only the measurements of stations close to $\boldsymbol{s}_i$ should be included. This methodology is summarized in Algorithm 1. An additional consideration is based on the results of [26], where the authors concluded that the best method for in-filling missing data is the Mean Top Bottom (MTB) method, which is a simple mean of the existing observation on the top and the bottom of the missing values. This method makes sense for small gaps. Thus, we implemented a two-stage in-filling method, in which the first stage corrects single missing gaps using MTB, and then in the second stage our proposed method is applied. Our approach consists of detrending the original time series by means of a 4th order polynomial model, and modeling the detrended time series using a Gaussian Process feed by the meteorological covariates taken from the NNRP data set (which do not contain missing values), and two features derived from the Date field (Day of week, and Day of year) to support seasonality.

Table 3. Goodness-of-fit metrics for the ozone predictions obtained through our tailored modeling approach applied on the stations with missing values.

MSE	RMSE	MAE	BIAS	R^2
0.000	0.005	0.004	0.000	0.854

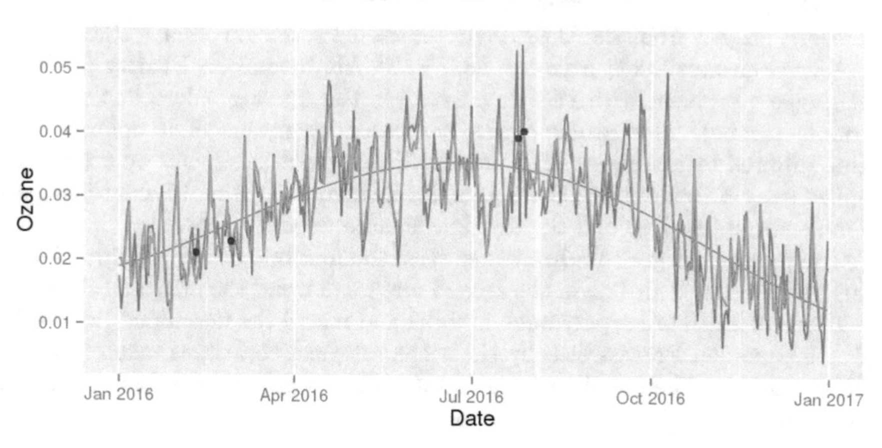

Fig. 6. Original and modeled time series for a single station. The modeled time series was used for data imputation. The trend curve is the polynomial linear model used to detrend the original time series. The red points represent the values in-filled using MTB in the first stage of the imputation algorithm. The highlighted yellow areas indicate zones where there are missing gaps of two or more contiguous records in the original time series that were not in-filled by the MTB.

For detrending the ozone time series we decided to use a polynomial of order 4 based on a qualitative evaluation. The decision regarding the number of neighbor stations that should be used for modeling a particular stations was taken from [9], where the authors state that the spatio-temporal model that they used was more precise when using 50 neighbor stations than only 10. Regarding which model to use, we chose GP based on the results of the previous section and based on the fact that it has been previously used for similar purposes [23]. The results of this procedure are presented in Table 3. Figure 6 illustrates the resulting time series and fitted trend on a station with single an multiple missing gaps. It is worth mentioning that although the result of our methodology are acceptable, the proposed approach was computationally expensive because the Gaussian Processes implemented through the *spTimer* library use random sampling. Training each model took around 2 and a half minutes, which multiplied for the 75 stations with missing values represented a considerable amount of time.

As can be seen in Table 3 and Fig. 6, the combined metrics and visual fitting of the imputed time series are better than those obtained from the methods in the previous section. The base trend has a major role in providing a foundation to support the Gaussian process.

6 Conclusions

Motivated by the general challenge of studying spatio-temporal modeling of air pollution concentrations measured from cheap sensor networks, we decided to assemble a real-life data set made up of ozone concentrations taken from the Air Data project from the U.S. Environmental Protection Agency (EPA) plus weather covariates taken from the NCEP/NCAR Reanalysis Project (NNRP). The biggest problem with this data set, was the missing values in the target variable (ozone). This problem was addressed by means of a tailored algorithm inspired by the hierarchical models. Throughout this study we extensively applied several Exploratory Data Analysis techniques and multivariate analysis to understand the data and the relationships between the features.

The proposed algorithm has two-stages: the first one corrects single missing gaps using MTB, and then in the second one we detrend the original time series by means of a 4th order polynomial model and model the detrended time series using a Gaussian Process feed by the meteorological covariates taken from the NNRP data set. The use of a base trend was the key factor to overcome the bias found when first using the Gaussian processes.

References

1. Allenby GM, Rossi PE, McCulloch RE (2005) Hierarchical Bayes models: a practitioners guide
2. Bakar KS, Sahu SK et al (2015) spTimer: spatio-temporal Bayesian modelling using R. J Stat Softw 63(15):1–32

3. Burke JA, Estrin D, Hansen M, Parker A, Ramanathan N, Reddy S, Srivastava MB (2006) Participatory sensing. Center for Embedded Network Sensing
4. Cameletti M, Lindgren F, Simpson D, Rue H (2013) Spatio-temporal modeling of particulate matter concentration through the SPDE approach. AStA Adv Stat Anal 97(2):109–131
5. Campozano L, Sánchez E, Avilés A, Samaniego E (2014) Evaluation of infilling methods for time series of daily precipitation and temperature: the case of the ecuadorian andes. Maskana 5(1):99–115
6. Cressie N, Wikle CK (2015) Statistics for spatio-temporal data. Wiley, New York
7. Finley AO, Banerjee S, Gelfand AE (2013) spBayes for large univariate and multivariate point-referenced spatio-temporal data models. arXiv preprint arXiv:1310.8192
8. Gelfand AE (2012) Hierarchical modeling for spatial data problems. Spat Stat 1:30–39
9. Gräler B, Pebesma E, Heuvelink G (2016) Spatio-temporal interpolation using gstat. R J 8(1):204–218
10. Hasenfratz D, Saukh O, Sturzenegger S, Thiele L (2012) Participatory air pollution monitoring using smartphones. Mob Sens 1:1–5
11. Kalnay E, Kanamitsu M, Kistler R, Collins W, Deaven D, Gandin L, Iredell M, Saha S, White G, Woollen J et al (1996) The NCEP/NCAR 40-year reanalysis project. Bull Am Meteorol Soc 77(3):437–471
12. Mukhopadhyay S, Sahu SK (2017) A Bayesian spatiotemporal model to estimate long-term exposure to outdoor air pollution at coarser administrative geographies in England and Wales. J R Stat Soc Ser (Stat Soc) 181(2):465–486
13. Pirani M, Gulliver J, Fuller GW, Blangiardo M (2014) Bayesian spatiotemporal modelling for the assessment of short-term exposure to particle pollution in urban areas. J Expo Sci Environ Epidemiol 24(3):319
14. R Core Team (2013) R: a language and environment for statistical computing. https://www.r-project.org/
15. S3L (2012) Matrix factorization as data imputation | S3l. http://s3l.stanford.edu/blog/?p=66
16. Sahu SK, Bakar KS (2012) Hierarchical Bayesian autoregressive models for large space-time data with applications to ozone concentration modelling. Appl Stoch Model Bus Ind 28(5):395–415
17. Sahu SK, Gelfand AE, Holland DM (2007) High-resolution space-time ozone modeling for assessing trends. J Am Stat Assoc 102(480):1221–1234
18. Samworth RJ et al (2012) Optimal weighted nearest neighbour classifiers. Ann Stat 40(5):2733–2763
19. Seo J, Youn D, Kim J, Lee H (2014) Extensive spatiotemporal analyses of surface ozone and related meteorological variables in south korea for the period 1999–2010. Atmos Chem Phys 14(12):6395–6415
20. Snyder EG, Watkins TH, Solomon PA, Thoma ED, Williams RW, Hagler GSW, Shelow D, Hindin DA, Kilaru VJ, Preuss PW (2013) The changing paradigm of air pollution monitoring. Environ Sci Technol 47(20):11,369–11,377. https://doi.org/10.1021/es4022602
21. Stocker M, Baranizadeh E, Portin H, Komppula M, Rönkkö M, Hamed A, Virtanen A, Lehtinen K, Laaksonen A, Kolehmainen M (2014) Representing situational knowledge acquired from sensor data for atmospheric phenomena. Environ Model Softw 58:27–47

22. US EPA (2016) Air data basic information | air data: air quality data collected at outdoor monitors across the US | US EPA. https://www.epa.gov/outdoor-air-quality-data/air-data-basic-information
23. Wen H, Xiao Z, Markham A, Trigoni N (2015) Accuracy estimation for sensor systems. IEEE Trans Mob Comput 14(7):1330–1343
24. WHO (2016) WHO global urban ambient air pollution database (update 2016). http://www.who.int/phe/health_topics/outdoorair/databases/cities/en/
25. Yanosky JD, Paciorek CJ, Laden F, Hart JE, Puett RC, Liao D, Suh HH (2014) Spatio-temporal modeling of particulate air pollution in the conterminous united states using geographic and meteorological predictors. Environ Health 13(1):63
26. Zakaria NA, Noor NM (2018) Imputation methods for filling missing data in urban air pollution data formalaysia. Urbanism. Arhitectura. Constructii 9(2):159

Author Index

M. Botto-Tobar et al. (Eds.): TICEC 2018, AISC 884, pp. 387–388, 2019.
https://doi.org/10.1007/978-3-030-02828-2

Printed in the United States
By Bookmasters